LAYOUT
BASICS

Überarbeitete Neuauflage

LAYOUT BASICS

Die wichtigsten Prinzipien für die Verwendung von Rastern

BETH TONDREAU

stiebner

Aus dem Englischen von der MCS Schabert GmbH, München, - www.mcs-schabert.de - unter Mitarbeit von Jürgen Brust, Stephanie Elsen und Hildegard Rudolph (Übersetzung).

Bibliografische Information der Deutschen Bibliothek
Die Deutsche Bibliothek verzeichnet diese Publikation in der Deutschen Nationalbibliografie; detaillierte bibliografische Daten sind im Internet über <http://dnb.ddb.de> abrufbar.

Printed in China

www.stiebner.com

ISBN-13: 978-3-8307-1450-7

MIX
Papier aus verantwortungsvollen Quellen
FSC® C016973
FSC
www.fsc.org

Wir produzieren unsere Bücher mit großer Sorgfalt und Genauigkeit. Trotzdem lässt es sich nicht ausschließen, dass uns in Einzelfällen Fehler passieren. Unter www.stiebner.com/errata/1450-7.html finden Sie eventuelle Hinweise und Korrekturen zu diesem Titel. Möglicherweise sind die Korrekturen in Ihrer Ausgabe bereits ausgeführt, da wir vor jeder neuen Auflage bekannte Fehler korrigieren. Sollten Sie in diesem Buch einen Fehler finden, so bitten wir um einen Hinweis an verlag@stiebner.com. Für solche Hinweise sind wir sehr dankbar, denn sie helfen uns, unsere Bücher zu verbessern.

Für

PATRICK JAMES O'NEILL

Er ist freundlich, lieb, lustig, fürsorglich, schwer in Ordnung, ironisch, geduldig, ein guter Koch und mindestens 92 andere Dinge.

INHALT

„So wir in der Natur Ordnungssysteme Wachstum und Struktur der belebten und unbelebten Materie regeln, so zeichnet sich auch die menschliche Aktivität seit dem Beginn durch ein Streben nach Ordnung aus."

—JOSEF MULLER-BROCKMANN

ARBEITEN MIT RASTERN

„Raster funktionieren nur dann wirklich gut, wenn sich der Designer nach Lösung der grundlegenden Probleme dann wieder über den struktur-bedingt uniformen Charakter des Rasters hinwegsetzt und auf dessen Basis die Elemente derart visuell und dynamisch erzählen lässt, dass keinerlei Langeweile aufkommt.“

—**TIMOTHY SAMARA**
Making and Breaking the Grid

EINLEITUNG

„Raster sind diejenigen Gestaltungswerkzeuge, die am häufigsten falsch verstanden oder falsch verwendet werden. Sie sind nur dann sinnvoll, wenn sie exakt zum Material passen.“

—**DEREK BIRDSALL**
Notes on Book Design

Ein Raster organisiert Räume und unterstützt zahlreiche Materialien für viele Arten der Kommunikation; es bestimmt und erhält die Ordnung, oft ohne selbst wahrgenommen zu werden. Ein Raster ist ein Plan und kein Kerker.

Auch wenn Raster bereits seit Jahrhunderten Verwendung finden, werden sie von vielen Grafikern mit den Schweizern assoziiert, deren Ordnungswut zu einer ausgeprägten Systematisierung in der Visualisierung so ziemlich aller Dinge führte. Zum Ende des letzten Jahrhunderts galten Raster als monoton und langweilig, aber in der heutigen Daten- und Bilderflut und bei den unterschiedlichsten Plattformen gelten sie bei Berufsanfängern wie bei erfahrenen Grafikern wieder als unverzichtbar.

Dieses Buch gibt einen kurzen Abriss über den Einsatz von Rastern. Jedes der 100 vorgestellten Gestaltungsprinzipien ist ein nützlicher Baustein für Layout- oder Kommunikationsprojekte. Und jedes wird mit einem Projekt in unterschiedlichen Medien illustriert, das Designer oder Designbüros aus aller Welt produziert haben.

Kein Prinzip funktioniert auf sich gestellt. Projekte oder Systeme umfassen oft zahlreiche Prinzipien. So zeigt diese neue Ausgabe, wie sich Prinzipien auf verschiedene Teile desselben Projekts oder auf ganz andere Projekte beziehen. In einem Fall basieren die Teile eines Kommunikationssystems auf mehreren Prinzipien. Wir zeigen auch mehr Beispiele von Designanwendungen für Print, Desktop, Tablet, Smartphone oder alles auf einmal.

Die Stärke dieses Buches liegt in den Arbeiten talentierter, großzügiger Designer, die inspirieren, unterstützen, aufmuntern, steuern und das Thema erweitern. Ich hoffe, die Beispiele werden Sie informieren, faszinieren und anregen und Sie auch an das erste Gebot der Kommunikation erinnern: Ihre Arbeit soll stets das widerspiegeln und aufwerten, was der Autor kommunizieren möchte.

DIE BASICS

1. Die wichtigsten Komponenten

Aller Anfang ist schwer. Beginnen Sie mit dem Inhalt und entwerfen Sie dann Ränder und Spalten. Da sind sicher Anpassungen nötig. Fangen Sie einfach an.

SPALTEN
sind vertikale Felder mit Texten und/oder Bildern. Breite und Anzahl der Spalten auf einer Seite oder dem Bildschirm variieren je nach Inhalt.

MODULE
sind Einzelfelder mit kompakter Fläche; damit generieren sie einen gleichmäßig strukturierten Raster. Kombination von Modulen ergibt Spalten und Reihen verschiedener Größen.

STEGE
sind Pufferzonen; damit werden die Flächen außerhalb des Satzspiegels zum Beschnitt hin bezeichnet. Auf den Stegen können auch Zusatzinformationen wie Anmerkungen oder Bildunterschriften stehen.

MODULGRUPPEN
sind mehrere Spalten oder Module, die als spezielle Flächen/Einheiten für Schrift, Werbung, Bilder oder sonstige Inhalte gestaltet werden.

HORIZONTALE ACHSEN
sind Geraden, die Flächen horizontal teilen. Selbst wenn sie nicht als Linien ausgeformt sind, können sie dazu verwendet werden, Flächen und Elemente so zu ordnen, dass sie als Wegweiser für den Leser dienen.

KOLUMNENTITEL X
sind Orientierungshilfen; Kopf- oder Fußzeilen mit inhaltlichen Angaben oder grafischen Symbolen heißen „lebender" Kolumnentitel, die Seitenzahl (Pagina) allein heißt „toter" Kolumnentitel.

2. Grundstrukturen von Rastern

Die Diagramme unten zeigen Grundstrukturen, die noch auf verschiedene Weise variiert werden können. Die Raster von Zeitungen und ihren Webseiten gehen über drei Spalten hinaus auf fünf oder mehr.

Der **EINSPALTIGE RASTER** wird für Fließtext verwendet, wie er in Abhandlungen, Berichten oder Belletristiktiteln vorkommt. Der Textblock ist das wichtigste Element auf der Seite, der Doppelseite oder dem Display.

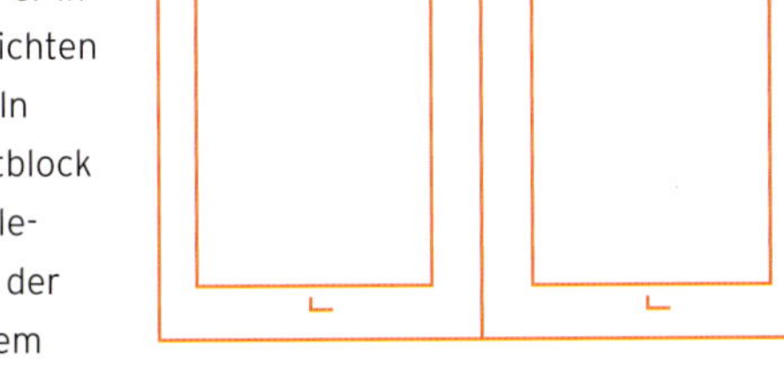

Der **ZWEISPALTIGE RASTER** eignet sich zur strukturierten Anordnung umfangreicher Textmengen oder zur Darstellung unterschiedlicher Informationseinheiten. Doppelspaltiger Raster kann sowohl aus gleichen wie unterschiedlich breiten Spalten bestehen. Spalten unterschiedlicher Breite sind ideal proportioniert, wenn die schmälere halb so breit ist wie die breite.

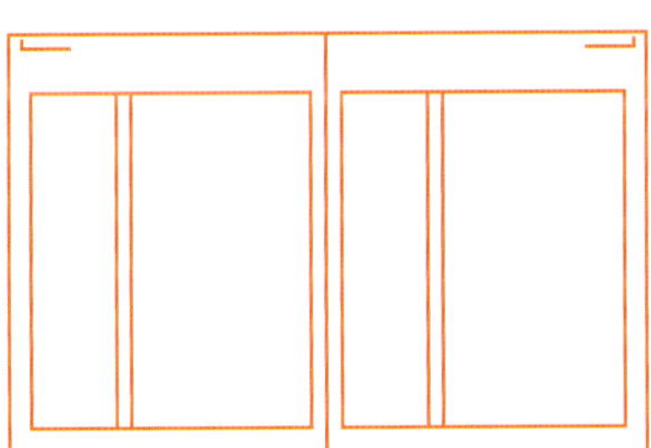

MEHRSPALTIGE RASTER erfordern eine größere Flexibilität. Damit lassen sich mehrere Spalten verschiedener Breite kombinieren; eignen sich gut für Magazine und Webseiten.

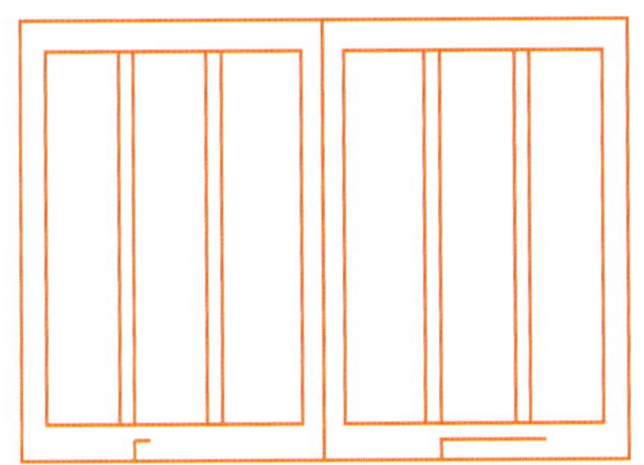

MODULRASTER sind das Mittel der Wahl für die übersichtliche Gestaltung komplexer Informationsmengen, wie man sie in Zeitungen, Kalendern, Schaubildern und Tabellen findet.

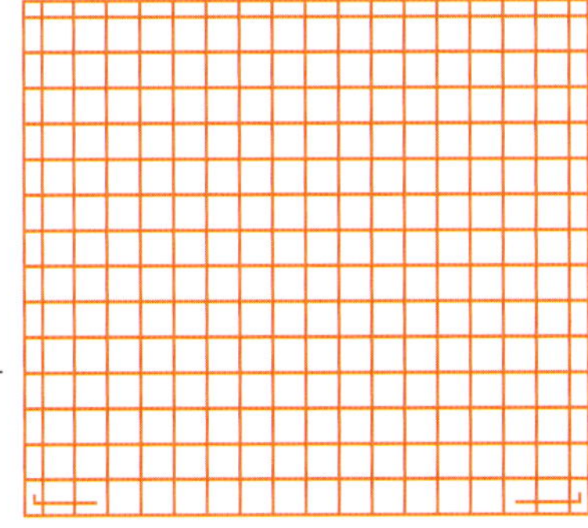

Mit **HIERARCHISCHEN RASTERN** lässt sich eine Seite in Zonen aufteilen. Hierarchische Raster sind oft in horizontale Einheiten untergliedert. Einige Magazine ordnen die Inhaltsseiten horizontal. Viele Geräte teilen das Material effizient in horizontale Bänder auf.

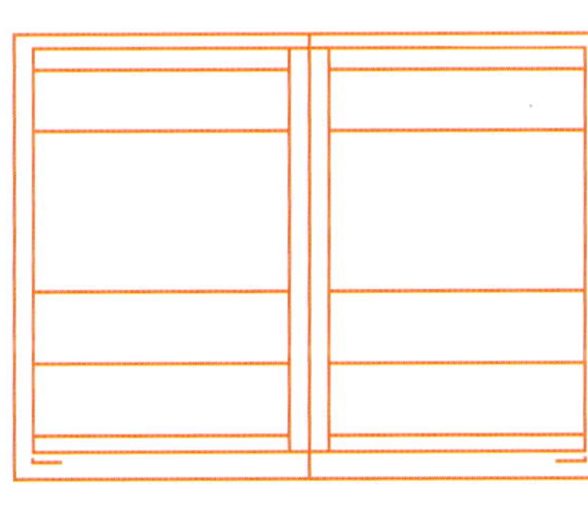

FRAGEN SIE SICH

- Welches Material? Ist es kompliziert?
- Wie viel Material?
- Welches Ziel?
- Wer ist der Leser/Betrachter/Benutzer?

---

3. Welcher Raster für welchen Inhalt?

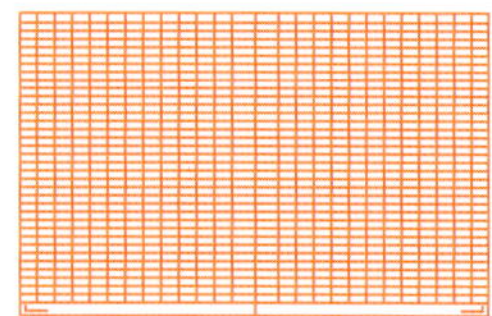

Inhalt, Stege, Anzahl der Bilder, gewünschte Seitenanzahl, Bildschirmdarstellungen und Rahmen sind Faktoren für die Auswahl des passenden Rasters. Das Hauptaugenmerk sollte auf dem Inhalt liegen. Jeder Raster bedingt seine eigenen gestalterischen Probleme, wobei jedoch einige grundlegende Richtlinien Beachtung finden können:

- Benutzen Sie für Fließtext **EINSPALTIGEN RASTER**, etwa für Aufsätze oder Bücher. Eine Spalte erscheint häufig weniger bedrohlich und eleganter als mehrere Spalten und eignet sich für Kunstbücher und Kataloge.

- Für komplizierteres Material bieten **ZWEI-** oder **MEHRSPALTIGE RASTER** mehr Flexibilität. Spalten, die zu einem späteren Zeitpunkt des kreativen Prozesses noch geteilt werden können, sind am variabelsten. Mehrspaltige Raster werden für Webseiten verwendet, um der Vielzahl an Informationen wie Texten, Videos und Werbung gerecht zu werden.

- Für große Informationsmengen etwa in Kalendern und Tabellen sind **MODULRASTER** ein geeignetes Mittel, um übersichtliche Informationseinheiten herauszuarbeiten. Modulraster haben sich auch bei der Gestaltung von Zeitungen mit zahlreichen Informationsfeldern bewährt.

- **HIERARCHISCHE RASTER** unterteilen Seiten oder Bildschirmansichten **HORIZONTAL** und eignen sich gut für einfache Webseiten, auf denen die Informationen systematisch angeordnet sind, damit sie beim Scrollen gut lesbar sind.

Alle Raster ermöglichen eine systematische Gestaltung, erfordern aber Planung und Rechenkenntnisse. Gleich ob mit Pixeln, Picas oder Millimetern gearbeitet wird, das Wichtigste beim Gestalten mit Rastern ist, dass Sie exakt rechnen.

PROJEKT
Good Magazine

KUNDE
Good Magazine, LLC

DESIGN
Open

DESIGNER
Scott Stowell

Diese Skizzen eines erfahrenen Designers machen deutlich, wie ein Raster entsteht.

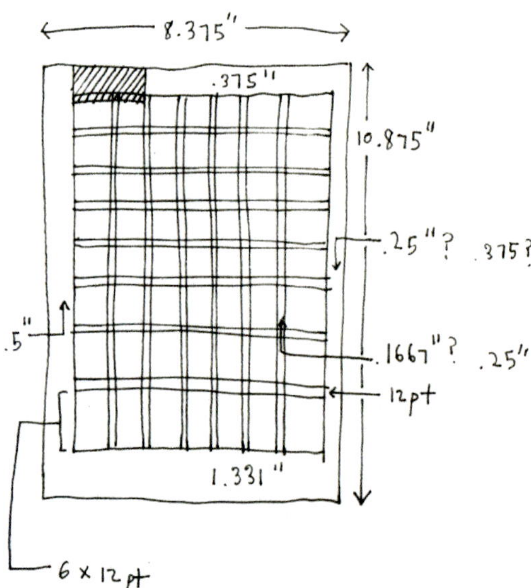

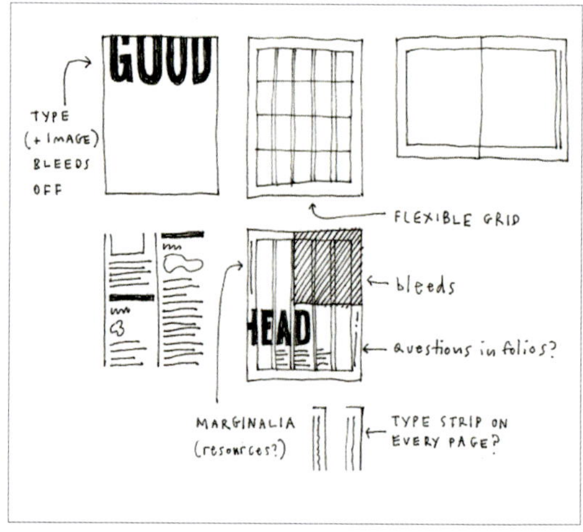

Die Entwürfe zeigen Variationsmöglichkeiten von Rastern für ein Magazin.

4. Rechnen Sie

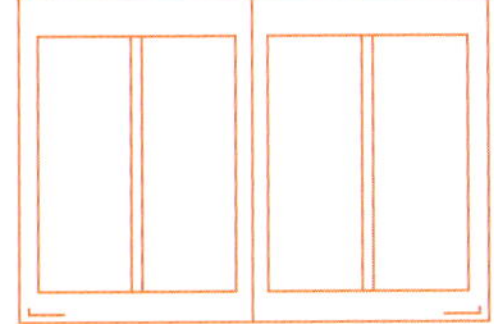

Räumen Sie dem Haupttext Priorität ein und beurteilen Sie die Komplexität des Projekts – meistens sind Größe, Seitenzahl und Farben bereits vorgegeben. Bei der Betrachtung des Inhalts sollten sämtliche rechnerischen Vorgaben des Projekts bereits berücksichtigt werden.

Sobald die Seitengröße bzw. der Bildschirmausschnitt und der hauptsächliche Text feststehen, bietet es sich an, die Anordnung der Elemente auf der Seite zu entwickeln. Wenn Sie ausschließlich mit Text arbeiten, teilen Sie den Text auf die verfügbaren Seiten auf. Sofern auch Bilder, Überschriften, Boxen oder Tabellen Platz finden müssen, legen Sie zunächst fest, wie viel Fläche der Text beanspruchen soll. Auf der verbleibenden Fläche können Bilder, Tabellen und andere Informationen platziert werden. Sie werden meistens die Größen für alle Elemente gleichzeitig ausrechnen müssen.

Wenn Sie sich über die Grundidee für das Material und dessen Anordnung im Klaren sind, können Sie sich eingehend und detailliert mit den Überschriften und Hierarchien befassen. (Siehe nächstes Beispiel.)

TYPOGRAFISCHE TIPPS

Schriften an sich sind eine wilde Ansammlung unterschiedlicher Größen, Zwischenräume, Breiten und Zeilenumbrüche. Ein gleichmäßiger Zeilenfluss erleichtert die Lesbarkeit und verleiht einem Artikel Kontinuität und Gleichmäßigkeit.

Bei größeren Textmengen sollte die Schrift sowohl funktional als auch augenfällig sein. Für Fließtext ist eine relativ große Schrift sowie ausreichend Zeilenabstand zu empfehlen, um den Leser nicht zu ermüden. Vermeiden Sie bei schmalen Spalten große Abstände zwischen den Wörtern durch die Wahl einer kleinen Schrift oder linksbündige Ausrichtung.

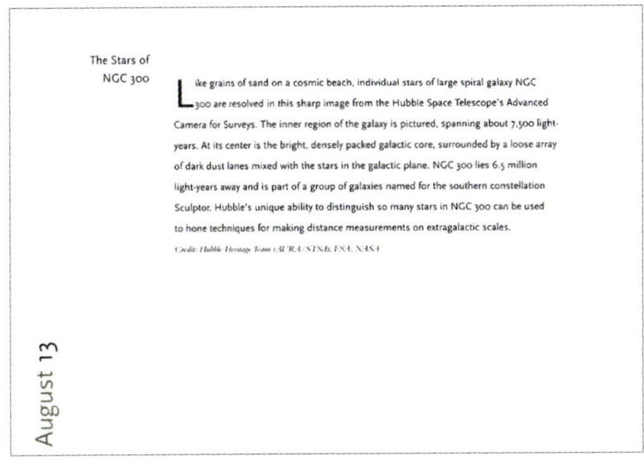

The Stars of NGC 300

Like grains of sand on a cosmic beach, individual stars of large spiral galaxy NGC 300 are resolved in this sharp image from the Hubble Space Telescope's Advanced Camera for Surveys. The inner region of the galaxy is pictured, spanning about 7,500 light-years. At its center is the bright, densely packed galactic core, surrounded by a loose array of dark dust lanes mixed with the stars in the galactic plane. NGC 300 lies 6.5 million light-years away and is part of a group of galaxies named for the southern constellation Sculptor. Hubble's unique ability to distinguish so many stars in NGC 300 can be used to hone techniques for making distance measurements on extragalactic scales.

August 13

In diesem Bildband über Astronomie lässt eine einspaltige Textseite an die Tiefe des unendlichen Weltraums erinnern.

Ein Beispiel für einen Katalog mit einer Unmenge an Text, in dem Zweispaltigkeit dafür sorgt, dass der Text angenehm wirkt und den Bildern einen Rahmen gibt.

PROJEKTE
Astronomy and *Symbols of Power*

KUNDE
Harry N. Abrams, Inc.

DESIGN DIRECTOR
Mark LaRivière

DESIGN
BTDnyc

DESIGNER
Beth Tondreau, Suzanne Dell'Orto, Scott Ambrosino (for *Astronomy* only)

Die Frage, ob ein ein- oder zweispaltiger Raster vorteilhafter ist, hängt vom Inhalt und von der Textmenge ab.

5. Es dem Leser leicht machen

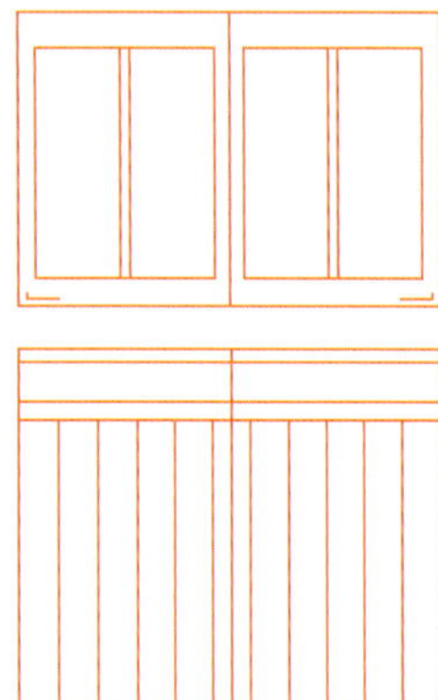

Sind im Material Überschriften, Zwischentitel, Auflistungen oder Bullets enthalten? Falls nicht, könnte es von Vorteil sein, das eine oder andere davon noch zu integrieren. Heben Sie die wichtigsten Informationen durch große, fette Lettern oder gleich eine andere Schrift hervor. Mit Variationen der Schrift, deren Größe und des Duktus können verschiedene Materialien voneinander abgehoben werden, wobei auch hier Schlichtheit vorherrschen sollte. Jeder Stil sollte unmissverständlich seinen Zweck erfüllen, da verschiedene Stilrichtungen verwirrend wirken könnten.

So wichtig Schriftgrade und Bildgrößen sind, darf auch die Weißfläche nicht vernachlässigt werden. Der freie Raum um eine Überschrift kann deren Bedeutung genauso gut unterstreichen. Teilen Sie unübersichtliche Materialansammlungen in leicht verdauliche Häppchen. Gut lesbare Texteinheiten sind das visuelle Gegenstück zu mundgerechten Happen. Lockern Sie das Layout mit Marginalien und Boxen auf, damit sich der Leser leicht orientieren kann. Anhand des Schrifttyps kann der Leser intuitiv auf die Art des Inhalts schließen.

PROJEKT (LINKS)
Symbols of Power

KUNDE
Harry N. Abrams, Inc.

DESIGN DIRECTOR
Mark LaRivière

DESIGN
BTDNYC

Eine klassische Typografie mit der Schrift Bodoni spiegelt, zum Thema passend, die napoleonische Zeit wider.

PROJECT (RECHTS)
Blueprint

KUNDE
Martha Stewart Omnimedia

DESIGN DIRECTOR
Deb Bishop

DESIGNER
Deb Bishop

Die zeitgemäße Typografie ist klar, informativ und deutlich.

120 • SYMBOLS OF POWER

35. Woman's Formal Dress

ANONYMOUS
c. 1805
Embroidered cotton plain weave (mull)
L. 112 cm (center front); 211 cm (center back)
Museum of Fine Arts, Boston; Gift in memory of Helen Kingsford Preston (1989.347)

This white cotton gown reveals some of the subtle changes that took place in women's fashions during the Empire period. The antique-inspired columnar silhouette, popular at the turn of the century, is still evident but has been modified by the gently flaring skirt (formed by triangular fabric inserts, known as *godets*, at the side seams), the squared neckline, the back tie closure, and the small puffed drawstring sleeves. Empire-period embroidery often achieved the grandeur of Ancien Régime embellishment; here, however, the delicate embroidered motifs that adorn the dress show restraint. A small vine pattern of alternating leaves and berries forms a discreet *J* filled with embroidered dots on the bodice, while the skirt bears a slightly different vine of alternating flower blossoms and leaves, again filled with dots, in a pattern that cascades down the center front of the skirt to form a scallop at the hem. The overall effect is one of delicacy and simplicity, underscored by the transparency of the finely woven, white cotton mull fabric imported from India.

The dress was probably purchased in Paris in 1805 by Sarah Bowdoin (1761–1826), the wife of the Hon. James Bowdoin III (1752–1811), who served under President Jefferson as U. S. Minister to Spain and Associate Minister to France. The Bowdoins lived in Paris from 1805 to 1808, and Sarah's diary, written between 1806 and 1808, offers an interesting glimpse into her daily life there.[1] A keen observer of the customs and fashions of the period, Sarah recorded her excitement at walking in the Bois de Boulogne and seeing "almost *all* the fashionables returning; therefore, we had the pleasure of facing them."[2] During one of her favorite walks in the gardens at the Tuileries, Sarah observed the Empress Josephine leaving her Palace, and made careful notes in her diary detailing Josephine's fashionable attire: "She was dressed in a flesh coloured sattin [sic] with long sleeves, short dress with a trimming of very wide lace festooned at the bottom, a small purple velvet bonnet with three small white feathers at the bottom, a small purple camel's hair shawl over her shoulders."[3]

• L.W.

NOTES

1. Sarah's busy schedule included nightly entertainment of dinner guests, daily walks, shopping for dress fabrics and trimmings with her niece, visiting exhibitions at the Louvre, and doing needlework at home. (Sarah Bowdoin Diary, 1806–08, Bowdoin Family Collection, George J. Mitchell Department of Special Collections and Archives, Bowdoin College, Brunswick, Maine.)
2. Ibid., February 15, 1807.
3. Ibid., February 13, 1807.

Bei Verwendung von nur einer Schrift gilt die Faustregel: Innerhalb des Schriftbilds mit Kapitälchen, Groß- und Kleinbuchstaben sowie Kursiven eine Hierarchie schaffen. Für komplexe Informationsdarstellung sollten auch andere Fonts einbezogen werden.

START HERE

★PLANNING

Mortgages, inspections, APRs—oh, my! If the idea of buying your own place ties you up in knots, just follow this road map and we'll walk you through every step of the process. Repeat after us: There's no place like home…

TEXT BY ESTHER HAYNES

Buying vs. renting: What makes more sense economically? In a nutshell: If you plan on living in the same city for at least five years, it's usually worth it to buy. **Advantages to owning:** Over the long term, real estate is rarely a bad investment. The recent market downturns in some areas just give you more choice and additional negotiating room. You'll build equity (read: grow your assets) and establish a good credit history, and you won't have to deal with a landlord (bye-bye, linoleum floors). **Disadvantages:** You'll have to take care of all your own maintenance and improvements (hello, hardwood floors) and pay property taxes. **Situations in which you should hesitate to buy:** If your current rent is unusually low; if you're considering a co-op or condo apartment that has high maintenance costs that aren't tax deductible (these will be noted in the listing); if you think you might get laid off or be relocated in the near future; if your credit is so bad that you can qualify only for loans with very high interest rates. **To help you make the right decision:** Check out the "Buy vs. Rent Calculator" at ginniemae.gov, where you can plug in all the numbers and see a comparison. The results can be shocking. For example, if you're paying $1,500 a month in rent and thinking of buying a place for $200,000 (with a 10% down payment), buying now could save you $110,000 over the next 10 years.

How much can you afford? A mortgage calculator (like the one at mortgage-calc.com) is the simplest way to get a ballpark idea. You punch in the loan amount, interest rate, and length, and it spits out your estimated monthly payment. (Just keep in mind that insurance, property taxes, and maintenance will add to your costs.) **If you want to find out for certain what your upper allowable limit is:** Get preapproved (not prequalified) for a loan. Find a certified mortgage broker (see page 6), who will have you fill out a loan application. Based on your income, debt, and credit score, you will receive a letter stating the maximum amount a bank is willing to lend you. **Bonus:** Preapproval will give you an advantage over other buyers because the seller will know you definitely have the money available to make a purchase. **Beyond the banks:** Calculators and lenders don't take your lifestyle into consideration. Do your own detailed budget that includes all nonessential expenses, from movies to mocha lattes, and talk to a tax pro to see how home ownership can add up for you. Then ask yourself how much you can comfortably spend on your mortgage each month. For peace of mind, be sure you stick with that figure, even if a bank will give you a bigger loan with a larger payment.

Down payments: Putting 20% down is great if you can swing it. You'll start with more equity, pay less total interest, and avoid some fees. But if you have high credit card debt, consider putting down less and using the rest of the money to pay off your cards—mortgage interest is tax deductible, but credit card interest isn't.

If you're broke: Don't panic! You are not a lost cause. Check to see if you qualify for a Housing Finance Agency program in your state that helps first-time buyers (ncsha.org). Another option is to ask a family member or friend to lend you some money interest free, or to cosign the mortgage. You can also save by negotiating for the seller to pay closing costs, which can add up to 3% to 6% of the house price.

CHECKLIST
Have these things handy when just starting out:
- **This guide**
- **Financial information** to begin loan preapproval: recent bank statements, tax forms, and pay stubs, plus employment history
- **Notebook and pen**
- **Digital camera** to document house visits
- **Map or GPS** to find and remember locations
- **The Department of Housing and Urban Development's checklist** of things to note, such as proximity to good restaurants (hud.gov/buying/checklist.pdf)

Mortgage banker: A lender, bank, or company that provides loans.

Mortgage broker: A person or firm that arranges loans from multiple lenders.

Preapproval: A lender's promise (subject to final guarantee) to supply you with a specific loan amount for a mortgage.

Prequalification: A lender's informal estimate of the maximum you'd be allowed to borrow.

Credit bureau score: A number showing how likely you are to default on a loan, based on your credit history (see yours at annualcreditreport.com). Check it and correct any mistakes, which can foil your mortgage plans.

Down payment: The amount of a home's price you pay in cash; usually from 0% to 20%.

85% OF HOME buyers used a REAL-ESTATE AGENT *to help them* during their SEARCH FOR A HOUSE.

Große Informationsmengen lassen sich anhand unterschiedlicher Schriften und Texteinheiten in Boxen übersichtlich gestalten.

6. Ordnung kreieren

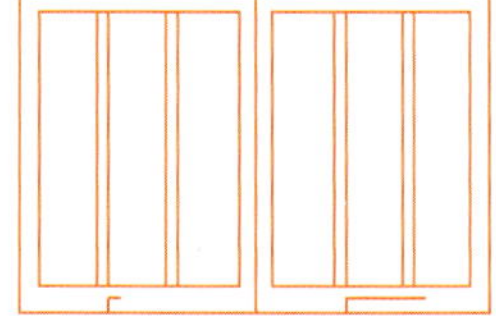

Selten sind alle Bildelemente einer Seite gleich groß. Wie über den Text Information kommuniziert wird, so ist die Größe von Bildern ein Indiz für deren Bedeutung. Einige Designbüros ordnen die Bilder vor dem Layouten nach Größe, andere überlassen es ihren Designern, mittels unterschiedlicher Bildgrößen Pfiff zu erzeugen. Komplexität bedingt allein schon ein größeres Format, Funktionalität und Größe erzeugen Wirkung und sind gegen Ermüdung des Lesers ebenso nötig wie gestalterische Abwechslung.

DUIPIT ALIQUAM CONSEQUIS

at. ɪpit atet praesse quisis alit vel iuscing ex estrud tet lobore feu feugiam consequiscil duisi.ɴa facidunt lutem velendre et, sed moluptat. ᴅuiscip suscidunt veliquat praeseq uatummodit, quatin hent num adio exercilla conse deliquis nos nonsed

des collaborations s'installent à nouveau. Et quand les projets échappent à la facilité du décor gratuit, du "graphique", quand ils réintroduisent la question du sens, c'est généralement autour de la typographie que ça se passe, en assumant quelque part une fonction d'information... Avec l'information comme alibi, mais pas uniquement...
Si on revient sur le début de notre discussion, à Bobigny, tu as fait quelquechose qui ressemble à une enseigne, mais c'est quand même le mot "Danse" que tu as écrit, et non pas "Centre national de la danse"! Et quand tu fais une proposition pour le tramway de Nice, tu ne produis pas une sculpture abstraite: c'est quand même d'un "T"!

PdS: Oui, mais il y a des gens qui vont passer devant pendant des années sans décrypter ce "T"... Il y a même quelque chose d'un peu plus subtil: j'ai toujours mis la face bleue dans un sens de circulation, et la face rouge dans l'autre: quand tu es entre deux stations, tu sais dans quel sens tu vas...

BG: Elle est donc vraiment très fonctionnelle, ton intervention!

PdS: Oui et non... Parce que c'est une chose qui ne se proclame pas. Et quand l'enseigne est de profil...

BG: On ne sait pas où on va?

56 : 2.2008

Bilder können halb-, ein- oder zweispaltig sein. Gelegentlich ist es sinnvoll, den Raster zu durchbrechen, um etwas Würze in das Layout zu bringen und Aufmerksamkeit zu erwecken. Die Bedeutung eines Bildes auf einer Seite lässt sich proportional zu seiner Größe steuern.

PROJEKT
étapes: magazine

KUNDE
Pyramyd/*étapes:* magazine

DESIGN
Anna Tunick

Bilder unterschiedlicher Größe lassen sich wie Orgelpfeifen anordnen.

7. Das Gleichgewicht der Elemente

Je nach Medium oder Projekt lassen sich mit Rastern Elemente isolieren, indem etwa der Text in einer anderen Spalte oder einem anderen Gestaltungsabschnitt als die Bilder dargestellt wird. Bei den meisten Rastern ist es möglich, Text und Bild gut zu verbinden und somit dem Leser beide Informationen gleichberechtigt zugänglich zu machen.

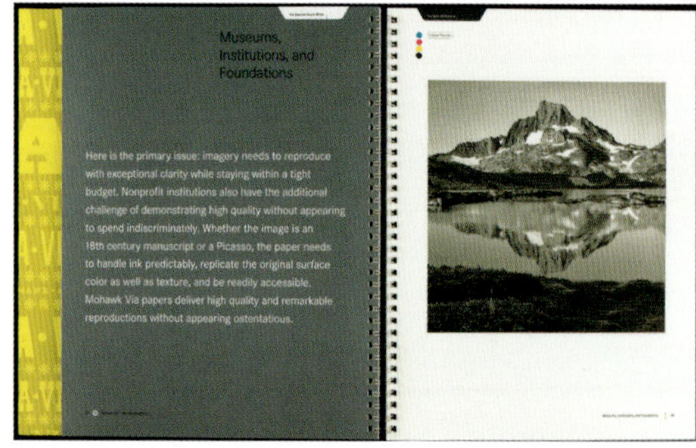

Hier wird der Text hervorgehoben und nimmt wie das Bild allein eine ganze Seite in Anspruch.

LINKS UND UNTEN: Raster bieten die Möglichkeit, Bilder horizontal über eine Seite wandern zu lassen, wobei die Bildunterschriften darunter Platz finden. Alternativ dazu die vertikale Anordnung der Bilder untereinander mit seitlich gesetzten Bildzeilen.

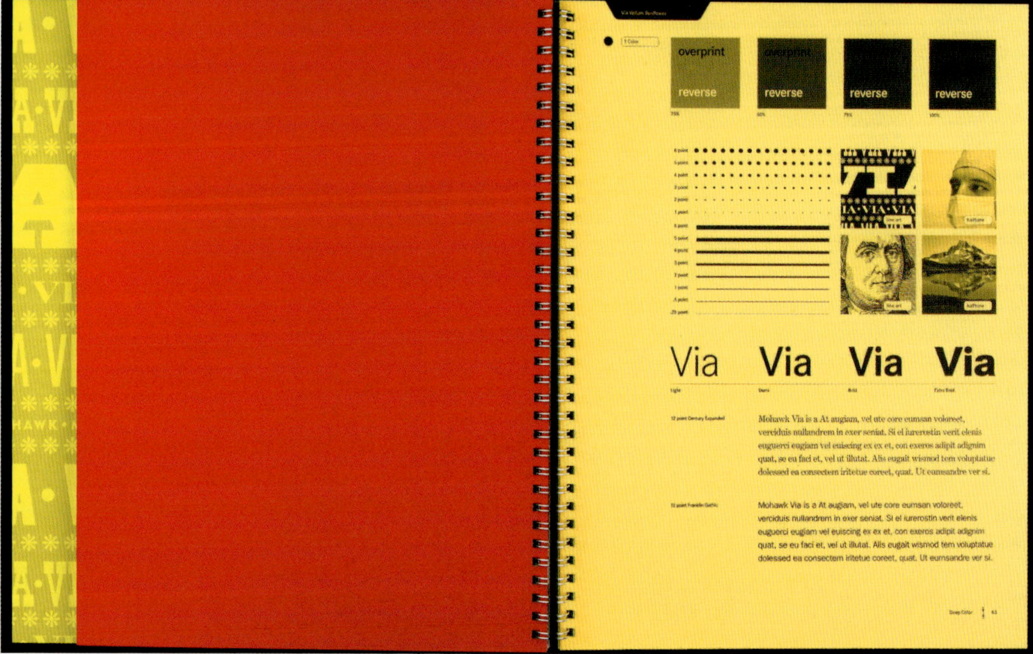

PROJEKT
Mohawk Via
The Big Handbook

KUNDE
Mohawk Fine Papers Inc.

DESIGN
AdamsMorioka, Inc.

DESIGNER
Sean Adams, Chris Taillon

Mit Rastern lässt sich bilderreiches Material übersichtlich darstellen.

FARBEN

8. Flächen mit Farbe definieren

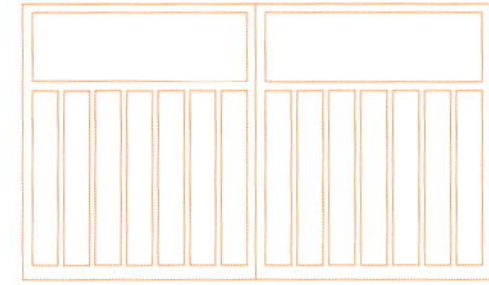

Module oder einzelne Abschnitte lassen sich mit Farben hervorheben. Sie definieren Flächen und bringen Struktur in die Elemente einer Seite bzw. Doppelseite. Farben machen das Layout lebendig und verleihen dem Inhalt psychologische Bedeutung. Denken Sie bei der Farbauswahl an die Zielgruppe. Gesättigte Farben erregen Aufmerksamkeit, ungesättigte lassen das Material subtiler sprechen. Zu viele Farben machen unruhig und irritieren den Blick.

FARBE AUF BILDSCHIRM UND AUF PAPIER

Wir leben in der Welt der RGB-Farben und sowohl Designer als auch Kunden sehen fast alles auf dem Bildschirm. Die Farben auf dem Monitor sind leuchtend, satt und schön. Der traditionelle Vierfarbdruck erfordert eine sorgfältige Papier- und Farbwahl, um die leuchtenden Bildschirmfarben annähernd wiederzugeben.

Auf bunten Flächen lassen sich Informationseinheiten anschaulich darstellen.

Mit Farben lassen sich Informationsblöcke gut absetzen, gleich, ob es sich um Module, Boxen oder Blöcke handelt. Module erfüllen so eine Funktion oder erhalten quasiornamentalen Charakter, wenn sie in Gegenüberstellung farbiger Kästen und Textboxen gestaltet werden.

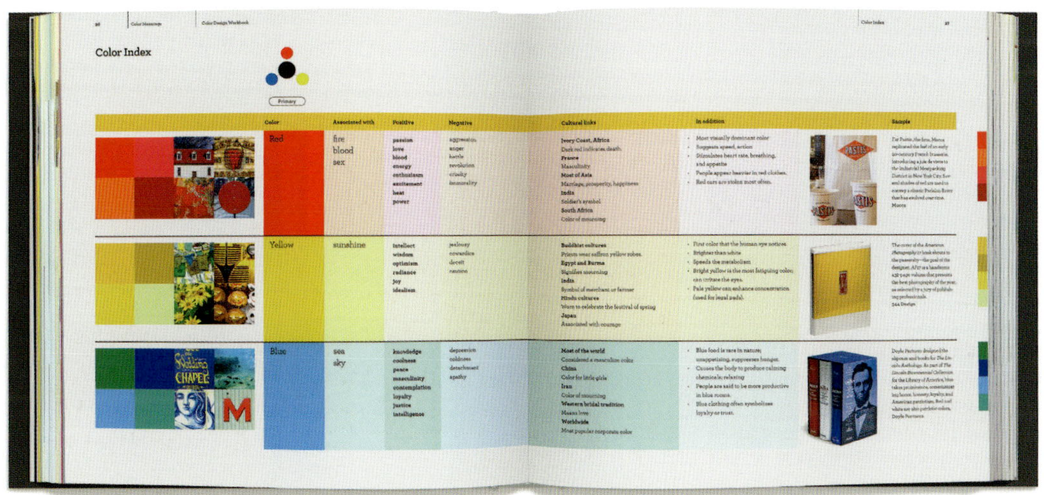

PROJEKT
Color Design Workbook

KUNDE
Rockport Publishers

DESIGN
Sean Adams

Dieses Buchbeispiel zeigt, wie Farben sowohl wesentlich zu der gewünschten Wirkung des Projekts als auch zu seiner positiven Präsenz beitragen können.

9. Freie Flächen als grafisches Element

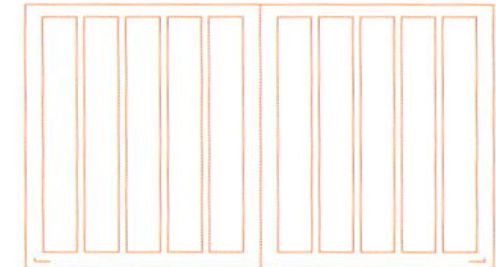

Raum vermittelt Größe. Auch wenn ein Raster solide und klar sein sollte, um große Informationsmengen aufzunehmen, muss er nicht überall ausgefüllt werden. Freie Flächen unterstützen den Informationsfluss und erleichtern Lesbarkeit und Textverständnis. Im Design erzeugen große freie Flächen Spannung und Konzentration. Weißfläche signalisiert Luxus. Leerräume vermitteln vollkommene Ästhetik.

Siehe auch Seite
176/177

Self-Portrait (5 Part), 2001.
Five daguerreotypes, each 8 1/2 x 6 1/2 in.
(21.6 x 16.5 cm)

Der Einsatz von Weißfläche ist eine bewusste Entscheidung, um dem Leser eine Pause zu gönnen oder seine Aufmerksamkeit zu steigern.

PROJEKT
Chuck Close | Work

KUNDE
Prestel Publishing

DESIGN
Mark Melnick

Wie in der Kunst geht es bei Design vor allem um Raum bzw. Fläche.

10. Der Rhythmus erhält die Aufmerksamkeit

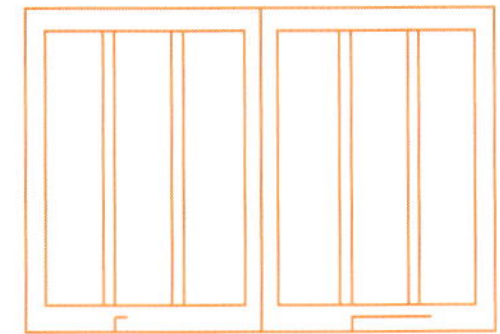

Es gibt Raster, die über Spalten mit mechanischem, klarem, sich wiederholendem bzw. kontinuierlichem Charakter verfügen. Das hat den Zweck, dass möglichst viel Information untergebracht werden kann. Trotzdem bieten die meisten Raster eine Möglichkeit, jeweils eine harmonische Verbindung von einem Informationsblock zum nächsten, von einer Doppelseite oder einer Bildschirmansicht zur nächsten herzustellen. Das Tempo eines Seitendesigns wirkt, trägt viel zur Attraktivität bei und weckt die Neugier des Lesers. Diese Wirkung lässt sich mit Größenvariationen und der räumlichen Anordnung der Bilder, der Typografie und mit der Weißfläche um die Bilder herstellen.

Eine Publikation als Meilenstein – Bobby Martin und sein Team bei OCD hängten die Doppelseiten mit ihren nachdenklichen Essays und historischen Fotos an die Wand, um Layout, Drama und Fluss zu prüfen und umzugestalten.

PROJEKT
King, eine Sonderausgabe zum 50. Jahrestag der Ermordung von Martin Luther King Jr.

KUNDE
The Atlantic

CREATIVE DIRECTOR
Paul Spella

ART DIRECTOR
David Somerville

DESIGNAGENTUR
OCD | Original Champions of Design

DESIGNER
Bobby C. Martin Jr., Jennifer Kinon

Die fließenden Layouts verdeutlichen die Geschichte zu 100 %.

ARBEITEN MIT RASTERN

11. Das Thema braucht ein Gesicht

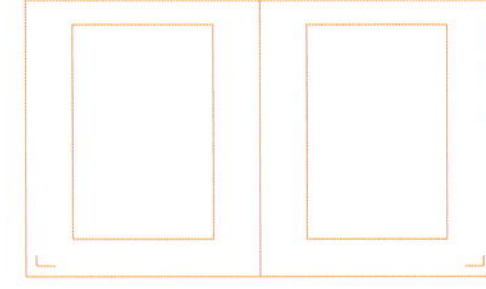

Bei der Wahl der Schrift für ein einspaltiges Layout sollte der Gegenstand des Projekts berücksichtigt werden. Einige der klassisch neutralen Schriften passen fast zu allem, wohingegen andere ihren eigenen Standpunkt vertreten und das Thema bereits durch ihren markanten Charakter formulieren. Schriften können eine Haltung ausdrücken oder sich diskret zurücknehmen. Die für den Text vorgesehene Fläche, die Schriftgröße und der Zeilenabstand sind alle bei der Gesamtkomposition des Textbilds relevant. Ungeachtet der Textmenge und des Materials spielen die Proportionen eine wichtige Rolle.

Siehe auch Seite
25

Bei diesem Projekt umgeht man mit der serifenlosen Geometric das Klischee einer hübschen Schrift für die Farbe Pink und sorgt stattdessen für Punk Power.

there were at least "two pinks" in the 1950s: a feminine pink and "young, daring—and omnisexual" pink.[69]

THE NAVY BLUE OF INDIA

Pink has long been a very popular color in India for both men's clothing and adornment. In Rajasthan, for example, it is still commo men wearing hot pink turbans. India's polychromatic sensibility h many Westerners. Already in 1956, Norman Parkinson did an infl

Bei einem schlichten Textdesign sind die typografischen Feinheiten umso wichtiger. Auch der Zeichenabstand und das Verhältnis der Schriftgrößen zueinander tragen zum gelungenen Gesamtbild bei.

Das Beispiel links zeigt die Schriften in der Abbildungsgröße, die Geometric-Kapitälchen in 10/15,75 pt und die Meridian in 10/15,75 pt. Die Serifenlose ist größer als die Schrift mit Serifen.

PROJEKT
Pink

KUNDE
Thames and Hudson

DESIGN
BTDNYC

Aus dem Text in sauberer Serifenschrift sticht die nüchterne Serifenlose heraus.

FEMININE DESIRE AND FRAGILITY: PINK IN EIGHTEENTH-CENTURY PORTRAITURE

A. Cassandra Albinson

A SURVEY OF OBJECTS MADE IN FRANCE AND BRITAIN IN THE EIGHTEENTH CENTURY REVEALS A REMARKABLE NUMBER OF PINK ITEMS: coverings for furniture; colored prints; porcelain; paint for interiors; paint for works of art; suits for men; and especially dresses and ribbons for women. Aside from its plentitude, what did the color pink represent in women's clothing? Portraiture provides us with a fruitful entryway because we often have indications about the sitter's life and circumstances that can provide further understanding of the choices they made in terms of costume and adornment.[1] In comparable portraits of Jeanne-Antoinette Poisson, Marquise de Pompadour, (1721–1764) and Frances Abington (1737–1815), each woman is depicted close to the picture plane and from the waist up. Both Pompadour and Abington were exceedingly famous at the time they were painted, and their fame and fortune rested in large measure on their physical beauty and prowess. And each woman had the means to be painted by the most famous artist of her day: François Boucher for Pompadour, Joshua Reynolds for Abington.[2] Each woman gestures toward herself and suggests a touch, either in the immediate future—in the case of Pompadour, who holds a rouge brush as if poised to apply color to her cheeks—or concurrently with being painted in the case of Abington in the role of Miss Prue. Hands in each portrait are as important as faces and are stressed by the inclusion of bracelets at the wrist. Both portraits also feature a second figure around the sitter's midriff. In the case of Pompadour we see a miniature portrait of her lover, the French king, Louis XV, while a fluffy white dog sits on

83. François Boucher, *Jeanne-Antoinette Poisson, Marquise de Pompadour*, 1750, with later additions.
Harvard Art Museum.

82. Joshua Reynolds, *Mrs. Abington as Miss Prue in "Love for Love" by William Congreve*, 1771.
Oil on canvas. Yale Center for British Art, Paul Mellon Collection.

without one." A few years later, it was back, "beautifully refreshed" in a variety of styles—"pink evening shirts, pink-shirt dresses, even a pink swimming shirt," not to mention "one of 1953's prettiest little-evening blouses."[67]

"Across the US, a pink peak in male clothing has been reached as manufacturers have saturated more and more of their output with the pretty pastel," reported *Life* magazine in 1955. "Sole responsibility lies with New York's Brooks Brothers," whose pink shirt "was publicized for college girls and caught on for men too." Gradually, pink neckties, dinner jackets, golf jackets, trousers, and other garments also became increasingly visible. "Like most male fashions, including the Ivy League Look, this pink hue and cry has taken some time to develop." But by 1955, the "traditionally feminine color" had become "a staple for [the] male."[68]

Elvis Presley not only wore pink suits, jackets, and trousers, he also drove a pink car and slept in a pink bedroom. Was he influenced by African-American style? His fans wore lipstick in Heartbreak Hotel Pink, and rock and roll extolled the color with songs like "Pink Pedal Pushers" (1958), "Pink Shoe Laces" (1959), and "A White Sport Coat (and a Pink Carnation)" (1957). Meanwhile, the warm carotenoid pink of flamingos was increasingly associated with newly affordable, warm-weather vacations in places like Florida and the Caribbean. So perhaps there were at least "two pinks" in the 1950s: a feminine pink and an emerging "young, daring—and omnisexual" pink.[69]

THE NAVY BLUE OF INDIA

Pink has long been a very popular color in India for both men's and women's clothing and adornment. In Rajasthan, for example, it is still common to see many men wearing hot pink turbans. India's polychromatic sensibility has influenced many Westerners. Already in 1956, Norman Parkinson did an influential photo shoot in India for British *Vogue*. One of his striking images juxtaposed a Western model in the latest fashion with an Indian girl in a hot pink sari. Another, shot in Jaipur, the "Pink City," posed the model in pale pink with a group of Indian men in bright pink coats and turbans. Diana Vreeland, then editor of *Harper's Bazaar*, saw the images and allegedly said, "How clever of you, Mr. Parkinson, also to know that pink is the navy blue of India."[70]

Traditionally, there were many rules about color in clothing related to age, region, caste, occasion, complexion, and time of day. More recently, the individual's personal taste has played an increasingly important role. In their book

43. Unknown artist, *Portrait of an Indian Prince Wearing a Wedding Sehra* [headgear], ca. 1920–40, Rajputana Photo Art Studio.
Gelatin silver print and watercolor. 14⅜ × 11 in. (36.5 × 28 cm). The Alkazi Collection of Photography.

12. Stege festlegen

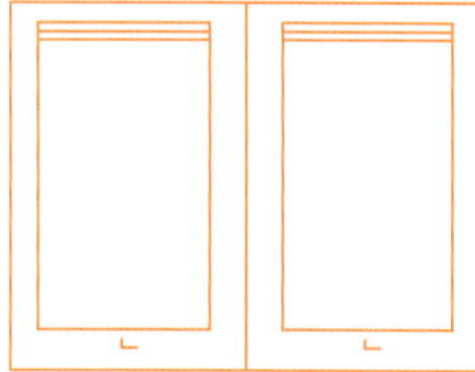

Siehe auch Seite
22–23

Bei Projekten mit großer Seitenzahl hat es sich bewährt, die Innenstege so breit zu belassen, dass sich der Text nicht im Bund verliert. Bei Buchprojekten können die Proportionen der Doppelseitenansicht auf dem Bildschirm oder im Farblaserausdruck ganz anders aussehen als im fertig gebundenen Buch. Der Bereich, der bei der Bindung verloren geht, hängt vom Umfang des Buches bzw. der Broschüre und der Bindetechnik ab. Ungeachtet der Bindemethode empfiehlt es sich immer, darauf zu achten, dass nichts verloren geht.

BINDEMETHODEN UND STEGE

Mit wachsendem Seitenumfang wächst auch die Gefahr, dass Text im Innensteg Richtung Bund verschwindet. Bei Faden- oder Klammerbindung kann das Buch oder die Broschüre weiter geöffnet werden als bei einer Klebebindung. Im letzten Fall traut sich der Leser vielleicht gar nicht einmal, das Buch ganz plan aufzuschlagen, da er es nicht beschädigen will. Bei Verwendung von Spiralbindung schließlich sollte am Innensteg ausreichend Platz für die Spirallöcher eingeplant werden.

15

BUTTER SAUCES

Butter sauces can be classified into four categories. In *beurre blanc-type sauces*, cold butter is whisked into a flavorful liquid base. *Broken butter sauces* are made by cooking whole butter in a sauté pan so that it breaks. These sauces are then usually finished with lemon juice or wine vinegar. *Compound butters* are prepared by working cold whole butter with flavorful ingredients, such as herbs or reduced vegetable purées. *Whipped butters* are prepared in almost the same way as compound butters, except that a hot flavorful liquid is also incorporated into the butter.

PROJEKT
Sauces

KUNDE
JOHN WILEY AND SONS

DESIGN
BTDNYC

Ein Kochrezept nach dem anderen, und das über 800 Seiten. Hier verlangt der Inhalt dringend nach breiten Stegen.

Die Beispiele sind aus *Sauces*, herausgegeben von John Wiley & Sons, © 2008 by James Peterson. Abgedruckt mit der Genehmigung von John Wiley & Sons Inc.

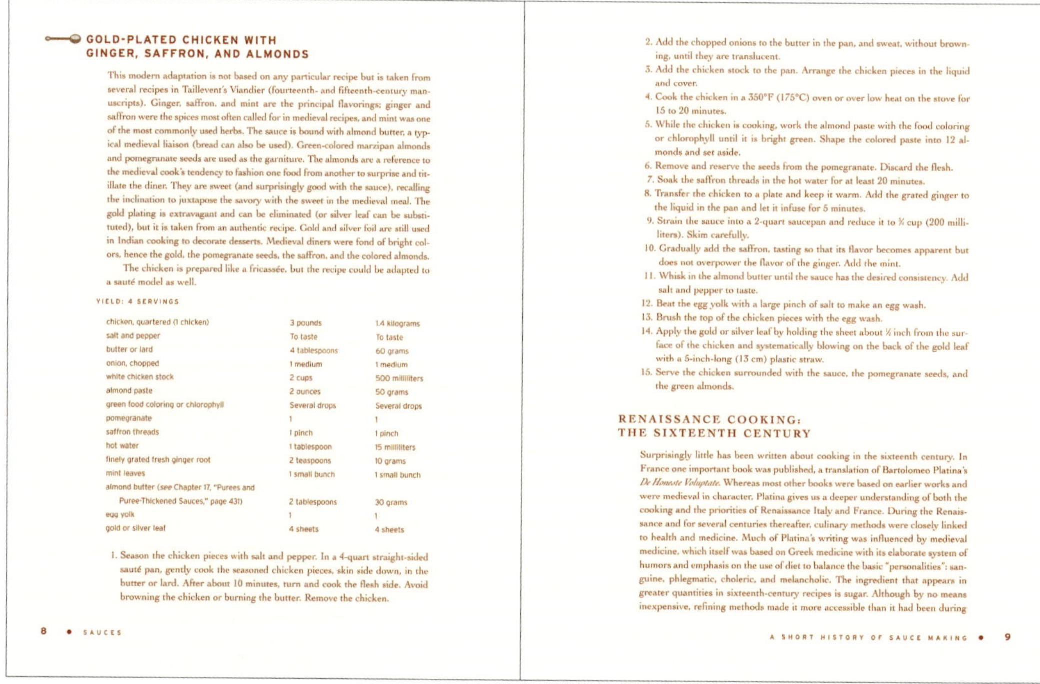

GOLD-PLATED CHICKEN WITH GINGER, SAFFRON, AND ALMONDS

This modern adaptation is not based on any particular recipe but is taken from several recipes in Taillevent's Viandier (fourteenth- and fifteenth-century manuscripts). Ginger, saffron, and mint are the principal flavorings; ginger and saffron were the spices most often called for in medieval recipes, and mint was one of the most commonly used herbs. The sauce is bound with almond butter, a typical medieval liaison (bread can also be used). Green-colored marzipan almonds and pomegranate seeds are used as the garniture. The almonds are a reference to the medieval cook's tendency to fashion one food from another to surprise and titillate the diner. They are sweet (and surprisingly good with the sauce), recalling the inclination to juxtapose the savory with the sweet in the medieval meal. The gold plating is extravagant and can be eliminated (or silver leaf can be substituted), but it is taken from an authentic recipe. Gold and silver foil are still used in Indian cooking to decorate desserts. Medieval diners were fond of bright colors, hence the gold, the pomegranate seeds, the saffron, and the colored almonds.

The chicken is prepared like a fricassée, but the recipe could be adapted to a sauté model as well.

YIELD: 4 SERVINGS

chicken, quartered (1 chicken)	3 pounds	1.4 kilograms
salt and pepper	To taste	To taste
butter or lard	4 tablespoons	60 grams
onion, chopped	1 medium	1 medium
white chicken stock	2 cups	500 milliliters
almond paste	2 ounces	50 grams
green food coloring or chlorophyll	Several drops	Several drops
pomegranate	1	1
saffron threads	1 pinch	1 pinch
hot water	1 tablespoon	15 milliliters
finely grated fresh ginger root	2 teaspoons	10 grams
mint leaves	1 small bunch	1 small bunch
almond butter (*see* Chapter 17, "Purees and Puree-Thickened Sauces," page 431)	2 tablespoons	30 grams
egg yolk	1	1
gold or silver leaf	4 sheets	4 sheets

1. Season the chicken pieces with salt and pepper. In a 4-quart straight-sided sauté pan, gently cook the seasoned chicken pieces, skin side down, in the butter or lard. After about 10 minutes, turn and cook the flesh side. Avoid browning the chicken or burning the butter. Remove the chicken.

8 • SAUCES

2. Add the chopped onions to the butter in the pan, and sweat, without browning, until they are translucent.
3. Add the chicken stock to the pan. Arrange the chicken pieces in the liquid and cover.
4. Cook the chicken in a 350°F (175°C) oven or over low heat on the stove for 15 to 20 minutes.
5. While the chicken is cooking, work the almond paste with the food coloring or chlorophyll until it is bright green. Shape the colored paste into 12 almonds and set aside.
6. Remove and reserve the seeds from the pomegranate. Discard the flesh.
7. Soak the saffron threads in the hot water for at least 20 minutes.
8. Transfer the chicken to a plate and keep it warm. Add the grated ginger to the liquid in the pan and let it infuse for 5 minutes.
9. Strain the sauce into a 2-quart saucepan and reduce it to ¾ cup (200 milliliters). Skim carefully.
10. Gradually add the saffron, tasting so that its flavor becomes apparent but does not overpower the flavor of the ginger. Add the mint.
11. Whisk in the almond butter until the sauce has the desired consistency. Add salt and pepper to taste.
12. Beat the egg yolk with a large pinch of salt to make an egg wash.
13. Brush the top of the chicken pieces with the egg wash.
14. Apply the gold or silver leaf by holding the sheet about ½ inch from the surface of the chicken and systematically blowing on the back of the gold leaf with a 5-inch-long (13 cm) plastic straw.
15. Serve the chicken surrounded with the sauce, the pomegranate seeds, and the green almonds.

RENAISSANCE COOKING: THE SIXTEENTH CENTURY

Surprisingly little has been written about cooking in the sixteenth century. In France one important book was published, a translation of Bartolomeo Platina's *De Honeste Voluptate*. Whereas most other books were based on earlier works and were medieval in character, Platina gives us a deeper understanding of both the cooking and the priorities of Renaissance Italy and France. During the Renaissance and for several centuries thereafter, culinary methods were closely linked to health and medicine. Much of Platina's writing was influenced by medieval medicine, which itself was based on Greek medicine with its elaborate system of humors and emphasis on the use of diet to balance the basic "personalities": sanguine, phlegmatic, choleric, and melancholic. The ingredient that appears in greater quantities in sixteenth-century recipes is sugar. Although by no means inexpensive, refining methods made it more accessible than it had been during

A SHORT HISTORY OF SAUCE MAKING • 9

Breite Innenstege gewährleisten bei umfangreichen Kochanweisungen gute Lesbarkeit, ohne dass der Text im Steg verschwindet. Dazu berücksichtigen sie auch Elemente wie Zwischentitel und Grafiken, die aus dem Textblock ausbrechen können. Großzügige Stege für Lauftitel in der Fußzeile und für Seitenzahlen vermitteln Ruhe und Entspannung.

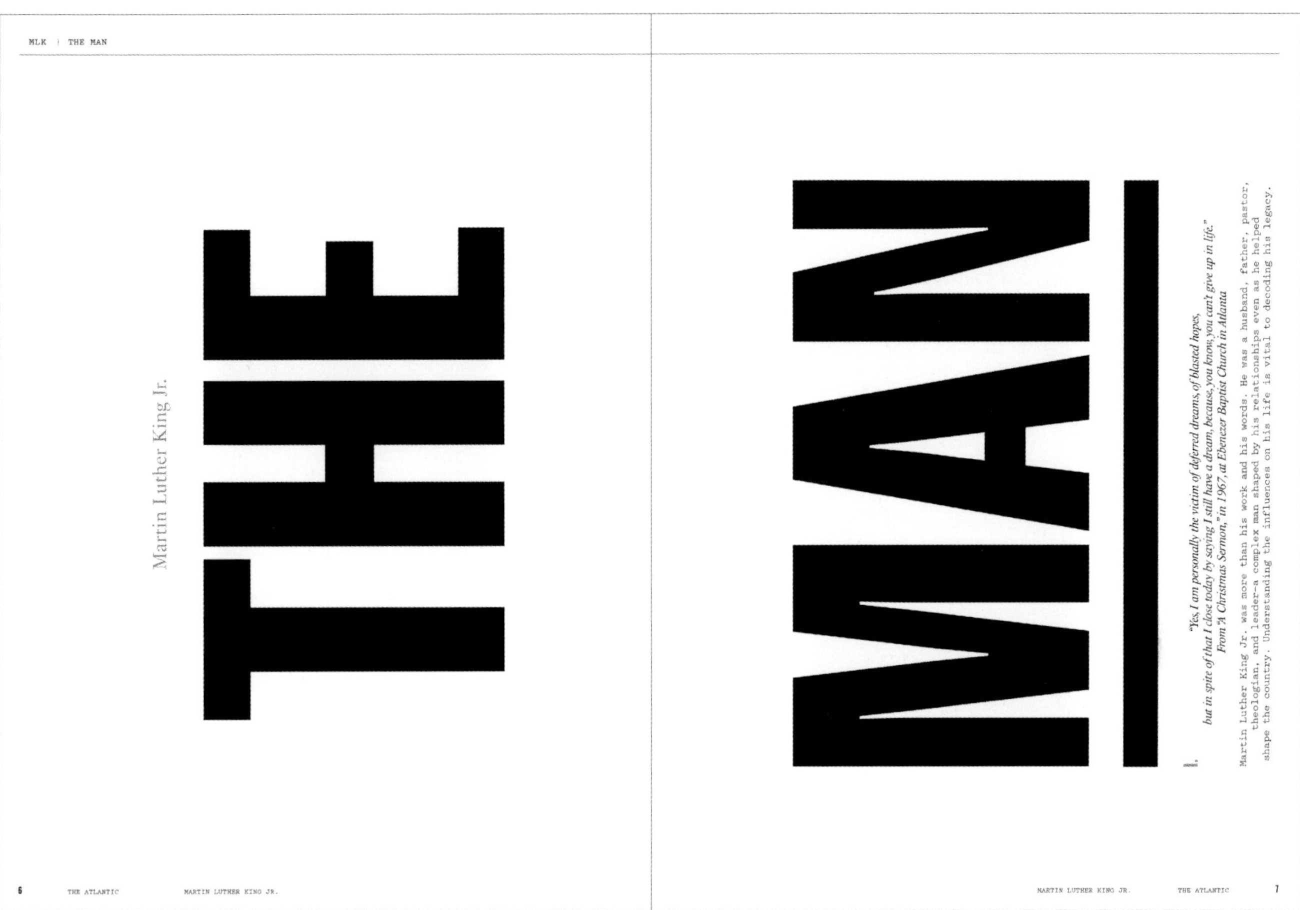

PROJEKT
King, eine Sonderausgabe zum 50. Jahrestag der Ermordung von Martin Luther King Jr.

KUNDE
THE ATLANTIC

CREATIVE DIRECTOR
Paul Spella

ART DIRECTOR
David Somerville

DESIGNAGENTUR
OCD | Original Champions of Design

DESIGNER
Bobby C. Martin Jr., Jennifer Kinon

Die Größe und die freien Flächen bringen Dynamik, die engen Stege Energie und Spannung.

Auf der anderen Seite – und das gilt nicht nur für einspaltige Layouts - sind manche Stege bewusst eng gehalten, um Spannung zu erzeugen und historische Umbrüche aufzuzeigen. Seitenzahlen und Kapiteltitel sind gefährlich nah am Rand und bilden auf dieser Doppelseite einen Kontrast zu den freien weißen Flächen. Auf den Doppelseiten auf Seite 85 rücken sie die Bedeutung des Materials in das Bewusstsein.

FAUSTREGEL

Ich werde häufig nach einer Faustregel für die Stege gefragt. Es gibt keine allgemeingültige Lösung. Ich rate dazu, mit 1,25 cm zu beginnen und dann in beide Richtungen zu experimentieren. Ein Außensteg von unter 6 mm kann zu Problemen beim Druck führen. Letztlich hängt das Maß von den Proportionen Ihrer Seite und Ihres Materials ab, beim Druck auch von den Möglichkeiten Ihres Anbieters. Beim Druck sieht man oft zu viel Text mit zu wenig Rand. Im Web, auf dem Tablet oder Smartphone sind Stege auch wichtig, aber mit kleineren Stegen geht weniger Information verloren.

Dies ist technisch ein einspaltiges Layout, aber das Prinzip der kleinen Stege zieht sich durch die gesamte Ausgabe dieses Magazins. Bei einem Beschnittformat von 19,5 x 26,4 cm gehen die 5 mm breiten äußeren Stege von der oberen Kante bis zum Lauftitel und von der unteren Kante bis zu den Seitenzahlen und Kapiteltiteln an die Grenze der technischen Möglichkeiten. Aber es funktioniert.

13. Mit Proportionen arbeiten

Vergessen Sie die Proportionen auch beim Fußsteg nicht und lassen Sie für die Seitenzahlen ausreichend Platz. Selbst eine auf den ersten Blick einfache Seite erfordert eine bewusste Nutzung des Platzes, die den Inhalt einer gedruckten Seite im Idealfall auch auf einem Bildschirm hervorhebt.

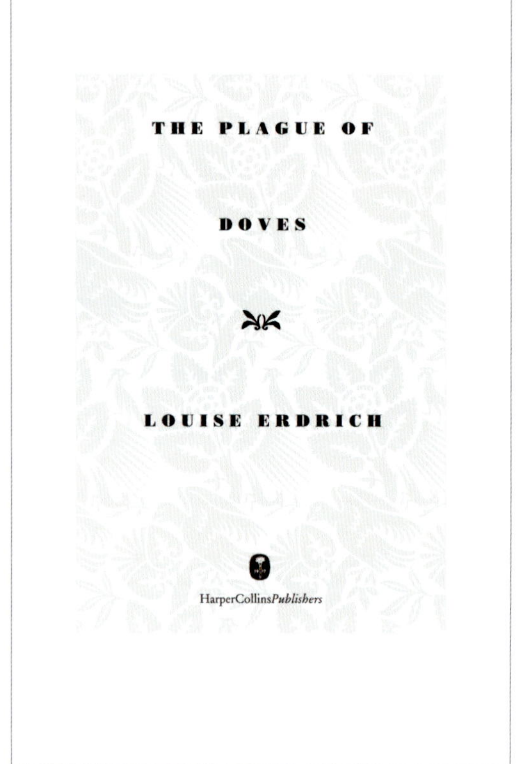

Der Fußsteg (der Rand am Seitenende) ist etwas breiter als der Kopfsteg. Das dezente Ornament legt eine zarte Betonung auf den Titel, der plakativ fett, aber bewusst bescheiden in kleinem Schriftgrad gesetzt wurde.

The Plague of Doves

IN THE YEAR 1896, my great uncle, one of the first Catholic priests of aboriginal blood, put the call out to his parishioners that they should gather at Saint Joseph's wearing scapulars and holding missals From that place they would proceed to walk the fields in a long sweeping row, and with each step loudly pray away the doves His human flock had taken up the plow and farmed among German and Norwegian settlers Those people, unlike the French who mingled with my ancestors, took little interest in the women native to the land and did not intermarry In fact, the Norwegians disregarded everybody but themselves and were quite clannish But the doves ate their crops the same When the birds descended, both Indians and whites set up great bonfires and tried driving them into nets The doves ate the wheat seedlings and the rye and started on the corn They ate the sprouts of new flowers and the buds of apples and the tough leaves of oak trees and even last year's chaff The doves were plump, and delicious smoked, but one could wring the necks of hundreds or thousands and effect no visible diminishment of their number The pole and mud houses of the mixed bloods and the bark huts of the blanket Indians were crushed by the weight of the birds They were roasted, burnt, baked up in pies, stewed, salted down in barrels or clubbed dead with sticks and left to rot But the dead only fed the living and each morning when the people woke it was to the scraping and beating of wings, the murmurous sussuration, the awful cooing babble, and the sight, to those who still possessed intact windows, of the curious and gentle faces of those creatures

5

Eine Pagina in der Mitte ist typisch für klassisches Design.

PROJEKT
The Plague of Doves

KUNDE
HARPERCOLLINS

DESIGN
Fritz Metsch

Ein Beispiel eines sogenannten Crystal-Goblet-Designs, in dem das schlicht gestaltete Textlayout ein literarisches Talent erglänzen lässt. In ihrem Buch *The Crystal Goblet* schreibt die Typografin Beatrice Warde, dass „der Druck gewissermaßen unsichtbar sein sollte", und merkte an, dass ruhiges Design einem kristallinen Glaskelch gleicht: „Er ist dazu gedacht, seinen schönen Inhalt in der ganzen Pracht zu zeigen, statt ihn zu verstecken."

The Plague of Doves

cousin John kidnapped his own wife and used the ransom to keep his mistress in Fargo Despondent over a woman, my father's uncle, Octave Harp, managed to drown himself in two feet of water And so on As with my father, these tales of extravagant encounter contrasted with the modesty of the subsequent marriages and occupations of my relatives We are a tribe of office workers, bank tellers, book readers, and bureaucrats The wildest of us (Whitey) is a short order cook, and the most heroic of us (my father) teaches Yet this current of drama holds together the generations, I think, and my brother and I listened to Mooshum not only from suspense but for instructions on how to behave when our moment of recognition, or perhaps our romantic trial, should arrive.

The Million Names

IN TRUTH, I thought mine probably had occurred early, for even as I sat there listening to Mooshum my fingers obsessively wrote the name of my beloved up and down my arm or in my hand or on my knee If I wrote his name a million times on my body, I believed he would kiss me I knew he loved me, and he was safe in the knowledge that I loved him, but we attended a Roman Catholic grade school in the early 1960's and boys and girls known to be in love hardly talked to one another and never touched We played softball and kickball together, and acted and spoke through other children eager to deliver messages I had copied a series of these second hand love statements into my tiny leopard print diary with the golden lock The key was hidden in the hollow knob of my bedstead Also I had written the name of my beloved, in blood from a scratched mosquito bite, along the inner wall of my closet His name held for me the sacred resonance of those Old Testament words written in fire by an invisible hand Mene, mene, teckel, upharsin I could not say his name aloud I could only write it on my skin with my fingers without cease until my mother feared I'd gotten lice and coated my hair with mayonnaise, covered my head with a shower cap, and told me to sit in the bathtub adding water as hot as I could stand.

The bathroom, the tub, the apparatus of plumbing was all new Because my father and mother worked for the school and in the tribal offices, we were hooked up to the agency water system I locked the bathroom door,

9

Gesperrte, fett gedruckte Kolumnentitel und Seitenzahlen setzen bei reinen Fließtextseiten einen hübschen Akzent. Großzügige Stege und ein breiter Zeilenabstand erleichtern das Lesen.

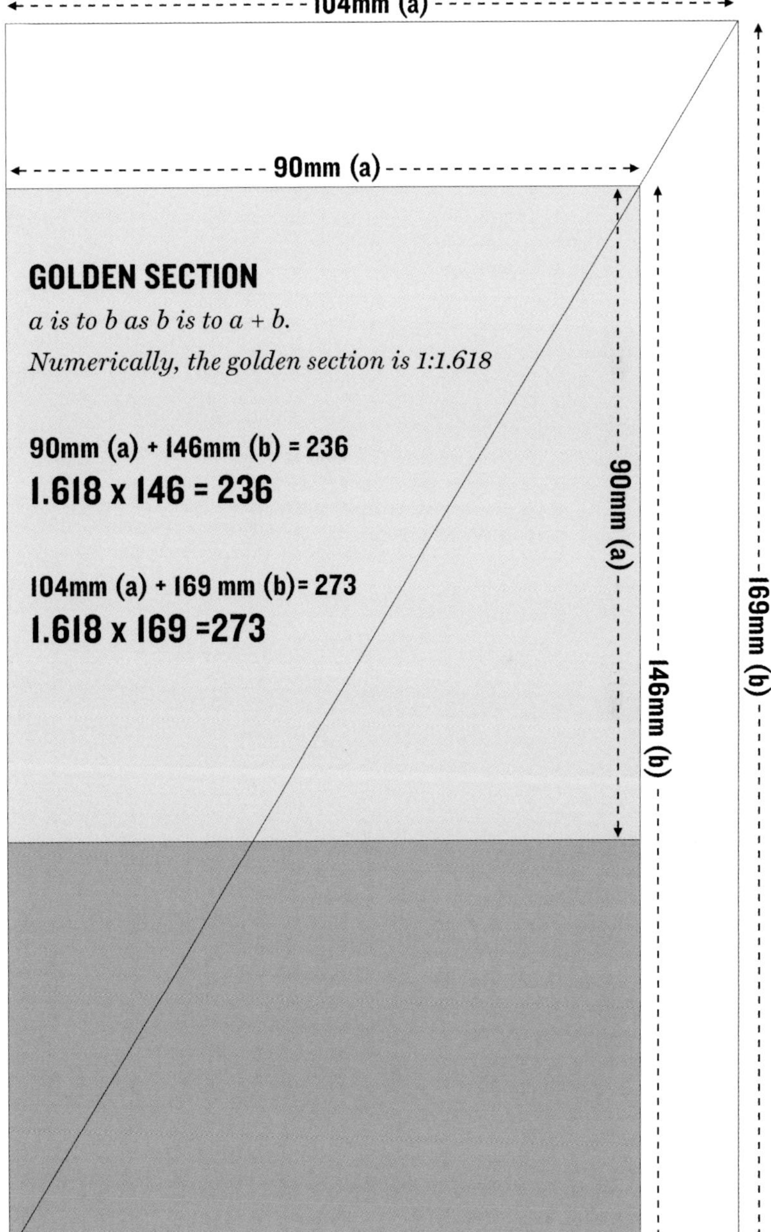

DER GOLDENE SCHNITT

Der Goldene Schnitt wird seit Jahrtausenden in Kunst und Architektur eingesetzt. Er gibt die Verhältnisse einzelner Elemente wie etwa Breite zu Höhe untereinander vor. Dieses Verhältnis beträgt etwa 0,618 oder 1 : 1,618. So verhält sich das kleinere Element, also die Breite (oder a) zum größeren Element, der Höhe (oder b) wie das größere zur Summe beider Elemente. Damit käme man bei einer Breite von 22 Pica auf eine Höhe von 35 Pica und 6 Punkten. Das Diagramm rechts zeigt zwei Rechtecke, bei denen der Goldene Schnitt angewandt wird. Designer arbeiten oft mit Instinkt und Augenmaß und schaffen dennoch angenehme Proportionen.

14. Für Gleichgewicht sorgen

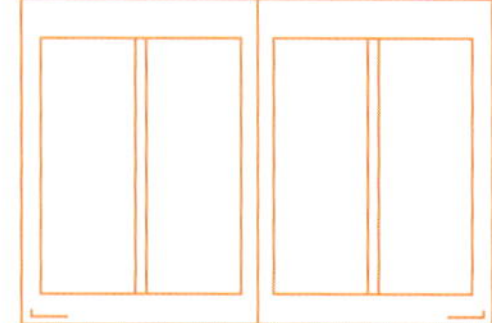

Mit einem zweispaltigen Raster lässt sich umfangreiches Material übersichtlich auf einer Seite anordnen. Symmetrische Spalten signalisieren Ordnung und wirken sich auf Bilder und Leerräume unterschiedlicher Größen beruhigend aus. Zweispaltige Publikationen, in denen für jede Sprache gleichberechtigt nebeneinander eine Spalte mit den gleichen Informationen Platz findet, sind ideal für internationales Publikum.

EVGENY CHUBAROV
ЕВГЕНИЙ ЧУБАРОВ

Für eher konservative Herausgeber und Leser bietet der traditionelle Blocksatz ein ästhetisch ordentliches Erscheinungsbild und Lesebequemlichkeit.

PROJEKT
Return to the Abstract

KUNDE
Palace Editions, for the Russian State Museums

DESIGN
Anton Ginzburg, Studio RADIA

In den zwei Spalten werden die Informationen parallel in zwei Sprachen auf Englisch und Russisch kommuniziert.

оказывается совершенно неспособным проявлять свои функции прозрачности и ясности. Разбросанные в пространстве произведения отвердевшие «знаки-образы художника», чем-то напоминающие «флаговые структуры», устойчивую эмблематику социальных систем, перефразированные элементы поп-арта, откровенно обнажают идеологию артдиверсий Евгения Чубарова, его открытость к языкам массовой культуры. Но в художественных измерениях картинного пространства они транспируются, скорее, как узоики, как внезапные описки, спотыкания о некое «что-то не так», превращаясь или возвращаясь в космическое сатори. Эти «случайные» ошибки дарят нам шок соприкосновения с неведомым при контакте с, казалось бы, заведомо освоенным. Они опровергают диктат идеи однородной абстракции над витальностью знакотворчества как выпад против здравого смысла, как выпадение в целительное безумие. В такой стратегии их образы утверждают новый принцип абстракции, освобожденный от власти личного монолога художника, но реализующий себя в контексте нового смыслового поля, «запаховывая» непосредственное чувство в интеллектуальную рефлексию. Она естественно возникает в своих сгущениях и пустотах, наплывах и разрывах как горизонтальная модель нового художественного сознания, прорывая гипноз знаковой поверхности через жест своеобразной «деконструкции».

Сама технология живописи Евгения Чубарова, ее способность комментировать и описывать саму себя порождает эффект картины как некого живописного объекта, где сама живопись раскрывается как чистая ностальгия по живописи, как воспоминание о картине, где в гуще информационного шума спрятан в коконе былой «абстрактный шедевр», «нетленка», по выражению Ильи Кабакова. Ее «почерк», ее многослойный ландшафт, блестяще выстроенный со всеми своими ассоциативными рядами, где подлинные слои художественной реальности просвечивают сквозь профанные, подбрасывая загадки – все это свидетельствует о новых глубинных ориентациях в искусстве абстракции. Они говорят о ветшании и абстрактных авангардных моделей и о рождении ее абсолютно новой телесности, отрефлексированной и генетически преображенной. Ее новые формы манипулируют следами и обломками ее исторического прошлого и последствиями собственного личного внутреннего опыта художника. В них проступают сознательные цитаты мирового культуронаследия, включающие целые

Closed and open-ended curves of Jackson Pollock exist in contemporary culture as translators, conveying the archaic world and that of the avant-garde art of the 1910s and 20s. They never disappear from view in our mythology. The sign symbolism in the art of Evgeny Chubarow corresponds to the archetypes of Shamanist texts, pulsing in his compositions and disclosing the meaning of their yet unexpressed messages.

Jackson Pollock, "Pages from Sketch-book", 1938

A search for original images related to the birth of human history, its archetypes inspired a paradoxical dialogue, unconditioned by any external influences. The English mystic, poet and artist William Blake and Evgeny Chubarow, a Moscow artist of our times, born in a Bashkirian village amidst Shamanist culture, both find a place in this dialogue. The same landscape, the same artistic style, the same extreme psychedelic states, it's as if they followed the same canon or had the same things before their eyes. These similarities have also been proclaimed by the independent American film director Jarmush in his film Dead Man, in which two men are confronted: a white man named Blake, as the famous English poet, and an Indian, who is immersed in his archaic mythology.

William Blake, "The Body of Abel Found by Adam and Eve," 1826

30

abstraction, one that is free from the pressure of the artist's monologue and one that realizes itself in the context of a new field of meaning, packaging spontaneous feelings into intellectual reflection. It emerges naturally as densities and empty spots, inflows and gaps, as a horizontal model of a new artistic consciousness, breaking through the hypnosis of the sign surface by way of a deconstructing gesture.

The technology of Chubarov's art, its capacity for self-commentary and self-description, creates paintings that have the effect of being objects of pure nostalgia for painting. A recollection of a painting where in the thick of the information noise, as in a cocoon, a former masterpiece of abstract art is concealed, "Imperishables" to quote Ilya Kabakov. Its style and complex landscapes, brilliantly structured with all their associations, where different layers of artistic reality show through the profane, suggesting all sorts of riddles – all this testifies to the new, deep-going orientations in abstract art. They demonstrate the withering of the abstract avant-garde models and the emergence of a new corporeality, carefully thought out and genetically transformed. These new forms manipulate with the traces and debris of history and the consequences of the artist's personal experience. Intentional quotation from the world cultural heritage is evident in this art, including whole movements and trends, skillfully woven into a new cultural context. Moreover, you find in its carpet-like continuity Chubarov's self-quotation and his mythologies existing in the collisions of dissimilar returns above the imagery and style of abstract expressionism, turning his heroic structures into archeological finds and ready-made objects. Both Jackson Pollack and Mark Toby as well as the German "New Wild" are impressed in Chubarov's intellectual energy much like film stars' names are on Hollywood plates. Post-historic handwriting reveals obvious legends in their contours of the remains of gilding, where respect borders on notions much broader than cultural memory, where irony alludes to the games in the labyrinths of time and space. In Einstein's shifted geometry with its "parallel" curvilinearity and relativity, these endless labyrinths bring to mind the abandoned caves and tunnels in Egyptian pyramids. Half-filled with crumbled stone, sand-drifts and excrescencies: they can be viewed both horizontally and vertically. Here you find forgotten and lost texts that were once declared revelations and prophecies. These multi-dimensional sign-bearing structures are being cleared and sorted out to be transformed into illuminations or oppositions like paradoxical tactile surfaces or jottings on the margins where the artist himself "archaeologizes" his mysterious verbalism weaving the fabric of a universal manuscript that

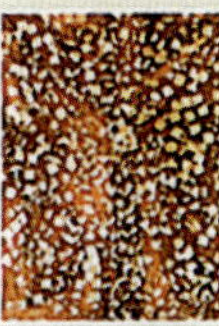

Pseudo-painting of Sigmar Polke, based on new technologies, returns us to pure "objectivity" and materiality of the text as such. It is precisely the problem of "space as text" that has become common for European and Russian cultures, marked by the fact of transition from the texture of meaning to the texture of space. In these textual layers Evgeny Chubarow acquires his own dimension.

Sigmar Polke, Yggdrasil, 1984

For Chubarow, the same as for A. Penk, address to the subconscious provided an outlet to other form of stability, distinct from those offered by the surrounding totalitarian reality. Penk's "Dinosaurs", symbolising absolute resistance to totalitarian sociality, are in fact the same snake-like creatures-signs, which infuse with dynamism Chubarow's compositions as well.

A.R. Penk, "G.B.I." 1988

31

Bei ausreichender Spaltenbreite und kleinem Schriftgrad erhält jede Spalte ein gleichmäßiges und gut lesbares Schriftbild. Eine aufgeräumte Schriftgestaltung bietet genug Gegengewicht für weitere Informationselemente wie Boxen, Tabellen oder Bilder.

15. Design folgt Funktion

Beim typischen Zweispalter sind die Spalten gleich breit. Natürlich geht es auch anders. Wenn für eine umfangreiche Informationseinheit ein offenes, leicht zugängliches und gut lesbares Layout erforderlich ist, kann diesen Anforderungen mit einem Raster entsprochen werden, der aus einer schmalen und einer breiteren Spalte besteht. Die breitere Spalte eignet sich gut für Fließtext und ermöglicht eine flüssige Erzählform, wohingegen in der schmalen Spalte Bildzeilen, Abbildungen oder Diagramme Platz finden.

Susan Brown

TEXTILES:
FIBER, STRUCTURE,
AND FUNCTION

With the Mars Exploration Rovers (MER) six miles off the surface of Mars and traveling at twelve thousand miles per hour, the NASA team experiences "six minutes of terror": a rapid-fire series of high-stress events known as entry, descent, and landing. The cruise stage, which provides support for the voyage, is discarded. A parachute opens, slowing the descent of the craft to about 250 miles per hour. The heat shield is jettisoned, and, for a moment, the lander is hanging from a narrow, braided tether—the world on a string. Five seconds before touchdown, braking rockets fire and explosive gas generators inflate the four clusters of airbags attached to the lander. The bags hit the jagged surface at fifty miles per hour and bounce a hundred feet back up in the air, crashing down dozens more times before rolling to a stop on the rocky surface of Mars (fig. 2).[1]

The airbag system was first developed for *Pathfinder* in 1996 as part of a series of low-cost Discovery program missions, and was further refined for the Mars Exploration missions in 2003. While the animations of the projected landings are both amazing and amusing to watch, the bags are highly engineered by any standard, and performance fibers and textiles play an indispensable role in the successful design of the system (fig. 3).

Die bequem lesbare Bildzeile in der schmalen Spalte fungiert hier als Pförtner am Kapitelanfang. Es ist zu beachten, dass am Kapitelanfang oft mehr Platz zur Verfügung steht, etwa im Kopfsteg, als auf einer normalen Textseite.

Fibers are considered high-performance if they have exceptional strength, strength-to-weight ratio, chemical or flame resistance, or range of operating temperatures. Advances in fiber strength were made throughout the twentieth century with the introduction of synthetic materials such as nylon in the 1930s and polyester in the 1950s, which still form the bulk of the technical-textiles market. But while polyester provides a 50% increase in strength over cotton, Kevlar delivers a 300% increase in strength and a 1,000% increase in stretch resistance.[2] Performance improvements of this magnitude are the factors leading this second textile revolution, a radical transformation in the way things are made.

The development of such high-performance fibers has caused engineers and designers to reexamine the structural capabilities of traditional textile techniques such as weaving, braiding, knitting, and embroidery. The qualities of textiles are dependent on the interaction between their material properties and their structural geometry, or on the fibers and the way in which those fibers are ordered. Each of the textile techniques represents a very specific architecture of fibers which can be used to create a wide variety of materials for design.

WEAVING

The textiles that protected the Mars Exploration Rovers on their descent and landing were made using the most fundamental textile technique: the "over-one under-one" interlacing of two perpendicular sets of threads that we learn as children to call weaving, also known as plain weave (fig. 4).

Bei diesem ausgewogenen und gelungenen Design ist die breitere Spalte doppelt so groß wie die schmälere. Für beide Spalten wurde die gleiche Schriftart gewählt, nur mit dem Unterschied, dass der Font in der schmalen Spalte etwas magerer ist. Die Verwendung unterschiedlicher Schriftgrade trägt zu einer facettenreichen Gestaltungsstruktur bei.

PROJEKT
Extreme Textiles

KUNDE
Smithsonian, Cooper-Hewitt, National Design Museum: Extreme Textiles Exhibition Catalog

DESIGN
Tsang Seymour Design

DESIGN DIRECTOR
Patrick Seymour

DESIGNER
Susan Brzozowski

In diesem Ausstellungskatalog werden nebeneinander verschiedene Formate entsprechend der jeweiligen Darstellungsbedürfnisse kombiniert.

38 EXTREME TEXTILES

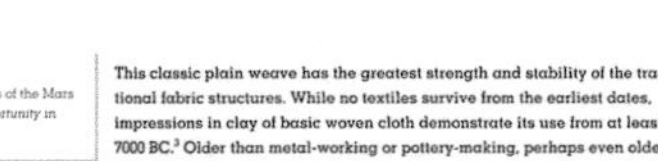

fig. 5
Impressions left by the airbags of the Mars Exploration Rover (MER) Opportunity in Martian soil, January 24, 2004

This classic plain weave has the greatest strength and stability of the traditional fabric structures. While no textiles survive from the earliest dates, impressions in clay of basic woven cloth demonstrate its use from at least 7000 BC.[3] Older than metal-working or pottery-making, perhaps even older than agriculture, cloth-weaving has a very primary relationship to the pursuits of humankind.[4]

It is fitting, then, that among the first marks made by man in the soil of Mars was that of a plain woven fabric: an impression made by the impact of the airbags (fig. 5).[5] Each bag has a double bladder and several abrasion-resistant layers made of tightly woven Vectran. Like most synthetic fibers, Vectran liquid crystal polymer is extruded from a liquid state through a spinneret, similar to a shower head, and drawn into filament fibers. The stretching of the fiber during the drawing process orients the polymer chains more fully along the fiber length, creating additional chemical bonds and greater strength. Vectran provides equal strength at one-fifth the weight of steel. Weight is of premium importance for all materials used for space travel, and Warwick Mills, the weaver of the fabric for the bags, achieved a densely woven fabric at a mere 2.4 ounces per square yard, but with a strength of 350 pounds per inch.[6]

The materials are also required to perform at severe temperatures. Because impact occurs two to three seconds after the inflation of the airbags, the fabrics endure their greatest stresses at both extremes of temperature: the explosive gasses that inflate the bags may elevate the temperature inside the

STRONGER 39

bladder layers to over 212°F, but the temperature on the Martian surface is −117°F. Retraction of the airbags to allow the egress of the rovers required that the fabrics remain flexible at these very low temperatures for an extended period of time—about ninety minutes for the deflation and retraction process. Two other fiber types, aramid fibers (Kevlar 29 and Technora T-240) and ultra-high molecular weight polyethylene (UHMWPE) Spectra 1000, were also considered during the development of the Pathfinder airbags. Spectra, a super-drawn fiber, is among the strongest fibers known—fifteen times stronger than steel. However, it performs poorly at extreme temperatures, and so was eliminated early in the development process. Vectran was ultimately selected for the best performance at low temperatures, but Kevlar 129 was used for the tethers inside the bags because of its superior performance at higher temperatures.

The rovers themselves are also textile-based; they are made from super-strong, ultra-lightweight carbon-fiber composites, which are being widely used for aerospace components as well as high-performance sports equipment.[7] As composite reinforcements, textiles offer a high level of customization with regard to type and weight of fiber, use of combinations of fibers, and use of different weaves to maximize the density of fibers in a given direction. Fiber strength is greatest along the length. The strength of composite materials derives from the intentional use of this directional nature. While glass fibers are the most commonly used for composites, for high-performance products the fiber used is often carbon or aramid, or a combination of the two, because of their superior strength and light weight.

One advantage of composite construction is the ability to make a complex form in one piece, called monocoque construction. A woven textile is hand-laid in a mold; the piece is wetted out with resin and cured in an autoclave. The textile can also be impregnated with resin and cured without a wet stage. The same drape or hand that makes twill the preferred weave for most apparel is also desirable for creating the complex forms of boats, paddles, bicycle frames, and other sports equipment. The weft in a twill, rather than crossing under and over each consecutive warp, floats over more than one warp, and with each subsequent weft the grouping is shifted over one warp, creating the marked diagonal effect typical of twills (fig. 8).

Boat builders were among the first to experiment with carbon-reinforced composites. One early innovator, Edward S. ("Ted") Van Dusen, began making carbon-fiber composite racing shells in the 1970s (fig. 7). The critical factor in shell design is the stiffness-to-weight ratio, with greater stiffness meaning that more of the rower's power is translated into forward motion. Van Dusen found that all of the standard construction materials had about the same specific stiffness, or stiffness per unit weight, and began experimenting with glass, boron, and carbon fiber–reinforced composites.[1]

For his Advantage racing shells, Van Dusen uses glass fiber in a complex twill commonly known as satin weave. In a satin, each weft may float over

Sind wenig oder keine Bilder vorhanden, ist es bei einem Raster mit zwei ungleichen Spalten aus ästhetischer Sicht nicht notwendig, die schmälere Spalte zu füllen.

Mit Schaubildern lässt sich der Gestaltungsraum teilen oder verbinden. Hier werden die blauen Schaubilder Bestandteil der Seite, ohne dass diese aufgrund der großen Informationsmenge überladen wirkt. Gleichzeitig betonen sie die Absätze.

74 EXTREME TEXTILES

The numbers in these tables represent typical values of some important fiber properties; the actual behavior of fibers may differ as variants are produced for diverse end uses. These numbers were compiled from many different sources and are meant for illustration purposes only.

COMPARISON OF YARN STRENGTH

MS
PBO
LCP
HMPE
P-Aramid
Carbon
Ceramic
Glass
Polyester
Nylon
Steel
0 1 2 3 4 5 6 7
Yarn strength based on area of fiber (GPa)
Yarn strength based on weight of fiber (N/tex)

COMPARISON OF MODULI

MS
PBO
LCP
HMPE
P-Aramid
Carbon
Ceramic
Glass
Polyester
Nylon
Steel
0 100 200 300 400 500
Modulus based on area of fiber (GPa)
Modulus based on weight of fiber (N/tex)

CARBON

Thomas Edison first used carbon fiber when he employed charred cotton thread to conduct electricity in a lightbulb (he patented it in 1879). Only in the past fifty years, however, has carbon developed as a high-strength, high-modulus fiber.[8] Oxidized then carbonized from polyacrylonitrile (PAN) or pitch precursor fibers, carbon's tenacity and modulus vary depending on its starting materials and process of manufacture.[9]

Less dense than ceramic or glass, lightweight carbon-fiber composites save fuel when used in aerospace and automotive vehicles. They also make for strong, efficient sports equipment. Noncorroding, carbon reinforcements strengthen deep seawater concrete structures such as petroleum production risers.[10] Fine diameter carbon fibers are woven into sails to minimize stretch.[11] In outer apparel, carbon fibers protect workers against open flames (up to 1000°C/1,800°F) and even burning napalm: they will not ignite, and shrink very little in high temperatures.[12]

ARAMIDS

Aramids, such as Kevlar (DuPont) and Twaron® (Teijin), are famous for their use in bulletproof vests and other forms of ballistic protection, as well as for cut resistance and flame retardance. Initially developed in the 1960s, aramids are strong because their long molecular chains are fully extended and packed closely together, resulting in high-tenacity, high-modulus fibers.[13]

Corrosion- and chemical-resistant, aramids are used in aerial and mooring ropes and construction cables, and provide mechanical protection in optical fiber cables.[14] Like carbon, aramid-composite materials make light aircraft components and sporting goods, but aramids have the added advantages of impact resistance and energy absorption.

LIQUID CRYSTAL POLYMER (LCP)

Although spun from different polymers and processes, LCPs resemble aramids in their strength, impact resistance, and energy absorption, as well as their sensitivity to UV light. Compared to aramids, Vectran (Celanese), the only commercially available LCP, is more resistant to abrasion, has better flexibility, and retains its strength longer when exposed to high temperatures. Vectran also surpasses aramids and HMPE in dimensional stability and cut resistance: it is used in wind sails for America's Cup races, inflatable structures, ropes, cables and restraint-lines, and cut-resistant clothing.[15] Because it can be sterilized by gamma rays, Vectran is used for medical devices such as implants and surgical-device control cables.[16]

HIGH-MODULUS POLYETHYLENE (HMPE)

HMPE, known by the trade names Dyneema (Toyobo/DSM) or Spectra (Honeywell), is made from ultra-high molecular-weight polyethylene by a special gel-spinning process. It is the least dense of all the high-performance

STRONGER 75

DECOMPOSITION TEMPERATURE

MS
PBO
LCP
HMPE (melts)
P-Aramid
Carbon
Glass (melts)
Polyester (melts)
Nylon 6.6 (melts)
Steel (melts)
0 500 1000 1500 2000 2500 3000 3500
Degrees Celsius

DENSITY

MS
PBO
LCP
HMPE
P-Aramid
Carbon
Ceramic
Glass
Polyester
Nylon 6.6
Steel
0 1 2 3 4 5 6 7 8
grams per cm³

fibers, and the most abrasion-resistant. It is also more resistant than aramids, PBO, and LCP to UV radiation and chemicals.[17] It makes for moorings and fish lines that float and withstand the sun, as well as lightweight, cut-resistant gloves and protective apparel such as fencing suits and soft ballistic armor. In composites, it lends impact resistance and energy absorption to glass- or carbon-reinforced products. HMPE conducts almost no electricity, making it transparent to radar.[18] HMPE does not withstand gamma-ray sterilization and has a relatively low melting temperature of 150°C (300°F)—two qualities that preclude its use where high temperature resistance is a must.

POLYPHENYLENE BENZOBISOXAZOLE (PBO)

PBO fibers surpass aramids in flame resistance, dimensional stability, and chemical and abrasion resistance, but are sensitive to photodegradation and hydrolysis in warm, moist conditions.[19] Their stiff molecules form highly rigid structures, which grant an extremely high tenacity and modulus. Apparel containing Zylon® (Toyobo), the only PBO fiber in commercial production, provides ballistic protection because of its high energy absorption and dissipation of impact. Zylon is also used in the knee pads of motorcycle apparel, for heat-resistant work wear, and in felt used for glass formation.[20]

PIPD

PIPD, M5 fiber (Magellan Systems International), expected to come into commercial production in 2005, matches or exceeds aramids and PBO in many of its properties. However, because the molecules have strong lateral bonding, as well as great strength along the oriented chains, M5 has much better shear and compression resistance. In composites it shows good adhesion to resins. Its dimensional stability under heat, resistance to UV radiation and fire, and transparency to radar expands its possible uses. Potential applications include soft and hard ballistic protection, fire protection, ropes and tethers, and structural composites.[21]

HYBRIDS

A blend of polymers in a fabric, yarn, or fiber structure can achieve a material better suited for its end use. Comfortable fire-retardant, anti-static clothing may be woven primarily from aramid fibers but feature the regular insertion of a carbon filament to dissipate static charge. Yarns for cut-resistant applications maintain good tactile properties with a wrapping of cotton around HMPE and fiberglass cores. On a finer level, a single fiber can be extruded from two or more different polymers in various configurations to exhibit the properties of both.

16. Linien ziehen

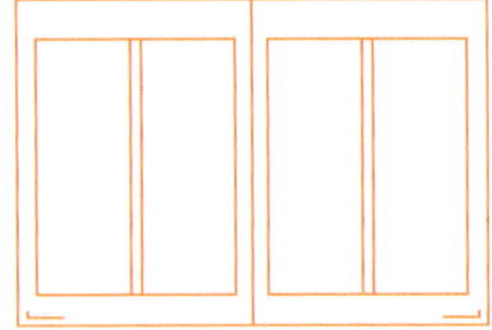

Manchmal bestehen didaktische Inhalte aus so vielen voneinander unabhängigen Informationseinheiten, dass die Weißfläche zwischen den Spalten für gute Lesbarkeit nicht ausreicht. Hier kann eine Vertikallinie als zusätzliche Trennung zwischen den Spalten fungieren. Mit Horizontallinien lassen sich innerhalb der Spalten Boxen vom Fließtext trennen oder auch der gesamte Satzspiegel von den Kolumnentiteln. Aber Vorsicht: Zu viele Linien können einer Seite einen langweiligen Charakter verleihen.

In diesem Seitenlayout werden die Informationseinheiten durch eine Vertikallinie getrennt. Innerhalb der Spalten kennzeichnen verschiedene Schriftvarianten (fett, Kapitälchen, kursiv oder Bruchzahlen) die Unabhängigkeit der Texteinheiten.

NONFAT ROASTED GARLIC DRESSING

MAKES about 1 ½ cups
PREP TIME: 10 minutes
TOTAL TIME: 2 hours (includes 1 ½ hours roasting and cooling time)

To keep this recipe nonfat, we altered our usual technique for roasting garlic, replacing the oil we typically use with water.

- **2 large garlic heads**
- **2 tablespoons water**
- **Salt**
- **2 tablespoons Dijon mustard**
- **2 tablespoons honey**
- **6 tablespoons cider vinegar**
- **½ teaspoon pepper**
- **2 teaspoons minced fresh thyme, or ½ teaspoon dried**
- **½ cup low-sodium chicken broth**

1. Adjust an oven rack to the upper-middle position and heat the oven to 400 degrees. Following the photos on page 000, cut ½ inch off the top of the garlic head to expose the tops of the cloves. Set the garlic head cut side down on a small sheet of aluminum foil, and sprinkle with the water and a pinch of salt. Gather the foil up around the garlic tightly to form a packet, place it directly on the oven rack, and roast for 45 minutes.

2. Carefully open just the top of the foil to expose the garlic and continue to roast until the garlic is soft and golden brown, about 20 minutes longer. Allow the roasted garlic to cool for 20 minutes, reserving any juices in the foil packet.

3. Following the photo on page 000, squeeze the garlic from the skins. Puree the garlic, reserved garlic juices, ¾ teaspoon salt, and the remaining ingredients together in a blender (or food processor) until thick and smooth, about 1 minute. The dressing, covered, can be refrigerated for up to 4 days; bring to room temperature and whisk vigorously to recombine before using.

LOWFAT ORANGE-LIME DRESSING

MAKES about 1 cup
PREP TIME: 10 minutes
TOTAL TIME: 1 hour (includes 45 minutes simmering and cooling time)

Although fresh-squeezed orange juice will taste best, any store-bought orange juice will work here. Unless you want a vinaigrette with off flavors make sure to reduce the orange juice in a nonreactive stainless steel pan.

- **2 cups orange juice (see note above)**
- **3 tablespoons fresh lime juice**
- **1 tablespoon honey**
- **1 tablespoon minced shallot**
- **½ teaspoon salt**
- **½ teaspoon pepper**
- **2 tablespoons extra-virgin olive oil**

1. Simmer the orange juice in a small saucepan over medium heat until slightly thickened and reduced to ⅔ cup, about 30 minutes. Transfer to a small bowl and refrigerate until cool, about 15 minutes.

2. Shake the chilled, thickened juice with the remaining ingredients in a jar with a tight-fitting lid until combined. The dressing can be refrigerated for up to 4 days; bring to room temperature, then shake vigorously to recombine before using.

Test Kitchen Tip: **REDUCE YOUR JUICE**

Wanting to sacrifice calories, but not flavor or texture, we adopted a technique often used by spa chefs in which the viscous quality of oil is duplicated by using reduced fruit juice syrup or roasted garlic puree. The resulting dressings are full bodied and lively enough to mimic full-fat dressings but without the chemicals or emulsifiers often used in commercial lowfat versions. Don't be put off by the long preparation times of these recipes—most of it is unattended roasting, simmering, or cooling time.

Salads 65

PROJEKT
America's Test Kitchen Family Cookbook

KUNDE
America's Test Kitchen

ART DIRECTION
Amy Klee

DESIGN
BTDnyc

Horizontallinien oben oder unten trennen Informationseinheiten oder rahmen eine ganze Box ein.

EASY JELLY-ROLL CAKE

MAKES an 11-inch log
SERVES 10
PREP TIME: 5 minutes **TOTAL TIME:** 1 hour

Any flavor of preserves can be used here. For an added treat, sprinkle 2 cups of fresh berries over the jam before rolling up the cake. This cake looks pretty and tastes good when served with dollops of freshly whipped cream (see page 000) and fresh berries.

- **¾ cup all-purpose flour**
- **1 teaspoon baking powder**
- **¼ teaspoon salt**
- **5 large eggs, at room temperature**
- **¾ cup sugar**
- **½ teaspoon vanilla extract**
- **1¼ cups fruit preserves**
- **Confectioners' sugar**

1. Adjust an oven rack to the lower-middle position and heat the oven to 350 degrees. Lightly coat a 12 by 18-inch rimmed baking sheet with vegetable oil spray, then line with parchment paper (see page 000). Whisk the flour, baking powder, and salt together and set aside.

2. Whip the eggs with an electric mixer on low speed, until foamy, 1 to 3 minutes. Increase the mixer speed to medium and slowly add the sugar in a steady stream. Increase the speed to high and continue to beat until the eggs are very thick and a pale yellow color, 5 to 10 minutes. Beat in the vanilla.

3. Sift the flour mixture over the beaten eggs and fold in using a large rubber spatula until no traces of flour remain.

4. Following the photos, pour the batter into the prepared cake pan and spread out to an even thickness. Bake until the cake feels firm and springs back when touched, 10 to 15 minutes, rotating the pan halfway through baking.

5. Before cooling, run a knife around the edge of the cake to loosen, and flip the cake out onto a large sheet of parchment paper (slightly longer than the cake). Gently peel off the parchment paper attached to the bottom of the cake and roll the cake and parchment up into a log and let cool for 15 minutes.

MAKING A JELLY-ROLL CAKE

1. Using an offset spatula, gently spread the cake batter out to an even thickness.

2. When the cake is removed from the oven, run a knife around the edge of the cake to loosen, and flip it out onto a sheet of parchment paper.

3. Starting from the short side, roll the cake and parchment into a log. Let the cake cool seam-side down (to prevent unrolling) for 15 minutes.

4. Unroll the cake. Spread 1¼ cups jam or preserves over the surface of the cake, leaving a 1-inch border at the edges.

5. Re-roll the cake gently but snugly around the jam, leaving the parchment behind as you go.

6. Trim thin slices of the ragged edges from both ends. Transfer the cake to a platter, dust with confectioners' sugar, and cut into slices.

658 THE AMERICA'S TEST KITCHEN FAMILY COOKBOOK

TYPE OF BEAN	AMOUNT OF BEANS	AMOUNT OF WATER	COOKING TIME
BLACK BEANS			
Soaked	1 pound	4 quarts	1½ to 2 hours
Unsoaked	1 pound	5 quarts	2¼ to 2½ hours
BLACK-EYED PEAS			
Soaked	1 pound	4 quarts	1 to 1¼ hours
Unsoaked	1 pound	5 quarts	1½ to 1¾ hours
CANNELLINI BEANS			
Soaked	1 pound	4 quarts	1 to 1¼ hours
Unsoaked	1 pound	5 quarts	1½ to 1¾ hours
CHICKPEAS			
Soaked	1 pound	4 quarts	1½ to 2 hours
Unsoaked	1 pound	5 quarts	2¼ to 2½ hours
GREAT NORTHERN BEANS			
Soaked	1 pound	4 quarts	1 to 1¼ hours
Unsoaked	1 pound	5 quarts	1½ to 1¾ hours
NAVY BEANS			
Soaked	1 pound	4 quarts	1 to 1¼ hours
Unsoaked	1 pound	5 quarts	1½ to 1¾ hours
PINTO BEANS			
Soaked	1 pound	4 quarts	1 to 1¼ hours
Unsoaked	1 pound	5 quarts	1½ to 1¾ hours
RED KIDNEY BEANS			
Soaked	1 pound	4 quarts	1 to 1¼ hours
Unsoaked	1 pound	5 quarts	1½ to 1¾ hours
LENTILS Brown, Green, or French du Puy *(not recommended for red or yellow)*			
Unsoaked	1 pound	4 quarts	20 to 30 minutes

Rice, Grains, and Beans 195

Die Weißfläche zwischen den Informationseinheiten trennt die horizontal angeordneten Elemente. Dadurch erscheint die Seite klar und übersichtlich.

Horizontallinien können ebenso hilfreich beim Trennen verschiedener Komponenten sein. Wenn im Satzspiegel viel Information untergebracht werden muss, lassen sich damit etwa Pagina und Kolumnentitel gut abheben.

17. Ordnung mit Fluss

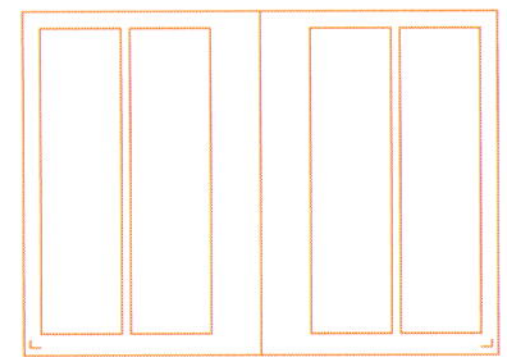

Ein zweispaltiger Raster ist ein solider Rahmen, der für einen guten Lesefluss sorgt. Die Bilder lassen sich bequem in die Spalten integrieren, wobei die Bildzeilen oben oder unten gesetzt werden können. Aber warum nicht noch mehr daraus machen? Sobald die Grundstruktur festgelegt ist, entsteht Raum, um einzelne Flächen auszudehnen. Bilder über zwei Spalten oder an die Ränder gesetzte Bildunterschriften können das ganze Projekt lebendiger machen und sowohl Rhythmus als auch Ordnung hineinbringen.

Man kann die Bilder breiter machen und verschiedene Schriftbreiten einsetzen.

DESIGN

TO KICKSTART INCOMES

PROJEKT
Design for the Other 90%, Ausstellungskatalog

KUNDE
Smithsonian, Cooper-Hewitt, National Design Museum

DESIGN
Tsang Seymour Design

DESIGN DIRECTOR
Patrick Seymour

ART DIRECTOR/DESIGNER
Laura Howell

Eine leicht „verdauliche" Reportage mit unterschiedlichen Bildformaten.

5. Felix Mururi with his MoneyMaker Hip Pump on his farm in Maragua District, Kenya

lawn mowers, and cell phones. They are made in large quantities in big factories. The economy of scale created by centralized manufacturing lowers the price, making the product affordable and ensuring higher quality and reliability. KickStart does the same thing. By centralizing our manufacturing in the most advanced factories available, we can produce high-quality, durable products at a lower cost (figs 7, 8). Wholesalers and middlemen move these goods from factory to marketplace, making a profit in the process. A network of more than 500 local retail shops in three countries stock and sell our pumps. This supply chain needs no artificial support, and will exist as long as there is consumer demand. KickStart also uses donor funds to market the new technologies and generate demand. As with any new product, this takes both time and money. When you are selling an expensive item to the poorest people in the world, it takes even longer and is more expensive (again, the KC) is a perfect example). But eventually, we will reach a point where we can end our marketing efforts and sell each pump at a profit, which we will then reinvest in developing new technology and expanding into new countries. This is a sustainable supply chain.

Third, there is a question of fairness. I have heard people say that it is not "fair" to ask poor people to invest in their own future, but is it fair to give one person or one village a gift when there are others just as needy? By making our products available through the marketplace, they are available to everyone, without patronage or favoritism. This is perhaps the hardest lesson for someone who wants to do good in the world. We see people in desperate need and want to alleviate their suffering. This spirit of generosity is human nature at its best. But as noble as this motivation is in the giving, it is demoralizing in the receiving. When people invest in themselves and their own futures, they have full ownership of their success, and that creates dignity.

INDIVIDUAL OWNERSHIP WORKS BEST

A good question to ask about any program is, Who will own the new technology? If the answer is unclear, or vague, then the program is unlikely to succeed in the long term. We have learned that individual ownership works better than group ownership. Africa is covered with failed community-owned technologies—tractors, water pumps, ambulances, water purification and irrigation systems, et cetera. The list goes on.

There is a common idea that poor people will come together for their collective benefit, or that "investing" in a community is more cost-effective or efficient than working with individuals. There are some situations where this works, like building roads or farmers' cooperatives. But it is much less likely to be effective with the joint ownership of a physical asset. The problem is that if everybody owns an asset, in reality, nobody owns it, and if nobody owns it, nobody will maintain it. Unless there is a way to extract a payment from everyone who uses the asset to cover the costs of maintenance, repair, and replacement, you have the classic free-rider problem.

It comes down to this: The poorest people in the world are just like you and me. No matter how community-minded we are, we will take care of the needs of our family first. And we value the most the items we had to work for.

DESIGN FOR AFFORDABILITY

Our best-selling Super MoneyMaker Pump can be used to irrigate more than two acres of land, and on average the users make $1,000 profit from selling fruits and vegetables in the first year of use. We continue to work to reduce the cost, but at $95, it is still too expensive for many families.

In response, we designed the Hip Pump, which can irrigate almost an acre and retails for less than $35. It looks like a bicycle-tire pump pivoted on a hinge at the end of a small platform. However, unlike a bicycle pump, it uses the operator's whole body. It is lightweight, portable, and extremely easy to use.

The Hip Pump has been a tremendous success: its initial production run of 750 units sold out almost immediately. One of them was bought by Felix Mururi, a young man from rural Kenya. He had a wife and three children to support, but they owned no land. Felix left his family to seek work in Nairobi, where he managed to earn $40 a month working in a restaurant in the city's slums, sending what he could home to his wife and children. When he saw the Hip Pump, he realized he could make more money farming back in his village. He saved his money, bought a pump, went home, and rented six small plots of land. He grew tomatoes, kale, baby corn, and French beans, which he sold to middlemen who took them to the city. Felix planted different crops on each of his small plots so he would have harvests at different times of the year. When we visited Felix three months after he started using his pump, he had already made $580 profit, and he and his wife were talking eagerly about buying land and building their own house. This small pump had enabled Felix to turn his own sweat and drive into cash, look after his family, and plan for his future (fig. 5).

MEASURE THE IMPACT OF WHAT YOU DO

Measuring real impact or outcome is where many would-be social entrepreneurs fail. The number of products you have sold or distributed tells the world nothing. You have to measure the change you are hoping to create with your invention. It is hard and expensive to do, but it is vital. We have learned a great deal from our impact-monitoring efforts. Not only does it enable us to measure ourselves against the goals we have set, it has also been hugely valuable in the design and improvement of our products and marketing efforts.

These are KickStart's core values, and they come together to create a very cost-effective and sustainable way to help people help themselves out of poverty. None of these principles are unique to KickStart or our technologies.

6. A farmer waters her French bean crop with water from a MoneyMaker pump, outside of Nairobi, Kenya.

7, 8. Kenya Vehicle Manufacturers (KVM), located in Thika, north of Nairobi, is one of the companies KickStart partners with to manufacture MoneyMaker Pumps.

They can be applied to many other technologies to make a real difference in the world. Each of these is important individually, but in our experience it is their combination that makes them truly effective.

Finally, for those people who are driven to innovate for the developing world (and also for those who are eager to fund such efforts), I offer this test. A truly successful program to develop and promote new technologies and/or business models needs to meet the following four criteria:

DOES THE PROGRAM CREATE MEASURABLE AND PROVEN IMPACT? This means that you need to carefully define the problem you are trying to solve, then carefully monitor and measure the actual impact you are having on that problem. In the case of KickStart, we are trying to bring people out of poverty by enabling them to earn more money. So we carefully measure how much more money the buyers of our technologies make as a result of owning them. If a program cannot create and prove real impact, then it is not worth implementing.

IS THE PROGRAM COST-EFFECTIVE? There are limited funds for developing and promoting new technologies, and we need to ensure that whatever is done uses these funds efficiently. "Cost-effective" is a subjective measure, so we offer this comparison: KickStart spends about $250 of donor funds to take an average family out of poverty, whereas a more traditional aid program claims on its Web site to do the same for $2,750.

IS THERE A SUSTAINABLE EXIT STRATEGY? One has to ensure that the benefits will continue to accrue for both the existing and new beneficiaries, even after the donor funds are depleted. Creating a program that continues to depend on donor funds forever is not a viable solution. There are four different ways that an effort can become sustainable: 1) build and leave in place a profitable supply chain to continue providing the goods/services; 2) hand over the program to a government which will fund it using tax money; 3) create a local situation that can continue to prosper without the injection of any new outside funds, for instance, establishing a local group savings and loan (merry-go-round) system; 4) completely eliminate the problem, such as eradicating a disease.

IS THE MODEL REPLICABLE AND SCALABLE? The problems we are trying to solve—poverty and climate change, among others—are very large in scale, and it is expensive to develop new technologies and new business models. So we want to ensure that the technologies themselves as well as the dissemination models are not too dependent on specific local conditions, and can be easily adapted to many different settings and locations.

Incorporating all of these guidelines into your work will be a challenge, but great inventors and designers enjoy a challenge. I can tell you that this experience has been an exciting, sometimes frustrating, often exhausting, and immensely satisfying journey. I wish you a fantastic journey of your own.

KICKSTART'S DESIGN PRINCIPLES:
Any tool or technology KickStart produces must meet all of the following design criteria:

INCOME-GENERATING—Any tool must have a profitable business model attached to it.

RETURN ON INVESTMENT—The business opportunity must be available to thousands of people, and the business must be profitable enough that the entrepreneur recoups his or her investment in six months or less.

AFFORDABILITY—We design our tools to retail at less than a few hundred dollars, ideally less than $100.

ENERGY-EFFICIENCY—All of our tools are human-powered, so they must be extremely efficient in converting human power into mechanical power.

ERGONOMICS AND SAFETY—Our products must be able to be used for long periods of time without injury.

PORTABILITY—Tools must be small and light enough to transport from store to home on foot, by bike, or by minibus.

EASE OF INSTALLATION AND USE—Tools must be easy to set up and use, without additional tools or training.

STRENGTH AND DURABILITY—Our tools are used in harsh conditions and will be pushed to their limits. They must be built to withstand abuse. We offer a one-year guarantee on all of our products.

DESIGN FOR AVAILABLE MANUFACTURING CAPACITY—Mass production keeps cost down, but locally available materials and processes can dictate the design.

CULTURAL ACCEPTABILITY—Local cultures will not change to adopt a new technology; the technology has to be adapted to local customs.

ENVIRONMENTAL SUSTAINABILITY—Our tools must not create a negative impact on the environment.

18. Übersicht durch Typografie

Gutes Design spiegelt das Material wider und bezieht sich darauf, d.h., der Leser wird angesprochen. Bei gelungener Typografie werden die einzelnen Gestaltungsabschnitte klar und verständlich voneinander abgehoben, gleich was für einem Zweck die Publikation dienen soll. Auch wenn die Abschnitte horizontal sowie vertikal angeordnet werden und sich über eine Doppelseite oder eine ganze Story erstrecken, wirken sie immer noch strukturierend und stützend. Ausschlaggebend ist auch hier die Korrespondenz mit dem Material. Besonders wichtig dabei: Der Leser muss die Grundinformation mit einem Blick erfassen können. Überschrift(-en) sollten deutlich abgehoben sein, Bildzeilen so gesetzt sein, dass sie eindeutig zuzuordnen und für den Leser hilfreich sind, besonders wenn es um anleitende Inhalte geht.

あの人に会って、聞きたかった、大切なこと。

佐橋慶女さん、教えてください。

「消えつつある日本の生活文化、日本各地で出会った暮らしの知恵」

木曽のお寺のお庫裏さんが、ゆべしを檀家の人たちと作り始めると、お寺が柚子の香りでいっぱいなんですって。

037

暮らしの知恵 036

Die Überschriften sind deutlich vom Text abgehoben, manchmal auch am rechten Seitenrand positioniert. Andere Überschriften haben ihren Platz in der Seitenmitte. Textblöcke sind durch Weißflächen oder Linien abgetrennt, wobei ausreichend Platz für die Bildzeilen gelassen wurde.

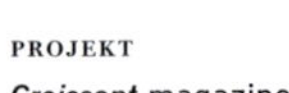

PROJEKT
Croissant magazine

KUNDE
Croissant magazine

ART DIRECTION AND DESIGN
Seiko Baba

Croissant, ein japanisches Magazin für Frauen ab 30, legt Wert darauf, dass die Anleitungen im Heft klar und gut verständlich sind. Die hier abgebildete Ausgabe heißt MOOK, eine Spezialausgabe der *Croissant*-Herausgeber. Der Titel lautet ***Mukashi nagara no kurashi no chie***, was frei übersetzt so viel heißt wie „Altehrwürdige Lebensweisheiten".

大根は、葉っぱから尻尾まで全部食べられるのよ。皮はキンピラにして、ね。

薬用酒なんて、台所の納戸にい〜っぱい！ 昔のものは、一つのものに効くんでなく、「効くんだとさ」、なんです。

Text in verschiedenen Gestaltungsbereichen kann unterschiedliche Informationen widerspiegeln. Hier wurden der normale Text und die Schritt-für-Schritt-Anleitungen in separate Bereiche gesetzt.

ZWEISPALTIGE RASTER

19. Laut und leise

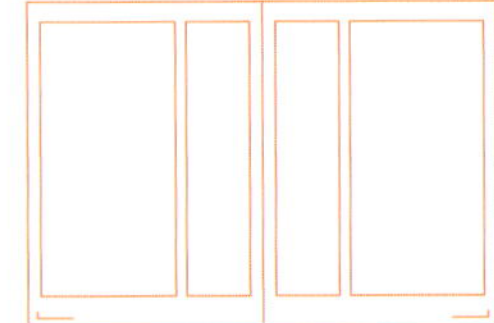

Die besten Raster verfügen über Kontinuität, Klarheit und eine solide Struktur, damit Experimente gewagt werden können. Zweispaltige Raster können unterschiedliche Spaltenbreiten haben, was die Gestaltung mit Spannung und Bewegung bereichert. Sollen ausgefallene Ideen ein Design beleben, bietet ein solider Raster Grundstock und klaren Rahmen für ungewöhnliche Dramatik.

Als beruhigende Elemente haben sich bewährt:

- eine Fläche nur für Überschriften am Seitenanfang
- eine wiederkehrende Box mit Text auf beiden Seiten an der gleichen Stelle, die als Orientierung für den Leser fungiert
- lebende Kolumnentitel unten; sie navigieren den Leser durch die Publikation

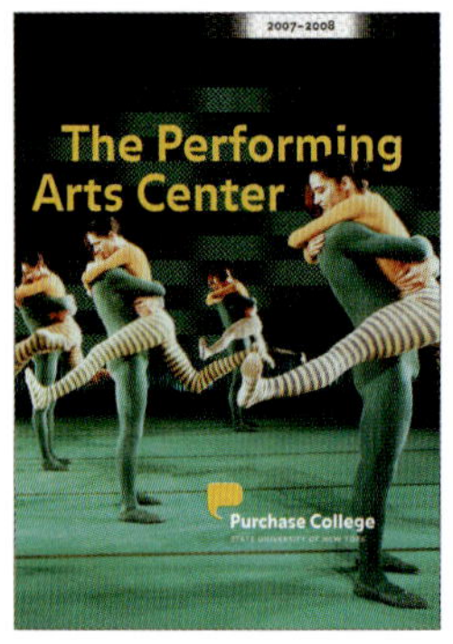

In diesem Projekt haben die Schlüsselinformationen in der gesamten Broschüre den gleichen Platz. Der Text mit zusätzlichen Informationen ist leicht zu finden. Der klare Strukturaufbau wird den dynamisch-ornamentalen Bildern entgegengesetzt.

DANCE

Grupo Corpo

"Brazil's best-known dance company starts with one inestimable advantage: its name, "Corpo" has multiple meanings in Portuguese (body of work, corps, and corporation). But the raw image of a group of bodies communicates in any language. It's punchy and exciting." – THE NEW YORK TIMES

FRIDAY & SATURDAY, APRIL 4 & 5, 2008, 8 PM
PEPSICO THEATRE

AFRICAN DIASPORA :::::::: A bold contemporary dance company that is as magical and endearing as its home country, **GRUPO CORPO** is the most successful dance troupe in Brazil. They combine the sensuality of Afro-Brazilian dance forms, the liquid swing of jazz, and the technical prowess of ballet with energy to burn. With deliriously captivating performances, Grupo Corpo presents an in-your-face spirit and sense of full-out dynamism that is visually riveting and viscerally charged. With its contagious energy, charm, wit and talent that have made the company a runaway worldwide hit, Grupo Corpo never fails to have audiences dancing out the door.

PROGRAM:
Seven or Eight Pieces for a Ballet
New Work

DANCE SERIES SUBSCRIPTION $207, 167, 122
SINGLE TICKETS $45, 35, 25
CYO $41, 32, 23

12 TEL 914.251.6200

MARVELS FROM AROUND THE GLOBE

The Shaolin Warriors

FRIDAY, NOVEMBER 16, 2007, 8 PM
CONCERT HALL

MARVELS SERIES SUBSCRIPTION $154, $120, $86
SINGLE TICKETS $50, 40, 30
CYO $45, 36, 27

This breathtaking show is back by popular demand!

In a fully choreographed theatrical production, the **SHAOLIN WARRIORS** bring the remarkable skill, stunning movement and spectacular imagery of Kung Fu to stages throughout the world. Performed by the Buddhist monks of the Shaolin Temple, a sect that has become known throughout the world for its disciplined spiritualism and deadly martial-arts prowess, the production features many forms of Shaolin Kung Fu as well as a look at the daily temple life of the monks and their Zen Buddhist philosophy.

WWW.ARTSCENTER.ORG 13

PROJEKT
Brochure for the Performing Arts Center, Purchase College

KUNDE
SUNY Purchase

DESIGN
Heavy Meta

ART DIRECTOR
Barbara Glauber

DESIGNER
Hilary Greenbaum

Eine solide Struktur erlaubt ungewöhnliche und erfrischende Variationen.

The **MARK MORRIS DANCE GROUP** was formed in 1980 and gave its first concert that year in New York City. In 1988, MMDG was invited to become the national dance company of Belgium and spent three years in residence at the Théâtre Royal de la Monnaie in Brussels. The company returned to the United States in 1991 as one of the world's leading dance companies, performing across the U.S. and at major international festivals. MMDG is noted for its commitment to live music, a feature of every performance on its full international touring schedule since 1996. The company's 25th Anniversary celebration included over 100 performances throughout 26 U.S. cities and ten U.K. cities.

PROGRAM:
The Argument
Sang-Froid
Italian Concerto
Love Song Waltzes

DANCE SERIES SUBSCRIPTION
$207, 167, 122

SINGLE TICKETS $65, 55, 45
CYO $59, 50, 41

8 TEL 914.251.6200

DANCE SERIES SUBSCRIPTION
$207, 167, 122

SINGLE TICKETS $45, 35, 25
CYO $41, 32, 23

Integrating China's traditional culture with influences from abroad and contemporary dance technique, **BEIJING LDTX** offers a unique and seamless blending of these three elements in a repertoire that shows off unsurpassed technical skill and choreographic excellence.

FRIDAY'S PROGRAM: *The Cold Dagger* is the company's new full-evening work, choreographed by Li Han-zhong and Ma Bo. Based on the traditional Chinese game of Weigi, this intricately choreographed look at human confrontation juxtaposes incredible acrobatics with paired movement that would be otherwise impossible on a normal stage.

SATURDAY'S PROGRAM: A rep program that includes *All River Red*, a striking piece performed to Stravinsky's classic, *The Rite of Spring*; coupled with the company's newest commissioned work *Pilgrimage*, featuring music by the "father of Chinese rock," Cui Jian.

WWW.ARTSCENTER.ORG 9

Oben: Solide strukturiert verfügt das Projekt über eine klare typografische Hierarchie. Die erste Überschrift ist größer, die nachfolgenden Überschriften wiederholen sich in Boxen gleicher Größe, aber in kleinerem Schriftgrad. Termine und Veranstaltungsorte sind in gleichfarbigen Boxen zu finden. Setzen Sie auch klare Bezüge und übersichtliche Hierarchien.

Farben harmonieren mit Informationen.

Linke Seite: Die meisten Bilder erstrecken sich horizontal über die ganze Seitenbreite. Texteinheiten, die die Bilder überlappen, und Farbriegel bringen Bewegung und Dramatik in die Gestaltung. Die Namen der Darsteller sind klar, aber an unterschiedlichen Stellen positioniert; das sorgt für Struktur und fungiert gleichzeitig als spielerisches Element.

Rechts: Silhouetten und Weißfläche variieren das Tempo.

20. Mut zur Fläche

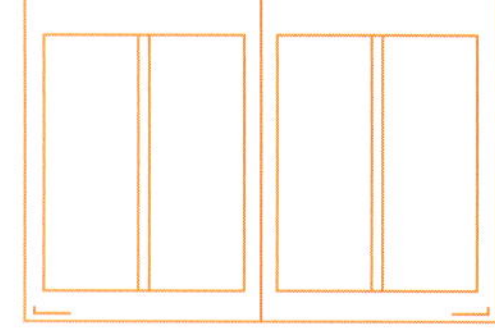

Auch innerhalb eines Projekts können verschiedene Raster- und Typografiesysteme miteinander kombiniert werden. Bei unterschiedlichen Informationseinheiten muss selbst ein Zweispalter noch etwas verändert werden, damit das Layout klar und harmonisch wirkt.

A John Macfarlane costume sketch for *Hansel and Gretel*

Engelbert Humperdinck
HANSEL *and* GRETEL

LIBRETTO
Adelheid Wette, after the fairy tale by the Brothers Grimm

ENGLISH VERSION
David Pountney

TUESDAY, JANUARY 1, 2008
1:30–3:50 PM ET

New Production

CAST & CREDITS

CONDUCTOR
Vladimir Jurowski

GRETEL
Christine Schäfer

HANSEL
Alice Coote

GERTRUDE
Rosalind Plowright

THE WITCH
Philip Langridge

PETER
Alan Held

PRODUCTION
Richard Jones

SET & COSTUME DESIGNER
John Macfarlane

LIGHTING DESIGNED BY
Jennifer Tipton

CHOREOGRAPHER
Linda Dobell

Production a gift of the Gramma Fisher Foundation, Marshalltown, Iowa, and Karen and Kevin Kennedy
Additional funding from Joan Taub Ades and Alan M. Ades

Hansel and Gretel was originally created for Welsh National Opera and Lyric Opera of Chicago.

18 • THE METROPOLITAN OPERA HD LIVE

SYNOPSIS

ACT I

Hansel and Gretel have been left at home alone by their parents. When Hansel complains to his sister that he is hungry, Gretel shows him some milk that a neighbor has given them for the family's supper. To entertain them, she begins to teach her brother how to dance. Suddenly their mother returns. She scolds the children for playing and wants to know why they have gotten so little work done. When she accidentally spills the milk, she angrily chases the children out into the woods to pick strawberries.

Hansel and Gretel's father returns home drunk. He is pleased because he was able to make a considerable amount of money that day. He brings out the food he has bought and asks his wife where the children have gone. She explains that she has sent them into the woods. Horrified, he tells her that the children are in danger because of the witch who lives there. They rush off into the woods to look for them.

19 • HANSEL AND GRETEL

Fließtexte wie bei fortlaufenden Geschichten oder Inhaltsangaben können in zwei Spalten angeordnet werden.

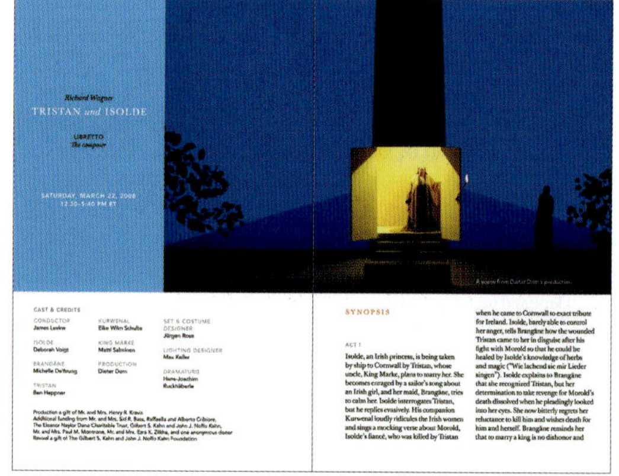

Richard Wagner
TRISTAN *und* ISOLDE

SYNOPSIS

Große dramatische Fotos stehen bei jeder Vorstellungsbeschreibung vornweg.

PROJEKT
2007–2008 HD Program Guide

KUNDE
The Metropolitan Opera

DESIGN
AdamsMorioka, Inc.

CREATIVE DIRECTORS
Sean Adams, Noreen Morioka

ART DIRECTOR
Monica Schlaug

DESIGNER
Monica Schlaug, Chris Taillon

Ein wohlgeordnetes und klassisches Design, das die zurückhaltende Grafik dennoch in einer zeitlosen Form mit jugendlicher Frische bereichert.

Alice Coote and Christine Schäfer sing the title roles.

GRETEL WAKES HANSEL,

and the two find themselves in front of a gingerbread house.

ACT II

Gretel sings while Hansel picks strawberries. When they hear a cuckoo calling, they imitate the bird's call, eating strawberries all the while, and soon there are none left. In the sudden silence of the woods, the children realize that they have lost their way and grow frightened. The Sandman comes to bring them sleep by sprinkling sand on their eyes. Hansel and Gretel say their evening prayer. In a dream, they see 14 angels protecting them.

ACT III

The Dew Fairy appears to awaken the children. Gretel wakes Hansel, and the two find themselves in front of a gingerbread house. They do not notice the Witch, who decides to fatten Hansel up so she can eat him. She immobilizes him with a spell. The oven is hot, and the Witch is overjoyed at the thought of her banquet. Gretel has overheard the Witch's plan, and she breaks the spell on Hansel. When the Witch asks her to look in the oven, Gretel pretends she doesn't know how: the Witch must show her. When she does, peering into the oven, the children shove her inside and shut the door. The oven explodes, and the many gingerbread children the Witch had enchanted come back to life. Hansel and Gretel's parents appear and find their children. All express gratitude for their salvation.

20 • THE METROPOLITAN OPERA HD LIVE

IN FOCUS

Engelbert Humperdinck

HANSEL *and* GRETEL

PREMIERE: HOFTHEATER, WEIMAR, 1893

Originally conceived as a small-scale vocal entertainment for children, *Hansel and Gretel* outgrew its original design to become the most successful fairy-tale opera ever created. Like so many children's classics, *Hansel and Gretel* achieved greatness because it resonates with both adults and kids. The composer Engelbert Humperdinck was a protégé of the musical titan Richard Wagner, and the score of *Hansel and Gretel* is flavored with the sophisticated musical lessons he learned from his idol while maintaining a charm and a light touch that were entirely Humperdinck's own. The ancient tale of the young brother and sister who get lost in a dark forest and almost get eaten by an old witch became a classic of German literature in the famous collected stories of the Brothers Grimm. The opera acknowledges the darker features present in the story, yet presents them within a frame of grace and humor. Humperdinck's fellow composer Richard Strauss was delighted with this score from the start and conducted its world premiere. *Hansel and Gretel* has been internationally popular ever since and must be one of the very few operas that can boast equal acclaim from such diverse and demanding critics as children and musicologists.

THE CREATORS

Engelbert Humperdinck (1854–1921) was a German composer who began his career as an assistant to Richard Wagner in Bayreuth in a variety of capacities, including tutoring Wagner's son Siegfried in music and composition. Humperdinck even composed a few minutes of orchestral music for the world premiere of Wagner's *Parsifal* (1882) when extra time was needed to effect a scene change. (This music is not included in the printed score of *Parsifal* and is no longer performed). *Hansel and Gretel* was Humperdinck's first complete opera and remains the foundation of his reputation. The world premiere of his later opera *Königskinder* took place at the Met and was one of the sensations of the company's 1910–11 season, following less than three weeks after the world premiere of Puccini's *La Fanciulla del West*. *Hansel and Gretel*, however, is the only one of Humperdinck's works to remain in the repertory. The libretto was written by his sister, Adelheid Wette (1858–1916), and is based on the famous fairy tale from the Grimms' collection. The brothers Jacob (1785–1863) and Wilhelm (1786–1859) Grimm were German academics whose groundbreaking linguistic work revolutionized the understanding of language development. Today, they are best remembered for editing and publishing collections of folk tales.

THE SETTING

In the libretto, the opera's three acts move from Hansel and Gretel's home to the dark forest to the witch's gingerbread house deep in the forest. Put another way, the drama moves from the real, through the obscure, and into the unreal and fantastical. In this production, which takes the idea of food as its dramatic focus, each act is set in a different kind of kitchen, informed by a different theatrical style: a D.H. Lawrence-inspired setting in the first, a German Expressionist one in the second, and a Theater of the Absurd mood in the third.

THE MUSIC

The score of *Hansel and Gretel* successfully combines accessible charm with subtle sophistication. Like Wagner, Humperdinck assigns musical themes to certain ideas and then transforms the themes according to new developments in the drama. Much of this development occurs in the orchestra, like the chirpy cuckoo, depicted by the winds in Act II, which becomes

21 • HANSEL AND GRETEL

Hier werden Informationen mit typografischen Mitteln voneinander getrennt, nämlich durch den Kontrast zwischen einer serifenlosen und einer serifenbetonten Schrift.

Bei Frage-und-Antwort-Formaten bieten sich ungleichmäßige Zweispalter an: Die Fragen stehen in der schmäleren, die Antworten in der breiteren Spalte.

that Tristan is simply performing his duty. Isolde maintains that his behavior shows his lack of love for her, and asks Brangäne to prepare a death potion. Kurwenal tells the women to prepare to leave the ship, as shouts from the deck announce the sighting of land. Isolde insists that she will not accompany Tristan until he apologizes for his offenses. He appears and greets her with cool courtesy ("Herr Tristan trete nah"). When she tells him she wants satisfaction for Morold's death, Tristan offers her his sword, but she will not kill him. Instead, Isolde suggests that they make peace with a drink of friendship. He understands that she means to poison them both, but still drinks, and she does the same. Expecting death, they exchange a long look of love, then fall into each other's arms. Brangäne admits that she has in fact mixed a love potion, as sailors' voices announce the ship's arrival in Cornwall.

ACT II

In a garden outside Marke's castle, distant horns signal the king's departure on a hunting party. Isolde waits impatiently for a rendezvous with Tristan. Horrified, Brangäne warns her about spies, particularly Melot, a jealous knight whom she has noticed watching Tristan. Isolde replies that Melot is Tristan's friend and sends Brangäne off to stand watch. When Tristan appears, she welcomes him passionately. They praise the darkness that shuts out all false appearances and agree that they feel secure in the night's embrace ("O sink hernieder, Nacht der Liebe"). Brangäne's distant voice warns that it will be daylight soon ("Einsam wachend in der Nacht"), but the lovers are oblivious to any danger and compare the night to death, which will ultimately unite them. Kurwenal rushes in with a warning: the king and his followers have returned, led by Melot, who denounces the lovers. Moved and disturbed, Marke declares that it was Tristan himself who urged him to marry and chose the bride. He does not understand how someone so dear to him could dishonor him in such a way ("Tatest Du's wirklich?"). Tristan cannot answer. He asks Isolde if she will follow him into the realm of death. When she accepts, Melot attacks Tristan, who falls wounded into Kurwenal's arms.

ACT III

Tristan lies mortally ill outside Kareol, his castle in Brittany, where he is tended by Kurwenal. A shepherd inquires about his master, and Kurwenal explains that only Isolde, with her magic arts, could save him. The shepherd agrees to play a cheerful tune on his pipe as soon as he sees a ship approaching. Hallucinating, Tristan imagines the realm of night where he will return with Isolde. He thanks Kurwenal for his devotion, then envisions Isolde's ship approaching, but the Shepherd's mournful tune signals that the sea is still empty. Tristan recalls the melody, which he heard as a child. It reminds him of the duel with Morold, and he wishes Isolde's medicine had killed him then instead of making him suffer now. The shepherd's tune finally turns cheerful. Tristan gets up from his sickbed in growing agitation and tears off his bandages, letting his wounds bleed. Isolde rushes in, and he falls, dying, in her arms. When the shepherd announces the arrival of another ship, Kurwenal assumes it carries Marke and Melot, and barricades the gate. Brangäne's voice is heard from outside, trying to calm Kurwenal, but he will not listen and stabs Melot before he is killed himself by the king's soldiers. Marke is overwhelmed with grief at the sight of the dead Tristan, while Brangäne explains to Isolde that the king has come to pardon the lovers. Isolde, transfigured, does not hear her, and with a vision of Tristan beckoning her to the world beyond ("Mild und leise"), she sinks dying upon his body.

44 • THE METROPOLITAN OPERA HD LIVE

CLOSE-UP

SCALING THE HEIGHTS

Deborah Voigt and **Ben Heppner** on how they'll ascend opera's Mount Everest—the title roles of *Tristan und Isolde*—with a little help from Maestro James Levine.

Debbie, you've only sung Isolde on stage once before, several years ago. Why the long interval?

Deborah Voigt: I first sang the part in Vienna five years ago. It came along sooner than I anticipated, but the circumstances were right and I decided to go ahead and sing it. When you sing a role as difficult as Isolde, people are going to want you to sing it a lot, and I didn't want to have a lot of them booked if it didn't go well. So I didn't book anything until the performances were over. The first opportunity I had after Vienna are the Met performances.

Ben, what makes you keep coming back to Tristan?

Ben Heppner: Before it starts, it feels like I'm about to climb Mount Everest. But from the moment I step on the stage to the last note I sing it feels like only 15 minutes have gone by. There is something so engaging about this role that you don't notice anything else. It takes all of your mental, vocal, and emotional resources to sing. And I like the challenge of it.

The two of you appear together often, and you've also both worked a lot with James Levine.

DV: Maestro Levine is so in tune with singers—how we breathe and how we work emotionally. I remember I was having trouble with a particular low note, and in one performance, he just lifted up his hands at that moment, looked at me and took a breath, and gave me my entrance. The note just landed and hasn't been a problem since.
BH: He has this wonderful musicality that is so easy to work with. As for Debbie, we just love singing together and I think that is really its own reward.

This *Tristan* will be seen by hundreds of thousands of people around the globe. How does that impact your stage performance?

DV: None of us go out to sing a performance thinking that it is any less significant than another, so my performance will be the same. But when you are playing to a huge opera house, gestures tend to be bigger. For HD, some of the operatic histrionics might go by the wayside.
BH: When the opera house is filled with expectant listeners—that becomes my focus. The only thing I worry about is that it's a very strenuous role, and I'm basically soaking wet from the middle of the second act on! ■

45 • TRISTAN UND ISOLDE

21. Einfach und gut

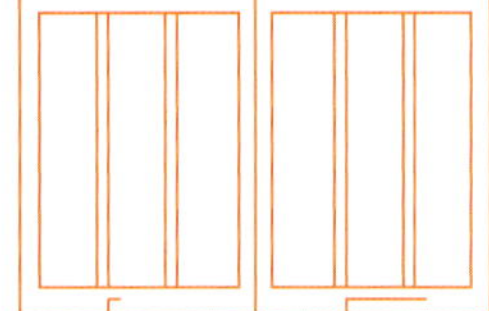

Erfolgreiches Design ist schlicht und vielseitig einsetzbar. Offenes, reduziertes Design, besonders in Büchern oder Katalogen, kann eine ganze Menge an Material aufnehmen. Bei Projekten mit Text und Bildern empfiehlt sich, zuerst das Verhältnis der Elemente zueinander festzustellen und danach erst festzulegen, wie viel Platz jeweils belegt werden soll. Falls die Bildzeilen besonders lang sind und viele zusätzliche Informationen und Namensnennungen beinhalten, können sie durch kleineren Schriftgrad oder Leerräume zwischen den Elementen deutlich vom Haupttext abgesetzt werden.

Ein dreispaltiger Raster, der wie ein ein- oder zweispaltiger wirkt, ist eine strukturelle Lösung. Verwenden Sie zwei Spalten rechts samt dem Raum zwischen den Spalten für Text. So erhalten Sie eine klare Struktur für den Fließtext und einen großzügigen linken Rand für die Bildzeilen.

Wenn es vom Material vorgegeben ist, können auch zwei Spalten Bildunterschriften eine Textspalte ersetzen, damit Bildzeilen und Bilder gut lesbar auf einer Seite Platz finden. Dreispaltige Raster erlauben es, die Bilder über ein, zwei oder drei Spalten oder den gesamten Satzspiegel zu ziehen.

PROJEKT
Beatific Soul

KUNDE
New York Public Library/
Scala Publishers

DESIGN
Katy Homans

In diesem Katalog zu einer Ausstellung über Leben, Karriere, schriftstellerische Arbeit, Veröffentlichungen und Manuskripte von Jack Kerouac wird sein großer Erfolg *On the Road* („Unterwegs") dargestellt. Der dreispaltige Raster erlaubt viele Variationen und hohe Flexibilität. Das Ergebnis ist eine Seite, die aufgeräumt und ruhig wirkt und durch gelungene Einfachheit überzeugt.

Dieser einfache, aber vielseitige Mehrspaltenraster nimmt alle Informationen problemlos auf. Die großzügige Serifenschrift im Fließtext garantiert gute Lesbarkeit. Die Bildzeilen links sind serifenlos gesetzt und heben sich deutlich ab. Diese Seitenstruktur ermöglicht problemlos Variationen im Text.

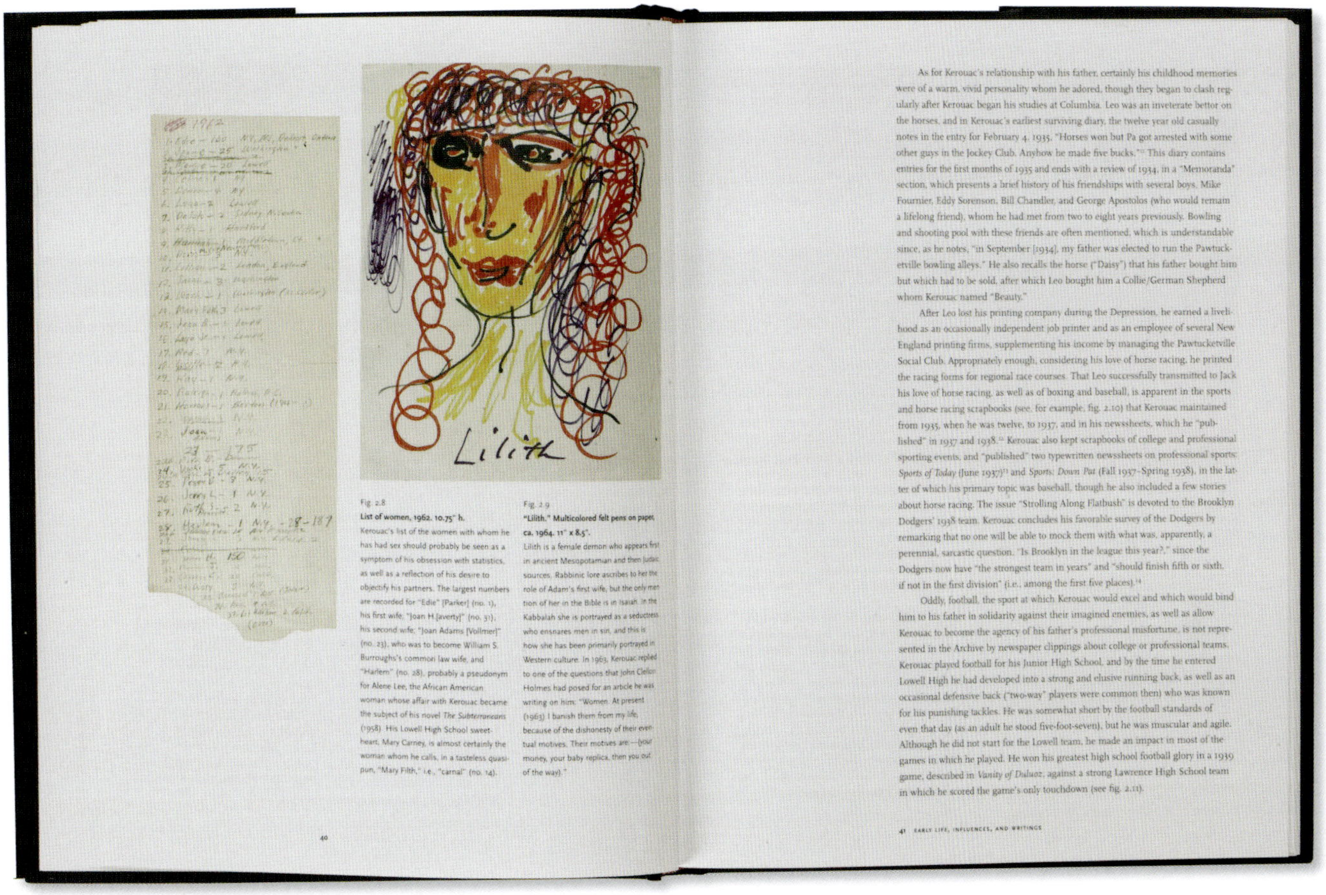

Hier bieten drei Spalten einen stabilen Rahmen für kleinformatige Kunstabbildungen und verschiedene Bildzeilen. Auf der linken Hälfte der Doppelseite ersetzen die Bildzeilen den Fließtext und in der linken Spalte findet sich eine schmale Abbildung. Die rechte Doppelseite ist ausschließlich für Fließtext reserviert.

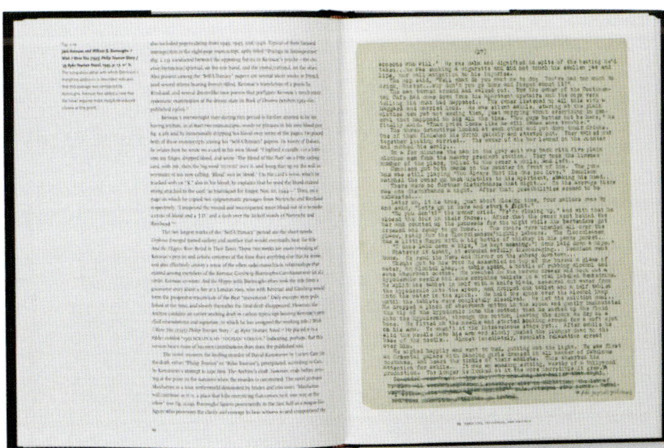

Um dem Projekt Klarheit zu geben, erhalten große Abbildungen gelegentlich eine ganze Seite allein. Hier steht ein handgeschriebenes Manuskript von Jack Kerouac einer ruhigen Textspalte auf der linken Seite gegenüber.

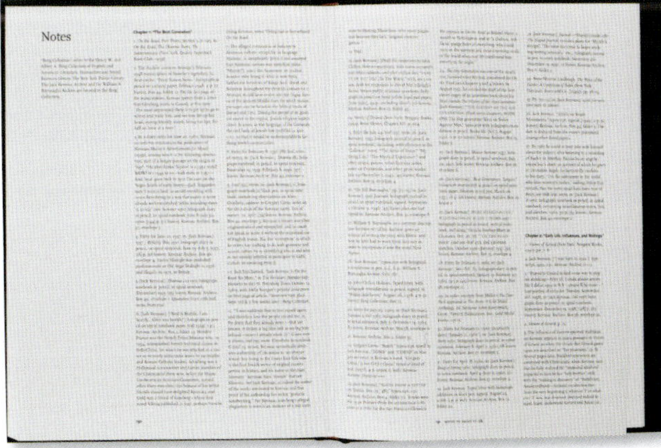

Für Referenzen wie Anmerkungen und Inhaltsverzeichnis eignet sich ein Raster mit drei Spalten besonders gut.

22. Spalten typografisch definieren

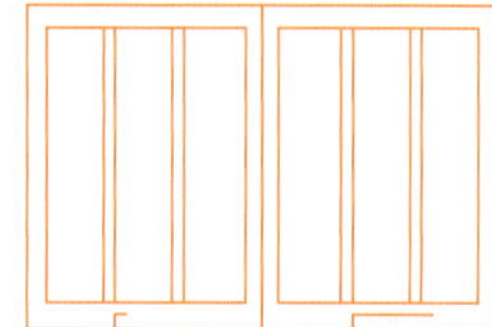

Mit typografischen Mitteln lassen sich die Spalten sehr gut voneinander abheben. Verschiedene Schriftstärken und -grade erlauben eine gute Anordnung der Informationen. Die Hierarchien können entweder horizontal (Überschrift, Beschreibung, Text) oder vertikal (in Spalten von links nach rechts) ausgeprägt sein. Mit unterschiedlichen Schriften, etwa serifenlos und mit Serifen, lässt sich ein Kontrast zwischen Informationseinheiten und Fließtext oder Anleitungen herstellen. Fette Überschriften oder Anleitungsnummerierungen erregen Aufmerksamkeit und bringen Würze in das Layout. Magere Buchstaben, vielleicht auch in einem anderen Font, eignen sich gut für Leitsätze und untergeordnete Texteinheiten. Der klar aufgeteilte Gestaltungsraum verleiht den typografischen Variationen Struktur.

Soft and Chewy Chocolate Chip Cookies

A sugar- and butter-rich batter is the foundation for these cookie-jar classics. Just baked, they make a perfect snack on a chilly winter night—or any time. MAKES ABOUT 3 DOZEN

2¼ cups all-purpose flour
½ teaspoon baking soda
1 cup (2 sticks) unsalted butter, room temperature
½ cup granulated sugar
1 cup packed light brown sugar
1 teaspoon coarse salt
2 teaspoons pure vanilla extract
2 large eggs
2 cups semisweet or milk chocolate chips, or a combination (about 12 ounces)

1. Preheat oven to 350°F. Whisk together flour and baking soda in a bowl. Put butter and sugars in the bowl of an electric mixer fitted with the paddle attachment. Mix on medium speed until pale and fluffy, about 2 minutes. Reduce speed to low. Add salt, vanilla, and eggs; mix until well blended, about 1 minute. Mix in flour mixture. Stir in chocolate chips.

2. Drop heaping tablespoons of dough onto baking sheets lined with parchment paper, spacing 2 inches apart. Bake cookies, rotating sheets halfway through, until edges turn golden but centers are still soft, 10 to 12 minutes. Let cool on sheets on wire racks 2 minutes. Transfer cookies to wire racks; let cool completely. Cookies can be stored between layers of parchment in airtight containers at room temperature up to 1 week.

58 • soft and chewy

Peanut Butter and Jelly Bars

This version of a well-loved combination from childhood concentrates the flavors into a sweet dessert that appeals to all ages. We like strawberry jam, but feel free to substitute any flavor you prefer. MAKES ABOUT 3 DOZEN

1 cup (2 sticks) unsalted butter, room temperature, plus more for pan
3 cups all-purpose flour, plus more for pan
1½ cups sugar
2 large eggs
2½ cups smooth peanut butter
1½ teaspoons salt
1 teaspoon baking powder
1 teaspoon pure vanilla extract
1½ cups strawberry jam, or other flavor
1 cup salted peanuts (5 ounces), roughly chopped

1. Preheat oven to 350°F. Butter a 9 by 13-inch baking pan, and line the bottom with parchment paper. Butter the parchment, dust with flour, and tap out excess.

2. Place butter and sugar in the bowl of an electric mixer fitted with the paddle attachment. Beat on medium speed until fluffy, about 2 minutes. With mixer running, add eggs and peanut butter; beat until combined, about 2 minutes. Whisk together flour, salt, and baking powder. Add to the butter mixture, and beat on low speed until combined. Add vanilla.

3. Transfer two-thirds of mixture to prepared pan; spread evenly with an offset spatula. Using offset spatula, spread jam on top of peanut-butter mixture. Crumble remaining third of peanut butter mixture on top of jam. Sprinkle evenly with peanuts.

4. Bake until golden, 45 to 60 minutes, rotating halfway through. Tent loosely with foil if bars are getting too dark. Transfer to a wire rack to cool. Run knife around edges and refrigerate, 1 to 2 hours. Cut into about thirty-six bars (about 1½ by 2 inches). Cookies can be stored in airtight containers at room temperature up to 3 days.

59

Die Zutaten werden in einer serifenlosen Schrift dargestellt, die Anleitungen in einer Serifenschrift. Eine fett gedruckte Variante der serifenlosen Schrift setzt Akzente.

PROJEKT
Martha Stewart's Cookies

KUNDE
MSL Clarkson Potter

DESIGN
Barbara deWilde

Anspruchsvolle Fotografien und erlesene Typografie spiegeln die Eleganz und den feinen Geschmack einer kulinarischen Expertin wider.

Coconut-Cream Cheese Pinwheels

Rich cream cheese dough, coconut-cream cheese filling, and a topper of jam make these pinwheels complex—chewy on the outside, creamy in the center. Create a variety of flavors by substituting different fruit jams for the strawberry. MAKES ABOUT 2½ DOZEN

for the dough:

- 2 cups all-purpose flour, plus more for work surface
- ⅔ cup sugar
- ½ teaspoon baking powder
- ½ cup (1 stick) unsalted butter, room temperature
- 3 ounces cream cheese, room temperature
- 1 large egg
- 1 teaspoon pure vanilla extract

for the filling:

- 3 ounces cream cheese, room temperature
- 3 tablespoons granulated sugar
- 1 cup unsweetened shredded coconut
- ¼ cup white chocolate chips

for the glaze:

- 1 large egg, lightly beaten
- Fine sanding sugar, for sprinkling
- ⅓ cup strawberry jam

1. Make dough: Whisk together flour, sugar, and baking powder in a bowl. Put butter and cream cheese into the bowl of an electric mixer fitted with the paddle attachment; mix on medium-high speed until fluffy, about 2 minutes. Mix in egg and vanilla. Reduce speed to low. Add flour mixture, and mix until just combined. Divide dough in half, and pat into disks. Wrap each piece in plastic, and refrigerate until dough is firm, 1 to 2 hours.

2. Preheat oven to 350°F. Line baking sheets with nonstick baking mats (such as Silpats).

3. Make filling: Put cream cheese and sugar into the bowl of an electric mixer fitted with the paddle attachment; mix on medium speed until fluffy. Fold in coconut and chocolate chips.

4. Remove one disk of dough from refrigerator. Roll about ⅛ inch thick on a lightly floured surface. With a fluted cookie cutter, cut into fifteen 2½-inch squares. Transfer to prepared baking sheets, spacing about 1½ inches apart. Refrigerate 15 minutes. Repeat with remaining dough.

5. Place 1 teaspoon filling in center of each square. Using a fluted pastry wheel, cut 1-inch slits diagonally from each corner toward the filling. Fold every other tip over to cover filling, forming a pinwheel. Press lightly to seal. Use the tip of your finger to make a well in the top.

6. Make glaze: Using a pastry brush, lightly brush tops of pinwheels with beaten egg. Sprinkle with sanding sugar. Bake 6 minutes. Remove and use the lightly floured handle of a wooden spoon to make the well a little deeper. Fill each well with about ½ teaspoon jam. Return to oven, and bake, rotating sheets halfway through, until edges are golden and cookies are slightly puffed, about 6 minutes more. Transfer sheets to wire racks; let cool 5 minutes. Transfer cookies to rack; let cool completely. Cookies can be stored in single layers in airtight containers at room temperature up to 3 days.

soft *and* chewy • 61

Die Elemente sind spielerisch und witzig angeordnet. Die Verwendung verschiedener Schriften wirkt belebend und damit unterhaltend und lehrreich zugleich.

23. Überfrachtung reduzieren

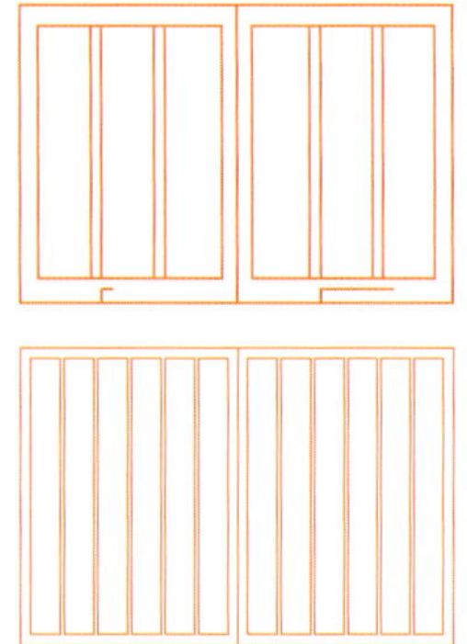

Beim Arbeiten mit einem Dreispalter ist es nicht unbedingt notwendig, jeden Winkel auszufüllen. Fläche weiß lassen ist hier die Devise. Weißfläche lädt den Leser dazu ein, mit den Augen auf dem Papier umherzuwandern und sich einzelne Storys, Bilder oder Logos herauszupicken. Linien unterschiedlicher Stärke wirken regulierend und betonen die einzelnen Informationseinheiten.

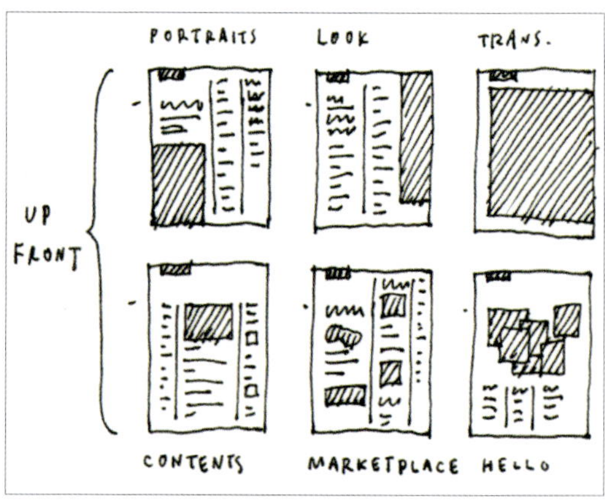

Vorbereitende Skizzen geben ein Gefühl für die Flächenaufteilung.

PROJEKT
Good magazine issue 008

KUNDE
GOOD MAGAZINE, LLC
Design Direction

Scott Stowell

DESIGN
Open

Viel Weißfläche und ein witziges, kantiges Design navigieren den Leser problemlos durch eine lebendige Ansammlung radikaler Ideen, die die Welt besser machen sollen.

CONTENTS

GOOD
Issue 008
Jan/Feb 2008

ON THE COVER It's our Big Ideas spectacular, starring Danny DeVito. Photo by Jeff Minton.

◄ GRAPHIC STATEMENT Each issue, GOOD asks an artist or group to set the tone for the magazine with a visual interpretation of the issue theme. Ilkka Halso's contribution is on pages 6 to 9. Learn more about him on page 16 and see more on page 82.

This Issue: Big Ideas!

64 Big Ideas!
Shooting for the Moon.

66 Big Ideas A to Z
GOOD presents an enormous, alphabetical list of people and ideas that will shape your future.

72 God, Without the Fuss
The head of the nation's largest church preaches a self-help gospel to millions of people. THOMAS GOLIANOPOULOS *finds out what's behind pastor Joel Osteen's smile.*

86 Buy Now, Pay Later
Can a new trend right our environmental wrongs? ADAM M. BRIGHT *scrutinizes the practice of carbon offsetting.*

92 Russian Gambit
Garry Kasparov tells CHRISTOPHER BATEMAN *about his stratagem to take over in the Kremlin.*

98 Most Likely to Secede
"Freedom and Unity" could take on new meaning if a faction of Vermonters has its way. CHRISTOPHER KETCHAM *dissects the possibility of modern secession.*

110 Jonathan and the Whale
In a remote Alaskan town, JONATHAN HARRIS *chronicles an Inupiat whale hunt in 3,214 photos.*

& Big Thinkers
Mario Batali, Laurence Lessig, Nicholas Negroponte, Samantha Power, Cameron Sinclair, and more...

CATEGORIES The symbols in the upper-right-hand corner of the page represent GOOD's content areas. Learn them. Know them. Love them. They'll be guiding you throughout the issue.

This is the INFOBAR. Check here for fun facts.

11
goodmagazine.com
Contents

Die Gestaltung von Inhaltsverzeichnissen erweist sich oft als schwierig. Hier hat das Chaos keine Chance und der Leser kann sich unkompliziert orientieren. Verschiedene Schriftgrade und -stärken machen die Seite interessant und harmonisch. Die Symbole oben rechts legen das im Magazin verwendete Format fest.

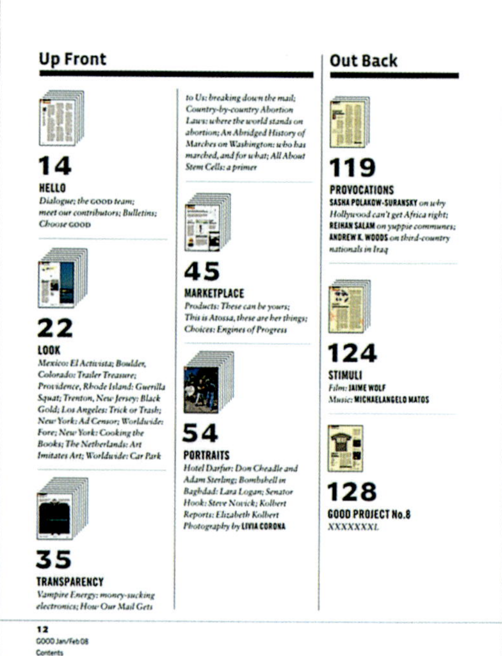

Up Front

14
HELLO
Dialogue; the GOOD team; meet our contributors; Bulletins; Choose GOOD

22
LOOK
Mexico: El Activista; Boulder, Colorado: Trailer Treasure; Providence, Rhode Island: Guerilla Squat; Trenton, New Jersey: Black Gold; Los Angeles: Trick or Trash; New York: Ad Censor; Worldwide: Fore; New York: Cooking the Books; The Netherlands: Art Imitates Art; Worldwide: Car Park

35
TRANSPARENCY
Vampire Energy: money-sucking electronics; How Our Mail Gets to Us: breaking down the mail; Country-by-country Abortion Laws: where the world stands on abortion; An Abridged History of Marches on Washington: who has marched, and for what; All About Stem Cells: a primer

45
MARKETPLACE
Products: These can be yours; This is Atossa, these are her things; Choices: Engines of Progress

54
PORTRAITS
Hotel Darfur: Don Cheadle and Adam Sterling; Bombshell in Baghdad: Lara Logan; Senator Hook: Steve Novick; Kolbert Reports: Elizabeth Kolbert
Photography by LIVIA CORONA

Out Back

119
PROVOCATIONS
SASHA POLAKOW-SURANSKY *on why Hollywood can't get Africa right;* REIHAN SALAM *on yuppie communes;* ANDREW K. WOODS *on third-country nationals in Iraq*

124
STIMULI
Film: JAIME WOLF
Music: MICHAELANGELO MATOS

128
GOOD PROJECT No.8
XXXXXXXL

12
GOOD Jan/Feb 08
Contents

Auf dieser Seite sind fünf Informationsebenen enthalten, die wegen ihrer ordentlichen Typografie im großzügig bemessenen Raum einfach und leicht zu lesen sind.

Mit Trennlinien und pfiffig strukturierter Typografie lassen sich eine ganze Menge Informationseinheiten nebeneinander unterbringen.

HELLO

: dialogue

Now you can find us in:

GUT

BOM

BUENO

GOOD

Ausdrucksstarke typografische Kaskaden mit viel Weißfläche können es gut mit einer ebenso ausdrucksstarken Illustration auf der gegenüberliegenden Seite aufnehmen.

BIG IDEAS!

QUANTUM HIPPIES

Q

Quantum mechanics is all about the relationship between matter and energy. So it's not hard to imagine why the science has been co-opted by a subculture of bong-toting academe—expand your mind, anyone? Scholars (cough) like Daniel Pinchbeck have struck mono-atomic gold with a prophetic philosophy that marries quantum theory and Ayahuasca-induced hallucination, all in an effort to come to grips with what rational materialism neglects: the inexplicable nature of being.

$\Delta x \Delta p \geq \frac{\hbar}{2}$ = whoa.

BIG THINKER:

WENDY KOPP

RUSSIAN DEMOCRACY

R

Russian Gambit

Garry Kasparov, the Russian presidential candidate and former chess grandmaster, is trying to keep Vladimir Putin in check.

interview by

CHRISTOPHER BATEMAN

illustrations by

DARREN BOOTH

Große Gedanken – große Buchstaben! Mit Initialen lassen sich Anfänge spielerisch gestalten. Hier beziehen sie sich auf den Anfangsbuchstaben der Überschriften. Grafische Symbole erscheinen je nach Bedarf an fester Position wie im Inhaltsverzeichnis.

24. Spalten absenken

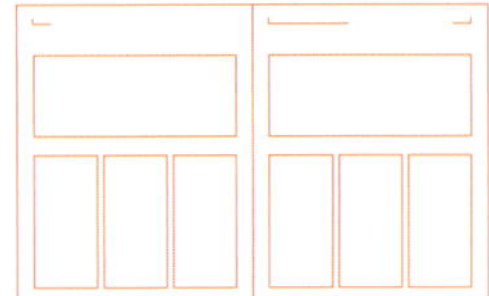

Bei einem Dreispalter und ausschließlich Text besteht die Gefahr, dass die Seiten zu dicht werden. Eine gute Möglichkeit, die Aufmerksamkeit des Lesers zu wecken und ihn nicht einzuschüchtern, ist eine Senkung, um damit klare Doppelseitenansichten und Bewegung zu schaffen. Senkungen schaffen auch Platz für Überschriften, Seitenzahlen, Kolumnentitel, Leitsätze und Fotos.

8

Pew Environment Group

Halloween 1948 was all trick and no treat in Donora, Pennsylvania. In the last week of October, this town of 14,000 in the western part of the state underwent a weather event called a "temperature inversion," trapping at ground level the smog from local metal factories.

Nearly half of Donora's residents experienced breathing problems, hundreds suffered permanent heart and lung damage, and some 50 deaths were attributed to the disaster. Sixty years ago, public policy gave Americans relatively little protection from industrial accidents. However, Donora and similar disasters helped focus national attention on the government's responsibility to protect the population from environmental hazards. Eventually, the Donora catastrophe led to the Air Pollution Control Act of 1955, the United States' first piece of federal legislation on this issue and an early step in what has become an ongoing effort to save the environment, for the sake of the natural world as well as public health.

A related development in 1948 produced no fatalities but was a harbinger of a situation that was ultimately even more serious. As energy demand and prices soared in the postwar boom and Western companies discovered vast oil fields in the Middle East, the United States for the first time became a net importer of oil.

Sixty years later, this country—indeed, the world—faces unprecedented environmental challenges. Changes to terrestrial and marine environments resulting from climate change, overfishing, agriculture, grazing and logging are already transforming the planet in ways that impair its ability to be hospitable to life—both ours and that of the countless other species that occupy it with us. The Pew Environment Group is focused on reducing the scope and severity of three major global environmental problems:

- Dramatic changes to the Earth's climate caused by the increasing concentration of greenhouse gases in the atmosphere;
- The erosion of large wilderness ecosystems that contain a great part of the world's remaining biodiversity;
- The destruction of the world's oceans, with a particular emphasis on global marine fisheries.

Climate change. To reduce the threat of climate change, we are urging the adoption of a mandatory national policy to reduce greenhouse gas emissions. While its centerpiece is a market-based cap and trade system, complementary measures are needed to create additional incentives to invest in less polluting technologies in key sectors, particularly transportation.

Early in 2007, we launched the Pew Campaign for Fuel Efficiency to promote legislation to increase fuel-efficiency standards for passenger vehicles to 35 miles per gallon by 2020. Nationwide, vehicles account for two-thirds of oil consumption and one-third of greenhouse gas emissions, with light-duty passenger vehicles—cars, pick-ups, minivans and SUVs—producing about 60 percent of transportation-related emissions. Globally, U.S. transportation accounts for about 8 percent of all greenhouse

Pew Prospectus 2008 IMPROVING PUBLIC POLICY
Pew Environment Group 9

Seeking protection from the smog in Donora, Pa., in 1948

Car emissions, a leading contributor to climate change

Sharks, sought for their fins, among the sea's endangered creatures

gas pollution and 17 percent of an increasingly tight and volatile world oil market. Higher standards would reduce our country's dependence on foreign oil, enhance national security, save consumers money and reduce global warming pollution.

Wilderness protection. Due to the spread of human civilization, habitat destruction and, increasingly, climate change, scientists estimate that we may be losing as many as 30,000 species each year. To slow or stop this loss, many conservation biologists say, we need to create new parks, wildlife refuges and protected areas where extractive activity and development are prohibited. Pew has played a critical role in the permanent protection of more than 200 million acres of wilderness in the United States and Canada since 1990. More recently, we have launched a joint initiative with The Nature Conservancy to establish new national parks and indigenous protected areas in Australia. Together, these three countries contain more than 30 percent of the world's remaining old growth forests and an even larger share of pristine wilderness areas.

Ocean conservation. Overfishing, chemical and nutrient pollution, habitat alteration, introduction of exotic species and climate change are taking what may be an irreversible toll on the world's marine environment. The Pew Environment Group has helped lead the way in bringing about many of the major improvements in fisheries management and marine conservation in the United States since the mid-1990s. In recent years, we have expanded our oceans work internationally and are working in various other regions of the world to curtail overfishing, protect critical marine habitat and reduce the amount of unintended bycatch—the fish, seabirds, sharks, whales and other species that are routinely thrown back into the sea, either dead or dying.

Pew today is in a stronger position to address all of these problems as the result of the merger of our Environment Program with the National Environmental Trust. The consolidated team has a domestic and international staff of more than 100, making us one of the nation's largest environmental scientific and advocacy organizations with a presence across not only the United States but also Australia, Canada, Europe, the Indian Ocean, Latin America and the Western Pacific.

Society has historically invested little time, thought or effort in protecting the environment for posterity. Sixty years ago, once the smog in Donora had cleared, most people simply assumed that things would return to the way they had been. We can no longer afford to make that mistake.

Joshua S. Reichert

Managing Director
Pew Environment Group

Variationen sind das Salz des Designers. In diesem Sinne entsteht der Kontrast zwischen dem breit angelegten Einführungstext und den schmäleren Textspalten. Noch interessanter wird es, wenn für die Überschriften eine optisch aus dem Rahmen fallende Schrift gewählt ist.

PROJEKT
Pew Prospectus 2008

KUNDE
The Pew Charitable Trusts

DESIGN
IridiumGroup

EDITOR
Marshall A. Ledger

ASSOCIATE EDITOR/
PROJECT MANAGER
Sandra Salmans

Die Arbeit einer gemeinnützigen Organisation erhält hier einen ernsthaften, aber nicht minder eleganten Rahmen.

Culture

Change was sweeping the arts scene in 1948, with an impact that would not be fully realized for years. American painters led the way into abstract expressionism, reshaping both the visual arts and this country's influence on the art world.

Meanwhile, technology was setting the stage for revolutions in music and photography. The LP record made its debut, and the Fender electric guitar, which would define the rock 'n roll sound in the next decade and thereafter, went into mass production. Both the Polaroid Land camera, the world's first successful instant camera, and the first Nikon went on sale.

In New York, the not-for-profit Experimental Theatre, Inc., received a special Tony honoring its path-breaking work with artists such as Lee Strasberg and Bertolt Brecht. But in April it was disclosed that the theatre had run up a deficit of $20,000—a shocking amount, given that $5,000 had been the maximum allocated for each play—and in October *The New York Times* headlined, "ET Shelves Plans for Coming Year."

Apart from its miniscule budget, there is nothing dated about the travails of the Experimental Theatre. The arts still struggle with cost containment and tight funds. But if the Experimental Theatre were to open its doors today, it might benefit from the power of knowledge now available to many nonprofit arts organizations in Pennsylvania, Maryland and California—and, eventually, to those in other states as well. Technology, which would transform music and photography through inventions in 1948, is providing an important tool to groups that are seeking to streamline a grant application process that, in the past, has been all too onerous.

That tool is the Cultural Data Project, a Web-based data collection system that aggregates information about revenues, employment, volunteers, attendance, fund-raising and other areas input by cultural organizations. On a larger scale, the system also provides a picture of the assets, impact and needs of the cultural sector in a region.

The project was originally launched in Pennsylvania in 2004, the brainchild of a unique collaboration among public and private funders, including the Greater Philadelphia Cultural Alliance, the Greater Pittsburgh Arts Council, The Heinz Endowments, the Pennsylvania Council on the Arts, Pew, The Pittsburgh Foundation and the William Penn Foundation. Until then, applicants to these funding organizations had been required to provide similar information in different formats and on multiple occasions. Thanks to the Pennsylvania Cultural Data Project, hundreds of nonprofit arts and cultural organizations throughout the state can today update their information just once a year and, with the click of a computer mouse, submit it as part of their grant applications. Other foundations, such as the Philadelphia Cultural Fund, the Pennsylvania Historical and Museum Commission and the Independence Foundation, have also adopted the system.

Long-playing records, enthralling the public in 1948

The Village of Arts and Humanities, revitalizing North Philadelphia

Development workshop for Bill Irwin's *The Happiness Lecture*

So successful has the project been that numerous states are clamoring to adopt it. In June, with funding from multiple sources, Maryland rolled out its own in-state Cultural Data Project. The California Cultural Data Project, more than five times the size of Pennsylvania's with potentially 5,000 nonprofit cultural organizations, went online at the start of 2008, thanks to the support of more than 20 donors. Both projects are administered by Pew.

As cultural organizations in other states enter their own data, the research will become exponentially more valuable. Communities will be able to compare the effects of different approaches to supporting the arts from state to state and city to city. And the data will give cultural leaders the ability to make a fact-based case that a lively arts scene enriches a community economically as well as socially.

The Cultural Data Project is not the first initiative funded by Pew's Culture portfolio to go national or to benefit from state-of-the-art technology. For example, the system used by Philly-FunGuide, the first comprehensive, up-to-date Web calendar of the region's arts and culture events, has been successfully licensed to other cities.

In addition to the Cultural Data Project, another core effort within Pew's Culture portfolio is the Philadelphia Center for Arts and Heritage and its programs, which include Dance Advance, the Heritage Philadelphia Program, the Pew Fellowships in the Arts, the Philadelphia Exhibitions Initiative, the Philadelphia Music Project and the Philadelphia Theatre Initiative. Since the inception of the first program in 1989, these six initiatives have supported a combined total of more than 1,100 projects and provided more than $48 million in funding for the Philadelphia region's arts and heritage institutions and artists.

Through its fellowships, Pew nurtures individual artists working in a variety of performing, visual and literary disciplines, enabling them to explore new creative frontiers that the marketplace is not likely to support. The center also houses the Philadelphia Cultural Management Initiative, which helps cultural groups strengthen their organizational and financial management practices.

Almost from the time it was established, Pew was among the region's largest supporters of arts and culture. While it continues in this role, committed to fostering nonprofit groups' artistic excellence and economic stability, and to expanding public participation, Pew—like the arts themselves—has changed its approach with the times.

Marian A. Godfrey

Managing Director
Culture and Civic Initiatives

2007 Milestones

Each year, we join with excellent organizations to produce work that exemplifies exactly what we mean in stating that Pew serves the public interest. On these pages, we highlight the results of some of the Pew-supported work that made a difference in 2007.

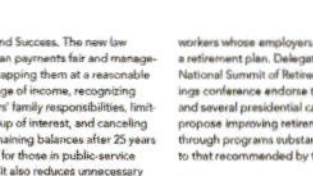

Environment

Health and Human Services

Pew Center on the States

25. Bildausschnitte variieren

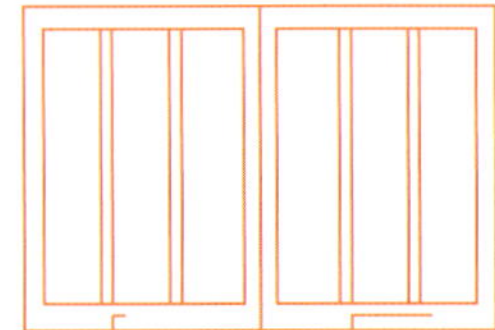

Ein Wechsel des Bildausschnitts bei Illustrationen oder Fotos für schrittweise Anleitungen macht das Layout lebendig. Wenn alles gleich groß ist, schafft das Übersicht, wirkt aber langweilig. Besser und durchaus möglich ist es hingegen, die Bilder zu variieren.

Handbook How-Tos

HOW TO
WASH, DRY, AND STORE LETTUCE

1. Fill a clean basin or a large bowl with cold water, and submerge the lettuce leaves completely. (For head lettuce, first discard the outer leaves; they're most likely to harbor bacteria. Chop off the end, and separate the remaining leaves.) Swish the leaves around to loosen dirt.

2. Once sediment has settled, lift out the lettuce, pour out the dirty water, and refill the bowl with clean water. Submerge the lettuce again, and continue swishing and refilling until there are no more traces of dirt or sand in the bowl. You may need to change the water 2 or 3 times.

3. Dry the lettuce in a salad spinner until no more water collects at the bottom of the bowl. Alternatively, blot the leaves between layered paper towels or clean dish towels until no water remains.

4. If you plan to store the lettuce, arrange the dry leaves in a single layer on paper towels or clean dish towels, roll up, and seal inside a plastic bag. Lettuce can be stored this way in the refrigerator for 3 to 5 days. To prevent it from browning rapidly, don't tear the leaves into smaller pieces until you're ready to use them.

SOAK AND SPIN THE LEAVES

STORE IN A TOWEL

1

2

3

4

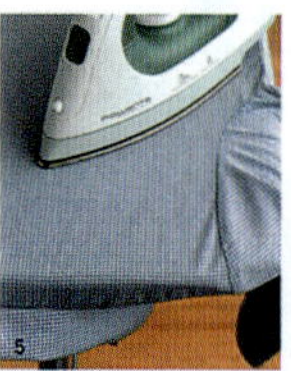

5

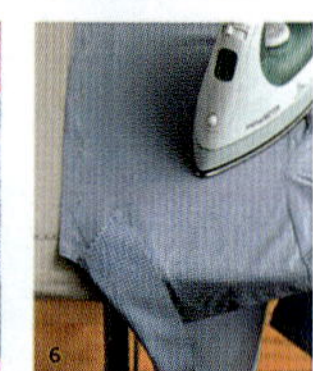

6

HOW TO
IRON A BUTTON-FRONT SHIRT

For easier ironing and the best results, start with a thoroughly damp shirt. Mist the shirt with water using a spray bottle, roll it up, and keep it in a plastic bag for 15 minutes or up to a few hours. (If you can't iron the shirt sooner, refrigerate it in the bag so the shirt won't acquire a sour smell.) Most of the ironing will be on the wide end of the board. If you're right-handed, position the wide end to your left; if you're left-handed, it should be on your right.

1. Begin with the underside of the collar. Iron, gently pulling and stretching the fabric to prevent puckering. Turn the shirt over, and repeat on the other side of collar. Fold the collar along seam. Lightly press.

2. Iron the inside of the cuffs; slip a towel under the buttons to cushion them as you work. Iron the inside of the plackets and the lower inside portion of the sleeves, right above the cuffs. Iron the outside of the cuffs.

3. Drape the upper quarter of the shirt over the wide end of the board, with the collar pointing toward the narrow end of the board, and iron one half of the yoke. Reposition, and iron the other half.

4. Lay 1 sleeve flat on the board. Iron from shoulder to cuff. (If you don't want to crease the sleeve, use a sleeve board.) Turn the sleeve over, and iron the other side. Repeat with the other sleeve.

5. Drape the yoke over the wide end of the board, with the collar facing the wide end, and iron the back of the shirt.

6. Drape the left side of the front of the shirt over the board, with the collar pointing toward the wide end; iron. Repeat with the right front side, ironing around, rather than over, buttons. Let the shirt hang in a well-ventilated area until it's completely cool and dry, about 30 minutes, before hanging it in the closet.

62

Um Text oder Anleitungen anschaulich zu gestalten, bietet es sich an, die einzelnen Schritte und das fertige Produkt mit Bildern oder Fotos zu illustrieren. Solche Bilder sind nützlich und lockern die Seite auf.

PROJEKT
Martha Stewart Living

KUNDE
Martha Stewart Omnimedia

DESIGN
Martha Stewart Living

CHIEF CREATIVE OFFICER
Gael Towey

Klar verständliche Bilder oder Fotos für eine schrittweise Anleitung erhalten ein solides, aber dennoch flexibles Format.

RECHTE SEITE: Die Typografie ist einerseits funktional und detailreich, andererseits exzellent, ohne gekünstelt zu wirken. In den separaten Boxen findet der Leser wichtige zusätzliche Tipps neben den Rezepten.

SAUTÉED SOLE WITH LEMON

SERVES 2

Gray sole is a delicately flavored white fish. You can substitute flounder, turbot, or another type of sole.

- ½ cup flour, preferably Wondra
- 1 teaspoon coarse salt
- ½ teaspoon freshly ground pepper
- 2 gray sole fillets (6 ounces each)
- 2 tablespoons unsalted butter
- 2 tablespoons olive oil
- 2 tablespoons sliced almonds
- 1½ tablespoons chopped fresh parsley
- Finely chopped zest and juice from 1 lemon, plus wedges for garnish

1. Combine flour, salt, and pepper in a shallow bowl. Dredge fish fillets in flour mixture, coating both sides, and shake off excess.

2. Melt butter with oil in a sauté pan over medium-high heat. When butter begins to foam, add fillets. Cook until golden brown, 2 to 3 minutes per side. Transfer each fillet to a serving plate.

3. Add almonds, parsley, zest, and 2 tablespoons juice to pan. Spoon over fillets, and serve with lemon wedges.

HARICOTS VERTS

SERVES 2

- Coarse salt, to taste
- 8 ounces haricots verts
- 2 tablespoons extra-virgin olive oil
- Freshly ground pepper, to taste
- 1 bunch chives, for bundling (optional)

1. Bring a pot of salted water to a boil. Add haricots verts, and cook until bright green and just tender, 3 to 5 minutes. Drain, and pat dry. Transfer to a serving bowl.

2. Toss with oil, salt, and pepper. Tie into bundles using chives.

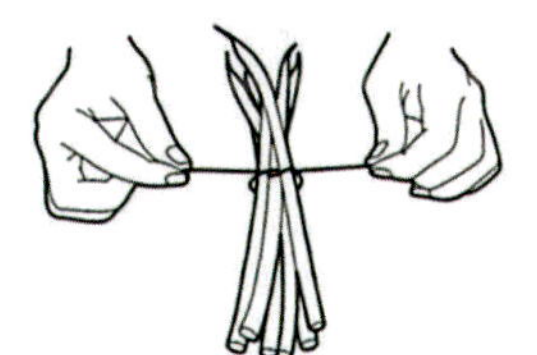

HOW TO BUNDLE GREEN BEANS

1. Cook haricots verts. Drain, and pat dry. Let stand until cool enough to handle.

2. Lay a chive on a work surface. Arrange 4 to 10 haricots verts in a small pile on top of chive. Carefully tie chive around bundle. Trim ends of chive if desired.

QUICK-COOKING CLASSIC Seared sole fillets glisten beneath a last-minute pan sauce made with lemon, parsley, and almonds. The resulting entrée, served with blanched haricots verts, is satisfyingly quick yet sophisticated.

26. Den Rhythmus variieren

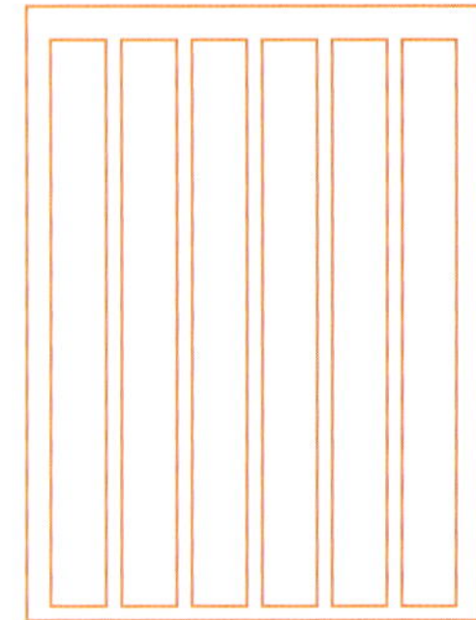

So wichtig ein klar strukturierter Aufbau eines Layouts ist, wenn andererseits die immer gleichen Elemente ohne Abwechslung wiederholt werden, versinkt der Leser dennoch in Langeweile.

Vermeiden Sie also eine sture und monotone Abfolge von Text und Bild. Mit Variationen lassen sich auch bei trockenen Fakten über die Gestaltung witzige Akzente setzen und Langeweile ausschalten.

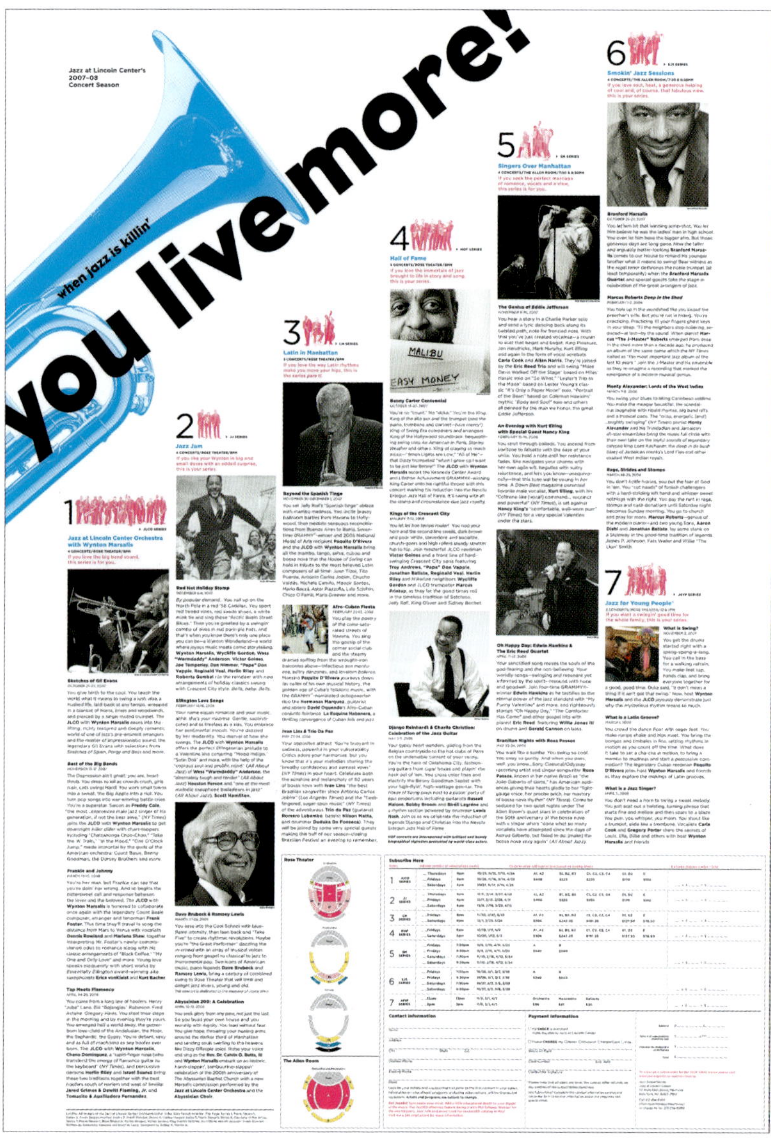

Dieser Raster fasst ein Unmenge an Informationen. Die abgestuften Spalten folgen der Schräge der Trompetengrafik und verleihen so dem bereits inhaltlich attraktiven Veranstaltungsprogramm noch einmal besonderen Pep. Typografisch ist das Programm in technischer, rhythmischer und harmonischer Hinsicht ein Meisterwerk.

PROJEKT
Program schedule

KUNDE
Jazz at Lincoln Center

DESIGN
Bobby C. Martin Jr.

Große Mengen an Information werden hier in einem kecken Layout aufgelockert geboten.

Dieser Spaltenraster schafft klare Rahmenbedingungen für Boxen, die mehrere Funktionen erfüllen: In ihnen sind die Informationseinheiten zu finden, sie signalisieren zugleich den zeitlichen Ablauf der Veranstaltungen und sind darüber hinaus in rhythmischer Vollendung über die Seite verteilt.

27. Die Mischung macht's

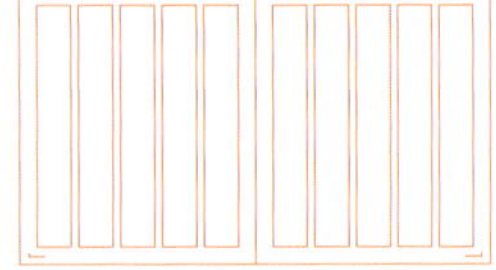

Die Gewichtung der einzelnen Elemente, der Aufbau der Seite, die Größenverhältnisse, die Fläche, die Farben – es ist durchaus möglich, viele Elemente in einem lebendigen Layout zu vereinen, das vielfältig und gleichzeitig dennoch in sich stimmig ist. Dank eines robusten Rasters als Basis kann bei einem Projekt mit ohnehin zahlreichen Bildern und Überschriften noch Platz für ein oder zwei Elemente mehr geschaffen werden. Unterschiedliche Schriftstärken und -grade sowie eine Dynamik bei den Bildgrößen schaffen Aufmerksamkeit ohne die leichte Lesbarkeit des Haupttexts zu stören.

Der mutige, über das gesamte Magazin verlaufende fünfspaltige Raster bietet die solide Grundmauer für eine ganze Vielfalt an Formen und Größen. Der Aufbau der Seite ist stabil, insbesondere durch die großzügige Weißfläche um die Abbildungen.

REGIONAL CRAFTS

A sampling of people and programs supporting the work of traditional artisans

REGIONAL CRAFTS GLASS WASHINGTON

PILCHUCK GLASS SCHOOL

Over three decades, the center has fostered a glass-blowing industry in the Pacific Northwest

A former New York designer returns home to create a unique partnership with gifted local artisans

REGIONAL CRAFTS HOUSEWARES • TEXTILES ALABAMA

ALABAMA CHANIN

PROJEKT
Metropolis magazine

KUNDE
Metropolis magazine

CREATIVE DIRECTOR
Criswell Lappin

In dem klar strukturierten Raster kommen diese traditionell hergestellten Arbeiten gut zur Geltung. Die solide Grundlage des mehrspaltigen Rasters über die gesamte Doppelseite bietet eine robuste Untermauerung für ein Sammelsurium an Elementgrößen, Strichbreiten und Farben.

REGIONAL CRAFTS CERAMICS MEXICO

FAST LEARNERS

The transformation of Mata Ortiz—renowned today for its pottery—is a recent phenomenon.

By MARTIN C. PEDERSEN

Obwohl die Typografie mit Ausnahme einiger Farbakzente schwarz gehalten ist, bringt sie durch den Kontrast der fetten, schablonenartigen und kleineren, zarteren Schrift Schwung und Struktur in das Layout.

RECHTE SEITE: Horizontallinien bilden die Basis für die Schaukelstuhlabbildungen.

REGIONAL CRAFTS HOUSEWARES + ARCHITECTURE NORTH CAROLINA

by BELINDA LANKS

HANDMADE HOME

A crafts group enlists local artisans to create a one-of-a-kind dwelling.

AKIRA SATAKE

CERAMICS

Satake produces functional ceramic pieces—from vases, platters, and bowls to decorative tiles—with a refined Japanese aesthetic.

FATIE ATKINSON

FURNITURE

Employing a steam-bending technique, Atkinson can make this chair out of any open-pored wood, including hickory, ash, and white or red oak (shown).

BARBARA ZARETSKY

TEXTILES

Zaretsky creates earth-toned patterns using natural fibers, plant dyes, and textile paints.

HandMade in America has been fervently promoting craft in Western North Carolina since 1993, but this year marks the nonprofit's first foray into real estate. In a novel collaboration, the group has partnered with private developer Biltmore Farms to construct the HandMade Home, a 3,700-square-foot model in Asheville showcasing the work of 100 local craftspeople. The house, which broke ground last September, is expected to meet the green-building standards of North Carolina's Healthy Built Homes program and fetch $2.25 million when it makes its debut in October as part of the city's annual "Parade of Homes."

Founding executive director Becky Anderson hopes the project will spur other developers, architects, and homeowners to tap the region's greatest resource: the 4,500 resident artisans making everything from furniture and lighting fixtures to tableware and rugs (examples shown above). "We want to become the center of handcrafted homes," she says. To make it easy, HandMade in America has produced directories featuring the work of and contact information for the craftspeople in its network. But Ben Brown, the project's publicist, recommends that people considering such an undertaking think smaller. "This is the first project of its kind, and it will probably be the last," Brown says. "With one hundred independent-minded artists involved, people are ready to shoot each other." ○

Courtesy HandMade in America

PEWABIC

The designs for Brookside School at Cranbrook (top), in Bloomfield Hills, and Detroit's Comerica Park stadium (above) were custom-made by the pottery's in-house team.

MOTAWI

The Frank Lloyd Wright Collection includes Avery (left) and Confetti (below). Also shown: Montrose (above) and Amaryllis (right), an adaptation of a Louis Sullivan design.

DAVID ELLISON

Sold by Country Floors, the Apollo Plaque (above) is a reinterpretation of historic details found on buildings in New York's Flatiron District.

Eastern Michigan is home to one of the most active crafts movements in the country.

REGIONAL CRAFTS TILE MICHIGAN

by EVA HAGBERG

MOTOR CITY GLAZE

"We'd start doing these tile shows that were just tile, and we'd think, How could anyone make a living at this?" says Marcia Hovland, part of a loose-knit group of Michigan-based tile-makers, reminiscing about the good old days before the tile industry took off. "And now everyone is doing really well."

Hovland is one of the artisans who came up through Detroit's famed Pewabic Pottery—a tile factory, exhibition space, and educational facility. She studied there with David Ellison—a name that comes up again and again in conversation with these eastern-Michigan tile fiends—and realized that she could turn her painting and design background into a whole new bag of [ceramic] chips.

Karim Motawi runs Motawi Tileworks out of Ann Arbor with his sister, Nawal. The company makes historically influenced pottery in line with the types of things that were produced in the earliest days of Pewabic in the 1900s. "We're literally plowing through the history books and the source books, the old catalogs," he says. "We're trying to re-create the lost craft." As the official Frank Lloyd Wright licensee, it's reproducing just fine.

Motawi Tileworks operates on a relatively tiny scale—it produces 16,000 square feet of tile a year, a drop in the bucket—and so do many of its local cohorts, which is why they're so happy to know Joseph Taylor, president of the Tile Heritage Foundation, which works to raise the historic craft's profile. "They are like tile cheerleaders," Motawi says. ○

Courtesy the manufacturers

REGIONAL CRAFTS SEATING NEW YORK

BROOKLYN'S OWN

A crafty, DIY-inspired furniture movement emerges in New York's most creatively vibrant borough.

ELUCIDESIGN

DANISH CHAIR

Inspired by the Scandinavian classics, this Chris Jondle-designed piece is made of maple and uses a hand-silkscreened linen for the back and seat.

WÖD

WÖD CHAIR

The dining-room chair, designed by Corey Springer and Eric Ervin in 2006, comes in a variety of woods, including cherry (shown), walnut, and maple.

UHURU DESIGN

OK METAL ARMCHAIR

Designed by Jason Horvath, this lounge chair consists of a one-inch-by-two-inch steel frame and upholstered cushions available in custom colors and patterns.

Courtesy the designers

CITY JOINERY

WEDGE CHAIR

This dining-room chair was designed by Jonah Zuckerman in 2007. Pictured in black walnut, it's available in a variety of woods.

SCRAPILE

PROTOTYPE I

Designed by Bert Bettencourt, the chair is made of repurposed wood scraps that were bound for a landfill. The process makes the materials unique to each piece.

PAUL SAMKO

ROCKING CHAIR

The walnut rocker is composed of 15 different pieces. Created by Samko in 2007, the chair can be customized using different types of wood or upholstery.

by EVA HAGBERG

Far from the maddening crowds of the contemporary-furniture scene, a small group of intrepid designers is sprouting like trees in Brooklyn. Aesthetically, they're all over the map. Scrapile (from Greenpoint) is known for the pun it's named after: a scrap pile of locally sourced wood that designers Bart Bettencourt and Carlos Salgado turn into a building material; each block incorporates everything from walnut to plywood and is then processed through a labor-intensive layering method. Uhuru, founded by Bill Hilgendorf and Jason Horvath, offers a line of sleek, multimaterial pieces, all of which, if viewed through a larger lens, are just as sustainable.

Those firms got started about four years ago, and they join the older guard Elucidesign, founded in 2001, and City Joinery, which set up shop in 1996. Elucidesign's Redpoint collection is a beautifully spare series of pared-down pieces; City Joinery's range and look is broader and heavier.

These firms may not share a look, but they do share a sensibility shaped by their size, scale, and voluntary outsider status in the design world. "We're in this straddling position," City Joinery's Jonah Zuckerman says. "We care a lot about design, but we also care a lot about craft." Horvath brings up a similar tension: "We don't want to be this big furniture company that does production overseas, but we don't want to be just building furniture in Red Hook." He shouldn't worry too much. His company and his compatriots are part of a new phenomenon—the rise of the artisan designer, Brooklyn division. ○

28. Komplexes vereinfachen

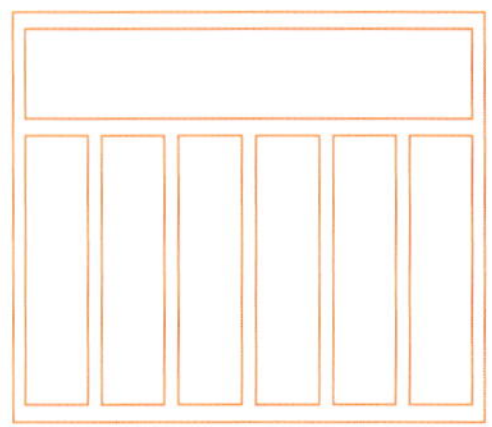

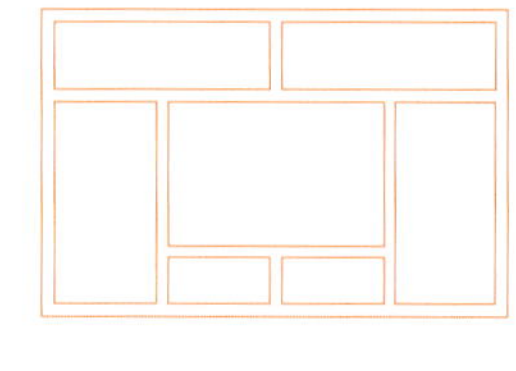

Mehrspaltige Raster sind ideal für übersichtliche Anordnung einer Vielzahl von Elementen eines wissenschaftlichen Berichts. Mit einem genauen Plan lassen sich große Informationsmengen gut in Einheiten aufteilen. Spalten, Linien, Text in verschiedenen Größen, Fonts und Farben lassen sich so kombinieren, dass technische Informationen verständlich werden.

In bzw. zwischen den breiten Horizontallinien sind Überschriften, Autoren, der Universitätsname und das Logo untergebracht. An einigen Stellen wird die Linie durchbrochen, um die einzelnen Spalten zu kennzeichnen.

Figure 2: The ICE probe is placed in the right heart for imaging during PFO closure and pulmonary vein isolation.

The ICE probe can be advanced into the inferior vena cava (IVC), enabling high quality imaging of the abdominal aorta (Figure 3).

Mit verschiedenen Größen und Zeilenabständen werden die allgemeineren Informationen über die Forschungen von den groß gesetzten Forschungsergebnissen abgehoben. In den kontrastreichen serifenlosen Bildzeilen wiederholen sich die Fakten. Die Texteinheiten werden mit Vertikallinien abgetrennt, um die Informationen noch deutlicher darzustellen.

PROJEKT (OBEN)
Poster

KUNDE
NYU Medical Center

DESIGN
Carapellucci Design

DESIGNER
Janice Carapellucci

Dieses Plakat für das NYU Medical Center ist ein Lehrbuchbeispiel für eine klar und eindeutig gestaltete Informationshierarchie. Die Fakten und Forschungsergebnisse sind leicht zu lesen. Die einzelnen Informationseinheiten heben sich gut voneinander ab, der Zeilenabstand und der Abstand zwischen den Elementen ist perfekt und optimal lesbar proportioniert. Trotz der großen Informationsdichte ist jeder Abschnitt, auch für Laien, gut lesbar.

PROJEKT (SEITE GEGENÜBER)
Tischset für SXSW-Workshop

KUNDE
smith + beta

DESIGN
Suzanne Dell'Orto

Lebendige Schritt-für-Schritt-Anleitungen für Macher mit verschiedenen Elementen, Symbolen, Tipps, Hierarchien, einer Checkliste und Wagemut.

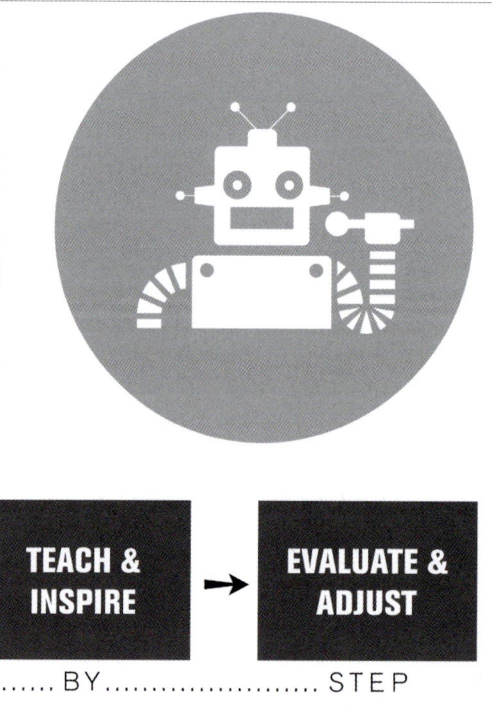

Der oben von links nach rechts laufende Text erläutert die übergeordnete Idee. Der mittlere Bereich platzt vor Informationen. Die weiße Schrift auf den horizontalen Bändern trennt den mittleren Bereich mit Listen, Flussdiagrammen und Quellen ab.

Evaluation of the Abdominal Aorta and its Branches Using an Intravascular Echo Probe in the Inferior Vena Cava

Carol L. Chen, MD
Paul A. Tunick, MD
Lawrence Chinitz, MD
Neil Bernstein, MD
Douglas Holmes, MD
Itzhak Kronzon, MD

New York University School of Medicine
New York, NY USA

Background

Ultrasound evaluation of the abdominal aorta and its branches is usually performed transabdominally. Not infrequently, the image quality is suboptimal. Recently, an intracardiac echocardiography (ICE) probe has become commercially available (Acuson, Mountain View CA, Figure 1). These probes are usually inserted intravenously (IV) and advanced to the right heart for diagnostic and monitoring purposes during procedures such as ASD closure and pulmonary vein isolation (Figure 2). Because of the close anatomic relation between the abdominal aorta (AA) and the inferior vena cava (IVC), we hypothesized that these probes would be useful in the evaluation of the AA and its branches.

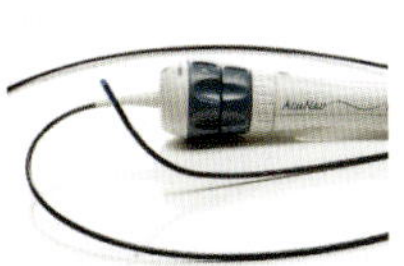

Figure 1: ICE probe (AcuNav, Acuson)

Figure 2: The ICE probe is placed in the right heart for imaging during PFO closure and pulmonary vein isolation.

The ICE probe can be advanced into the inferior vena cava (IVC), enabling high quality imaging of the abdominal aorta (Figure 3).

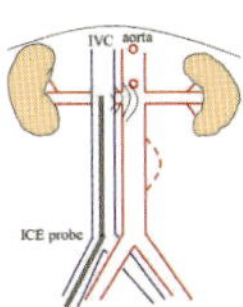

Figure 3: The position of the ICE probe in the IVC allows for excellent imaging and Doppler flow interrogation of the abdominal aorta and its branches (renal arteries, SMA, celiac axis) and the diagnosis of diseases such as renal artery stenosis and abdominal aortic aneurysm.

Methods

Fourteen pts who were undergoing a pulmonary vein isolation procedure participated in the study. In each pt, the ICE probe was inserted in the femoral vein and advanced to the right atrium for the evaluation of the left atrium and the pulmonary veins during the procedure. At the end of the procedure, the probe was withdrawn into the IVC.

Results

High resolution images of the AA from the diaphragm to the AA bifurcation were easily obtained in all pts. These images allowed for the evaluation of AA size, shape, and abnormal findings, such as atherosclerotic plaques (2 pts) and a 3.2 cm AA aneurysm (1 pt). Both renal arteries were easily visualized in each pt. With the probe in the IVC, both renal arteries are parallel to the imaging plane (Figure 4), and therefore accurate measurement of renal blood flow velocity and individual renal blood flow were possible.

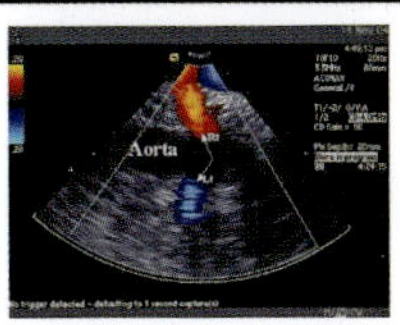

Figure 4: Two-dimensional image with color Doppler, of the abdominal aorta at the level of the right (Rt) and left (Lt) renal ostia. Note visualization of the laminar renal blood flow in the right renal artery, toward the transducer (red) and the left renal artery, away from the transducer (blue).

Calculation of renal blood flow:

The renal blood flow in each artery can be calculated using the cross-sectional area of the artery ($\pi r2$) multiplied by the velocity time integral (VTI, in cm) from the Doppler velocity tracing, multiplied by the heart rate (82 BPM in the example shown).

Figure 5

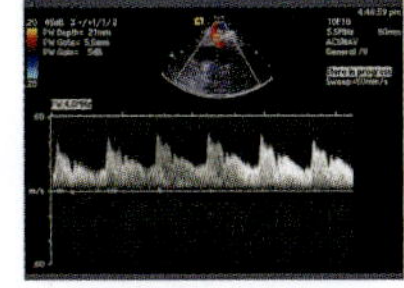

Figure 6: Pulsed Doppler of the right renal artery blood flow. The diameter of the right renal artery was 0.65 cm, and the VTI of the right renal blood flow was 0.19 meters (19 cm). Therefore the right renal blood flow was calculated as 516 cc/minute.

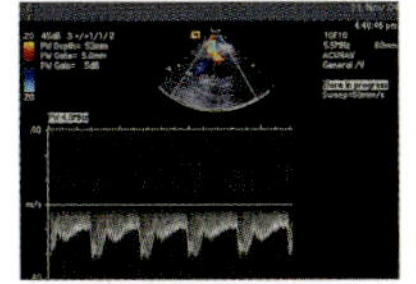

Figure 7: Pulsed Doppler of the left renal artery blood flow. The diameter of the left renal artery was 0.51 cm, and the VTI of the left renal blood flow was 0.2 meters (20 cm). Therefore the left renal blood flow was calculated as 334 cc/minute.

The total renal blood flow (right plus left) in this patient was therefore 850 cc/min. (average normal = 1200 cc/min.)

Conclusions

High resolution ultrasound images of the AA and the renal arteries are obtainable using ICE in the IVC. The branches of the abdominal aorta can be visualized and their blood flow calculated. Renal blood flow may be calculated for each kidney using this method. This may prove to be the imaging technique of choice for intra-aortic interventions such as angioplasty of the renal arteries for renal artery stenosis, fenestration of dissecting aneurysm intimal flaps, and endovascular stenting for AA aneurysm.

Why making? *Are you a maker? We hope so.*

It is a particularly critical time to put intelligent, ethical thought into "things." Perhaps you are shaping products that move markets…or, knitting fluffy hats. Do you recognize, in an antique chair, its narrative… The hand of the artisan who reshaped a tree to offer comfort?

Have you ever optimistically pulled apart broken electronics with hope of resuscitation? Confused by new car engines? *(You are not alone.)* Since the onset of the Anthropocene Age, humans have been obsessed with things. We have allowed them to help us, crowd us, amuse us, comfort us, etc. ***Joy, sustainability, curiosity and purpose*** are some of the keywords for 21c making manifestos. One must have trust in invisible electronic worlds yet remember the many paths we have traveled.

Why make makers?

Since making can be manifested in so many ways—software, toys, or an epicurean meal, it is essential to recognize the elements of a making processes that transcend media.

Materials, meaning-making, and mastery come together as a guide for companies who value creative processes and courageous individuals.

“I HEAR AND I FORGET. I SEE AND I REMEMBER. I DO AND I UNDERSTAND.”
—*Confucius*

Making Makers Who Fearlessly Make

SXSW 2015

Lori Kent, Ed. D.
Allison Kent-Smith
Catherine McGowan

10 tips

1. Know what your team makes. Know their skills.
2. Design learning experiences that engage the senses…have emotional meaning and connect to everyday work.
3. Define common terminology around making. Acknowledge team's existing knowledge.
4. Manage people so that their inner imaginations soar. *Tell them that what they know recombines as "creativity."*
5. Encourage everyone around you to have pride in their craft and continue to grow over a lifetime.
6. Design learning experiences that support multiple learning styles and configure complementary teams.
7. Making EXPLICIT a vision and your provisional goals.
8. AND…create a work (making) process that is shared and iterative.
9. Get people to connect with their inner child to lift creative blocks. Take makers to unexpected places.
10. This workshop is a beginning. You have a specific culture, individual needs…***take a first step.***

PROGRAM .... BUILDING .... STEP .... BY .... STEP

Resources

Texts

Shopcraft as Soulcraft: An Inquiry into the Value of Work by Matthew B. Crawford
The Courage to Create by Rollo May
Spark: How Creativity Works by Julie Bernstein (Studio 360)
Makers: The New Industrial Revolution by Chris Anderson
The Craftsmen by Richard Sennett
Ten Faces of Innovation by Kelley & Littman (IDEO)

WWW

http://aeon.co/magazine/being-human/
https://dschool.stanford.edu/groups/dhandbook/
http://edge.org/
http://makerfaire.com/
http://dx.cooperhewitt.org/lesson-plans/
http://www.fixerscollective.org/
http://www.techshop.ws/

Thanks to

Strawbees, SparkFun, Sally Oettinger, Meredith Olsen, Grace Borchers, and the s & b teacher collective.
Designed by Suzanne Dell'Orto.

3 ELEMENTS OF MAKING : MATERIALS, MEANING-MAKING, & MASTERY

Materials

- Materials tell you what to do.
- "Functional fixedness" is seeing a "thing" or material as having a specific use…rethink.
- Ordinary materials can inspire, transform….

Meaning-Making

- Be a generative thinker…able to sort, filter, bifurcate, combine and expand.
- Your experience gives you an incredibly rich "well" for making.
- Develop wonder. Think too much.

Mastery

- Mastery? What do you do best?
- How can you deconstruct process to teach mastery?
- How do you support individual and team mastery?

www.smithandbeta.com

SXSW Evaluation Link: sxsw.feedogo.com/fdbk.do?sid=IAP36301

29. Anleitungen verständlich gestalten

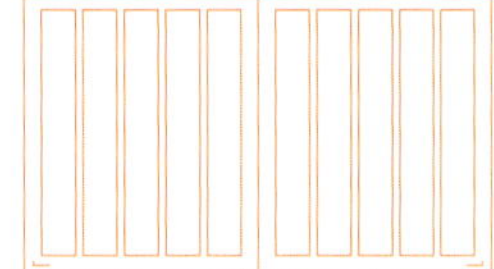

Schritt-für-Schritt-Anleitungen müssen leicht erfassbar sein. Der Leser kann einem klar gestalteten Layout (bis zu einem gewissen Grad) noch folgen, selbst wenn Sprachkenntnisse fehlen. Klarheit kann außerdem mit der Nummerierung einzelner Schritte und Bilder erreicht werden. Gut ausgewählte Fotos, die für sich allein sprechen, und ein leicht erfassbares Layout sind ideal.

PROJEKT

Kurashi no techo (Everyday Notebook) **magazine**

KUNDE

Kurashi no techo (Everyday Notebook) **magazine**

DESIGNER

Shuzo Hayashi, Masaaki Kuroyanagi

In dieser Schritt-für-Schritt-Anleitung wird ein westlicher Cartoon (Charlie Brown mit seiner Frühstückstüte) mit östlichem Raumgefühl kombiniert.

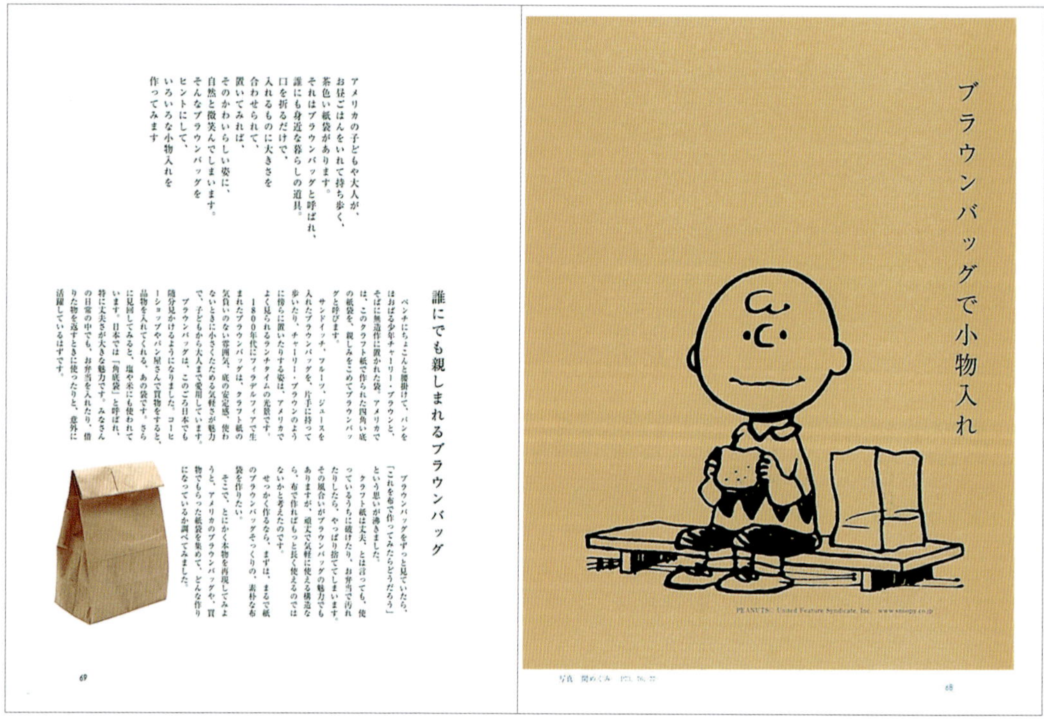

ブラウンバッグで小物入れ

アメリカの子どもや大人が、
お昼ごはんをいれて持ち歩く、
茶色い紙袋があります。
それはブラウンバッグと呼ばれ、
誰にも身近な暮らしの道具。
口を折るだけで、
入れるものに大きさを
合わせられて、
置いてみれば、
そのかわいらしい姿に、
自然と微笑んでしまいます。
そんなブラウンバッグを
ヒントにして、
いろいろな小物入れを
作ってみます

誰にでも親しまれるブラウンバッグ

ベンチにちょこんと腰掛けて、パンをほおばる少年チャーリー・ブラウンと、そばに無造作に置かれた袋。アメリカでは、このクラフト紙で作られた四角い底の紙袋を、親しみをこめてブラウンバッグと呼びます。

サンドイッチ、フルーツ、ジュースを入れたブラウンバッグを、片手に持って歩いたり、チャーリー・ブラウンのように傍らに置いたりする姿は、アメリカでよく見られるランチタイムの光景です。

1800年代にフィラデルフィアで生まれたブラウンバッグは、クラフト紙の気負いのない雰囲気、底の安定感、使わないときに小さくたためる気軽さが魅力で、子どもから大人まで愛用しています。

ブラウンバッグは、このごろ日本でも随分見かけるようになりました。コーヒーショップやパン屋さんで買物をすると、品物を入れてくれる、あの袋です。さらに見回してみると、塩や米にも使われています。日本では「角底袋」と呼ばれ、特に丈夫さが大きな魅力です。みなさんの日常の中でも、お弁当を入れたり、借りた物を返すときに使ったりと、意外に活躍しているはずです。

ブラウンバッグをずっと見ていたら、「これを布で作ってみたらどうだろう」という思いが沸きました。

クラフト紙は丈夫、とは言っても、使っているうちに破けたり、お弁当で汚れたりしたら、やっぱり捨ててしまいます。その風合いがブラウンバッグの魅力でもありますが、頑丈で気軽に使える構造なら、布で作ればもっと長く使えるのではないかと考えたのです。

せっかく作るなら、まずは、まるで紙のブラウンバッグそっくりの、素朴な布袋を作りたい。

そこで、とにかく本物を再現してみようと、アメリカのブラウンバッグや、買物でもらった紙袋を集めて、どんな作りになっているか調べてみました。

69

68

Durch eine freie Fläche lässt sich eine Einleitung ersetzen. Ein Cartoon wird von Angehörigen verschiedener Kulturen verstanden.

Deutlich abgehobene Boxen erleichtern den Weg von einem Anleitungspunkt zum nächsten. Auf dieser Seite hat jede Komponente ihren Platz in einem festen Raster.

Die Nummerierung kennzeichnet jeden Schritt, die Unterschritte erhalten kleine, umringelte Zahlen. Alle Elemente sind deutlich dargestellt und die Diagramme so verständlich, dass ein handwerklich begabter Mensch auch ohne Sprachkenntnisse dieser Anleitung mit gutem Ergebnis folgen kann. Lockere Raumaufteilung und angenehme Abbildungsgrößen sowie gelungene Fotos lassen auch die detailreichsten Anleitungen nicht bedrohlich erscheinen.

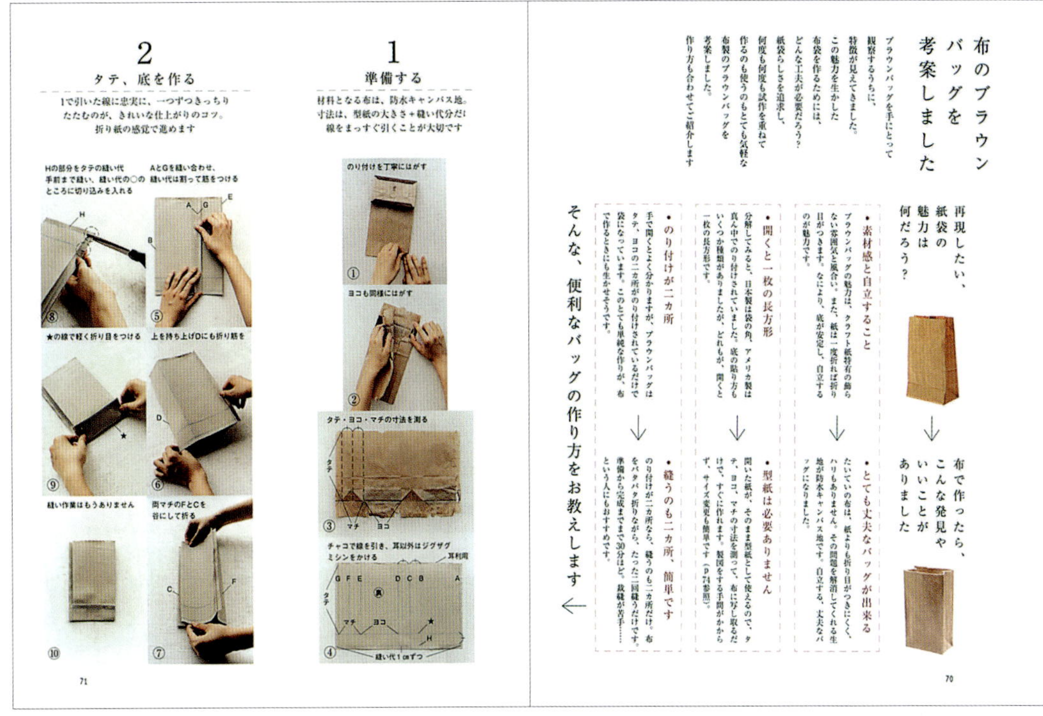

30. Grundlagen für Webseiten

Umfangreiche Webseiten benötigen Raster, die großen Datenmengen gerecht werden. So können die Informationseinheiten übersichtlich angeordnet werden. Verschaffen Sie sich zunächst einen Überblick über das Material und den Platz. Berücksichtigen Sie dabei Bildschirmränder, Symbolleisten (etwa die Menüleiste) und den Browser. Wie bei Printmedien muss beim Webdesign auch alles berücksichtigt werden, was Platz beansprucht. Dazu gehören bei den meisten Webseiten auch Elemente wie Werbung, Videos, die komplexe Gestaltung von Überschriften, Untertitel, Bildzeilen, Auflistungen und Links. Daher ist die Wahl klarer und deutlicher Fonts von großer Bedeutung.

BILDSCHIRMGRÖSSEN

Bildschirme sind unterschiedlich groß. Daher setzen viele Designer auf einen Satzspiegel von bestimmter Pixelzahl und -tiefe, der auch auf einem kleinen Display lesbar bleibt. Die Monitore werden zwar immer größer, aber seit dem Aufkommen von Tablets und Smartphones gibt es die unterschiedlichsten Displays und einfachere Hierarchien.

SECTIONS SEARCH SUBSCRIBE NOW LOG IN

ENGLISH 中文 (CHINESE) ESPAÑOL

The New York Times

Friday, July 6, 2018 | Today's Paper | Video | 77°F | Nasdaq +1.18% ↑

World U.S. Politics N.Y. Business Opinion Tech Science Health Sports Arts Books Style Food Travel Magazine T Magazine Real Estate ALL

Trade War With China Becomes Reality as Tariffs Take Effect

By ANA SWANSON 9:21 AM ET

- The Trump administration followed through with its threat to impose tariffs on $34 billion worth of Chinese products, setting up a clash between the world's two largest economies.
- The penalties prompted quick retaliation by Beijing. China said it immediately put its own similarly sized tariffs on an unspecified clutch of American goods.

322 Comments

China Strikes Back at Tariffs, but Its Consumers Worry

By RAYMOND ZHONG
20 minutes ago

While many maintain a defiant tone, Chinese shoppers could get hit as the trade war hurts exports and makes even basic types of food more

David Moskowitz

Who's Afraid of the Big Bad Wolf Scientist?

Rob Wielgus was one of America's pre-eminent experts on large carnivores. Then he ran afoul of the enemies of the wolf.

By CHRISTOPHER SOLOMON

Your Friday Briefing

By CHRIS STANFORD 8:34 AM ET

Here's what you need to know to start your day.

Listen to 'The Daily': A Reunification Story

Since President Trump ended the practice of

The Daily

Opinion

We'll All Be Paying for Scott Pruitt for Ages

By THE EDITORIAL BOARD

A hotter planet, shortened life spans and diminished trust in public service are the cost of the E.P.A. chief's service.

Pruitt's Resignation Is Just the Beginning

By NORMAN L. EISEN and NOAH BOOKBINDER

The damage he caused to the E.P.A., the federal government and the Trump administration will linger for years.

- Bruni: The Moist Mystery of Scott Pruitt's Resignation

OP-ED COLUMNISTS

- Brooks: Fred Rogers and the Loveliness of the Little Good
- Cohen: America Never Was, Yet Will Be
- Krugman: Big Business Reaps Trump's Whirlwind

READER CENTER »

The Times at Gettysburg, July 1863: A Reporter's Civil War Heartbreak

What Nelson Mandela Lost

By TAYARI JONES

The South African freedom fighter's letters from prison remind us that the separation of families is the ultimate expression of state power.

'Hope Is a Powerful Weapon': Unpublished Mandela Prison Letters

By THE EDITORS

A new volume offers insight into the personal and political life of one of the 20th century's most influential freedom fighters.

- Egan: Down and Out in San Francisco on $117,000
- Making America Unemployed Again
- The Déjà Vu of Mass Shootings
- Adrian Piper Speaks! (for Herself)
- Sign Up for Our World Cup Newsletter »

THE CROSSWORD »

Play Today's Puzzle

PROJEKT
nytimes.com

KUNDE
The New York Times

DESIGN
The New York Times

CREATIVE DIRECTOR
Tom Bodkin

Mit Liebe zum Detail und dem Kontrast zwischen einer Serifenschrift und einer serifenlosen werden auf dieser Webseite journalistische Fakten mit einer klassisch klaren und schönen Typografie kombiniert. Zeitangaben und Titel sind farbig abgehoben.

Diese dichte Struktur schafft Raum für ein Navigationsmenü, Storys, Bilder verschiedener Größen, Werbung und Videos.

Khoi Vinh, ehemaliger Design Director der ***New York Times:*** „Die Informationseinheiten sind die Grundsteine eines Rasters, die in den Spalten zusammengefasst werden und so die optische Struktur einer Seite bilden. Gut sind Einheiten, die ein Vielfaches von 3 oder 4 bilden, 12 ist ideal." Solche Überlegungen sind zwar nicht direkt sichtbar, verleihen der Seite aber einen starken Unterbau, der Einheiten und Spalten diszipliniert kommunizieren lässt.

Bei der Gestaltung der Spalten ist es wichtig, auf beiden Seiten der Schrift Platz zu lassen, sodass die Ausrichtung einheitlich vorgenommen werden kann, gleich ob in der Spalte Bilder, Text oder eine Box mit Text untergebracht werden sollen.

World »

Thai Cave Rescue Will Be Murky, Desperate Ordeal, Divers Say

English City, Stunned, Tries to Make Sense of New Poisonings

Japan Executes Cult Leader Behind 1995 Sarin Gas Subway Attack

Business Day »

U.S. Hiring Stayed Strong in June Despite Trade Strains

The Unemployment Rate Rose for the Best Possible Reason

China Strikes Back at Trump's Tariffs, but Its Consumers Worry

Opinion »

What Nelson Mandela Lost

'Hope Is a Powerful Weapon': Unpublished Mandela Prison Letters

We'll All Be Paying for Scott Pruitt for Ages

U.S. »

A Black Oregon Lawmaker Was Knocking on Doors. Someone Called the Police.

Migrant Shelters Are Becoming Makeshift Schools for Thousands of Children

Trump Administration in Chaotic Scramble to Reunify Migrant Families

Technology »

Tech Giants Win a Battle Over Copyright Rules in Europe

State of the Art: Employee Uprisings Sweep Many Tech Companies. Not Twitter.

The New New World: Why Made in China 2025 Will Succeed, Despite Trump

Arts »

If It's on 'Love Island,' Britain's Talking About It

Four Musicals on Three Continents: An Australian Company's Big Bet

The Art of Staying Cool: 10 Can't-Miss Summer Shows in New York

Politics »

Amy McGrath Set Her Sights on the Marines and Now Congress. Her Mother Is the Reason.

Trump Assails Critics and Mocks #MeToo. What About Putin? 'He's Fine'

Brett Kavanaugh, Supreme Court Front-Runner, Once Argued Broad Grounds for Impeachment

Fashion & Style »

Modern Love: This Is What Happens When Friends Fall in Love

The Secret Price of Pets

Fashion Review: A Declaration of Independence at Valentino and Fendi

Movies »

Lakeith Stanfield Is Playing Us All

Review: 'Ant-Man and the Wasp' Save the World! With Jokes!

Review: 'Sorry to Bother You,' but Can I Interest You in a Wild Dystopian Satire?

New York »

The Rise of the Stressed-Out Urban Camper

A City Founded by Alexander Hamilton Sets the Stage for Its Next Act

Culture of Fear and Ambition Distorted Cuomo's Economic Projects

Sports »

Neymar and the Art of the Dive

Garbiñe Muguruza and Marin Cilic Join the Wimbledon Exodus

On Pro Basketball: Finally Free From LeBron's Reign, the N.B.A. East Has No Reason to Change

Theater »

Four Musicals on Three Continents: An Australian Company's Big Bet

Review: 'The Royal Family of Broadway,' This Time in Song

Critic's Notebook: Orlando Bloom and Aidan Turner Are Drenched in Blood in London

Science »

Trilobites: Never Mind the Summer Heat: Earth Is at Its Greatest Distance From the Sun

The Lost Dogs of the Americas

Rhino Embryos Made in Lab to Save Nearly Extinct Subspecies

Obituaries »

Ed Schultz, Blunt-Spoken Political Talk-Show Host, Dies at 64

Michelle Musler, Courtside Perennial in the Garden, Is Dead at 81

Claude Lanzmann, Epic Chronicler of the Holocaust, Dies at 92

Television »

If It's on 'Love Island,' Britain's Talking About It

'Sharp Objects,' a Mesmerizing Southern Thriller, Cuts Slow but Deep

On Comedy: A Netflix Experiment Gives Deserving Comics Their 15 Minutes

Health »

Global Health: In a Rare Success, Paraguay Conquers Malaria

Trilobites: Lots of Successful Women Are Freezing Their Eggs. But It May Not Be About Their Careers.

Voices: When a Vegan Gets Gout

Travel »

The Getaway: Looking for a Weekend Escape? Here Are 5 Family-Friendly Options

Carry On: What W. Kamau Bell Can't Travel Without

The Rise of the Stressed-Out Urban Camper

Books »

Profile: Attention, Please: Anne Tyler Has Something to Say

Books of The Times: When It Comes to Politics, Be Afraid. But Not Too Afraid.

Captain America No. 1, by Ta-Nehisi Coates, Annotated

Education »

'Access to Literacy' Is Not a Constitutional Right, Judge in Detroit Rules

Colleges and State Laws Are Clamping Down on Fraternities

In the Age of Trump, Civics Courses Make a Comeback

Food »

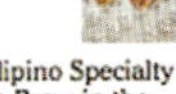

Wines of The Times: American Rosés Without Clichés

Hungry City: A Filipino Specialty Best Paired With a Brew in the East Village

Australia Fare: Yatala Pies Has Served Nostalgia for More Than 130 Years. Arguably.

Magazine »

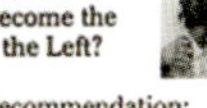

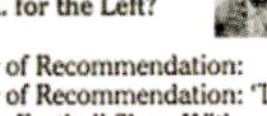

Feature: Can the A.C.L.U. Become the N.R.A. for the Left?

Letter of Recommendation: Letter of Recommendation: 'The Totally Football Show With James Richardson'

Feature: Who's Afraid of the Big Bad Wolf Scientist?

Real Estate »

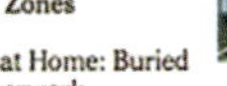

New Buildings Rise in Flood Zones

Right at Home: Buried in Paperwork

The Hunt: Trading Chelsea Clatter for Greenpoint Calm

The Upshot »

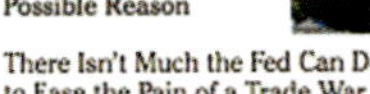

The Unemployment Rate Rose for the Best Possible Reason

There Isn't Much the Fed Can Do to Ease the Pain of a Trade War

Americans Are Having Fewer Babies. They Told Us Why.

Times Insider »

Outsmarted by a Smart TV? Not This Reporter.

With Our World Cup App, Fans Are Part of the Action

The Times at Gettysburg, July 1863: A Reporter's Civil War Heartbreak

REAL ESTATE »

THE HUNT

Trading Chelsea Clatter for Greenpoint Calm

By JOYCE COHEN

Living on Eighth Avenue was fun, but after six years Emery Myers wanted some peace and quiet — not to mention a garden. Walls were optional.

- Search for Homes for Sale or Rent
- Mortgage Calculator

MOST EMAILED | MOST VIEWED | **RECOMMENDED FOR YOU**

1. What Can You Do About a Hammertoe? 
2. Mom, I Need a Break 
3. Countdown to Retirement: A Five-Year Plan 
4. A Cult Show's Recipe for Success: Whiskey, Twitter and Complex Women 
5. Facebook Removes a Gospel Group's Music Video 
6. A Cult Show's Recipe for Success: Whiskey, Twitter and Complex Women 
7. Airline Crew Have Higher Cancer Rates 
8. London Mayor Allows 'Trump Baby' Blimp for President's Trip to U.K. 
9. When a Vegan Gets Gout 
10. Statue of Liberty Stamp Mistake to Cost Postal Service $3.5 Million 

Log in to discover more articles based on what you've read.

LOG IN | REGISTER NOW

What's This? | Don't Show

MODULE

31. Mit Linien Klarheit schaffen

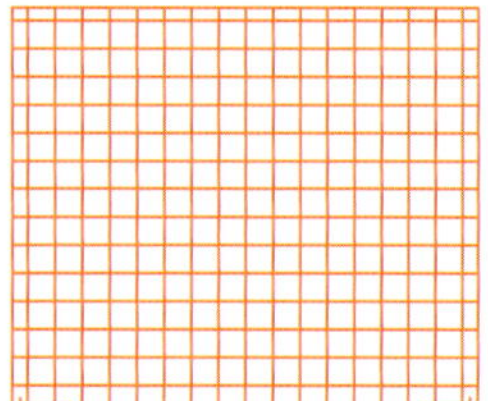

Manchmal besteht der Informationsfluss aus einer Kombination von Tabellen und Modulen. Bei der Darstellung komplexer Information sind Klarheit, gute Lesbarkeit, Großzügigkeit und Abwechslung wichtige Faktoren. Das Aufbrechen komplizierter Sachverhalte in überschaubare Einheiten ergibt ein klares Layout.

Hierfür sind Modulraster zweckmäßig:

- bei so vielen voneinander unabhängigen Informationseinheiten, dass zusammenhängendes Lesen entweder nicht nötig oder nicht möglich ist
- wenn Sie das gesamte Material in gleich großen Blöcken darstellen möchten
- wenn Sie ein einheitliches, oder nahezu einheitliches Format gestalten wollen
- wenn die Informationseinheiten gleich groß sind und ihnen Nummern oder Zeitangaben vorangestellt sind

Klarheit sollte ebenso bei der Typografie gegeben sein. Sie muss inhaltlich zum Projekt passen. Der Kontrast zwischen Schriftgrad und -größe sowie den erklärenden Einheiten fördert die leichte Erfassbarkeit einer Seite. Wie bereits bei anderen Gestaltungsprinzipien erwähnt, lässt sich mit dem souveränen Einsatz unterschiedlicher Fonts die Gratwanderung zwischen klaren aber langweiligen und grenzwertig skurrilen Seiten meistern.

RECHTE SEITE: In dieser Auflistung von Tipps erhält jede Texteinheit gleich große Ränder, wobei sich die Größe der Boxen nach der Textmenge richtet. Das Material in den Boxen wird nicht vom Duktus der unterteilenden Linien erdrückt. In den Seitenleisten können so noch zusätzliche Informationen untergebracht werden.

In jeder Sprache fungieren Aufzählungszeichen als Warnsignale in Überschriften und, wie praktisch immer, signalisieren Schriftstärke und -größe die Informationshierarchie.

Bei diesen nummerierten Einheiten sorgen in der Typografie nur Schriftstärke und -größe für etwas Abwechslung. Arabische Ziffern und Kanji-Schriftzeichen versehen die nützlichen, jedoch teilweise merkwürdigen Informationen mit einem anheimelnden Touch. Die Übersetzung von Nummer 7 lautet etwa: „Es wird trocken. Gurgeln Sie möglichst, wenn Sie wieder nach Hause kommen. Ein Glas neben dem Waschbecken vereinfacht die Sache."

PROJEKT
Zeitschrift ***Kurashi no techo (Notizbuch für jeden Tag)***

KUNDE
Zeitschrift ***Kurashi no techo (Notizbuch für jeden Tag)***

DESIGNER
Shuzo Hayashi, Masaaki Kuroyanagi

In diesem Ratgeber werden übersichtlich Tipps für den Haushalt gegeben.

●暮らしのヒント集

ここにならんでいる
いくつかのヒントのなかで、
ふと目についた項目を
読んでみてください。
たぶん、ああそうだったと
いうことになるでしょう

今日はなにを

1 テーブルにコップを置くときは、静かに置くことを心がけましょう。やさしいしぐさが気持ちをやわらげます。

2 組み立て式の椅子やテーブルのネジは、意外とゆるんでいるものです。締めなおしておきましょう。

3 暮らしには笑顔が大事です。いろいろあっても、にっこり笑顔を忘れずに。

4 一年使った枕を新しいものに替えてみましょう。新しい気持ちで眠りにつけるでしょう。

5 今日こそゆるんだ水道のパッキンを取替えましょう。家中の蛇口をチェックします。

6 毎日の暮らしのなかで見て見ぬふりはやめましょう。そういう癖を身につけてはいけません。

7 空気が乾燥してきます。外から帰ったらすぐにうがいができるように、洗面所のコップをきれいにしておきましょう。

8 朝、目が覚めたら、ベッドの中で今日一日、何をするかを考えます。することがたくさんあれば、うかうかしていられず、すぐ起きるでしょう。

9 どんなことでもまずはお金を使わずにできるかを考えてみましょう。それが工夫の一歩になります。

10 言いたいことを言った後は、笑顔で接することが大事です。険悪にならないように、まわりに気を使いましょう。

11 日曜日の朝、天気が良かったら、外でご飯にしませんか。ごく簡単なお弁当を近所の公園などで食べるのです。散歩もかねて気分も変わります。

12 風邪をひいて、お風呂に入れないときは、足だけでも洗って、温めましょう。さっぱりして気分がよくなります。

13 今日は一歩ゆずってみましょう。その一歩がそのまま新しい一歩を進めるちからになるものです。

14 裁縫箱を整理しましょう。さびた針やよれた糸は処分して、新しいものに取替えます。

15 今夜は粗食デーにしましょう。味噌汁にお漬物とか、ありあわせのおかずで間に合わせます。明日は今夜の分もごちそうにしましょう。

16 冷蔵庫が夏の設定になっていませんか。気温も下がったし、あけての回数も減ってきたので、調節しておきます。

17 虫歯があったら、いますぐ治しておきましょう。年末年始のお医者さんが休みのときに痛くなったら大変です。

18 手紙ばさみを買ってみましょう。とても便利なので、毎日届く郵便をさっさと片づけられます。

19 今日は一日、お年寄りのお相手をつとめましょう。お茶を飲みながら、ゆっくりと昔話を聞いてあげたり、一緒に出かけたりします。

20 毎日を心地よく過ごすには、あまりに潔癖すぎてもいけません。よごれやけがれも受け入れてこそ暮らしがあるのです。人との関係も同様です。

21 きびしい肌寒さをおぼえる夜になりました。ことにお年寄りにはひざ掛けか、肩掛けを一枚、早めに用意してあげましょう。

22 しめきりの窓をあけて、敷居のゴミを払いましょう。アルミサッシの溝など、ほこりがつまっているものです。

23 洋服ダンスの防虫剤は大丈夫でしょうか。においはしていても、中身はもうなくなっていることが案外多いものです。

24 新しいチャレンジは自分で決めるものです。ひとに惑わされて後悔しないように。

25 ガス台の下やすきまを掃除しましょう。汚れているものです。意外にきれいになると気持ちよく料理ができるでしょう。

MODULE

32. Luft zum Atmen

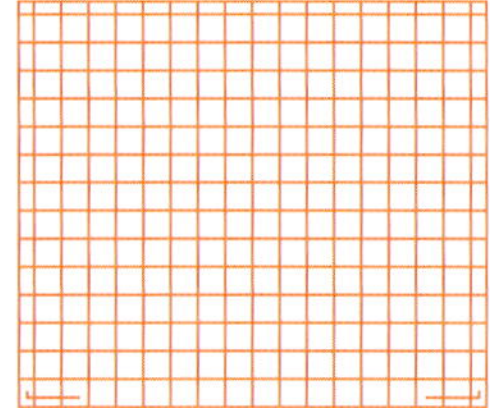

Nicht alle Module eines Rasters müssen auch aufgefüllt werden. In einem Modulraster werden die Schriftgrößen exakt festgelegt, wobei der Designer verschiedene Details entwickeln und darstellen kann. Die Module können sichtbar oder unsichtbar, groß oder klein sein. Dank ihrer stabilen Struktur nehmen sie Lettern, Farben und Ornamente zuverlässig auf – oder sie bleiben einfach leer.

Diese Schaugröße wurde speziell für Überschriften und Titel, aber nicht für Fließtext gestaltet. Bei kleineren Formaten sind Schaugrößen oft schlecht lesbar, da ihre differenzierenden Eigenschaften verschwinden.

PROJEKT
Restraint Font

KUNDE
Marian Bantjes

DESIGN
Marian Bantjes, Ross Mills

Handgezeichnete Typografie in digitaler Form.

DISPLAY

RESTRAINT

Restraint is a display face

&

is best used at sizes above 36 point.

It is not a full typeface,
in that it does not contain a
complete character set.
Think of it as an ornament
font that happens
to include letterforms.
It contains 26 letters of the
alphabet, a hyphen, question mark
and an ampersand.
The rest of the font is ornamentals
for creating extensive borders
and shapes.

Restraint works on a grid of squares.
Creating a design in Restraint
requires thinking along a grid.
Every glyph that you insert will move
the following glyphs over
by a full square.

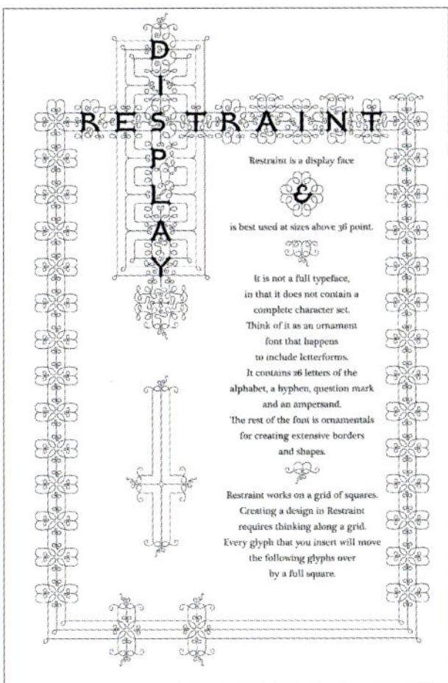

Durch das Auffüllen der Module in der Mitte der Seite und das Freilassen der Ecken lässt sich eine Art Rahmenoptik erzeugen.

USE
RESTRAINT

In most programs, Restraint sets solid.
This means that if your typesize
is 36, your leading should also be 36.
Or 60/60, 72/72 … etc.
This works best in InDesign, but also works
in MS Word, although in Word there is a
very slight gap between the lines,
and if you can, you could close it
a couple of points.

VERY
IMPORTANT:

In InDesign, make sure you use the
kerning setting for 'Metrics', *not* 'Optical'
(in the Character palette).
If you use Optical kerning, the glyphs
will not match up from line to line. Also,
make sure that the paragraph alignment
does not effect spacing (eg. justifying)
Restraint requires no manual kerning.

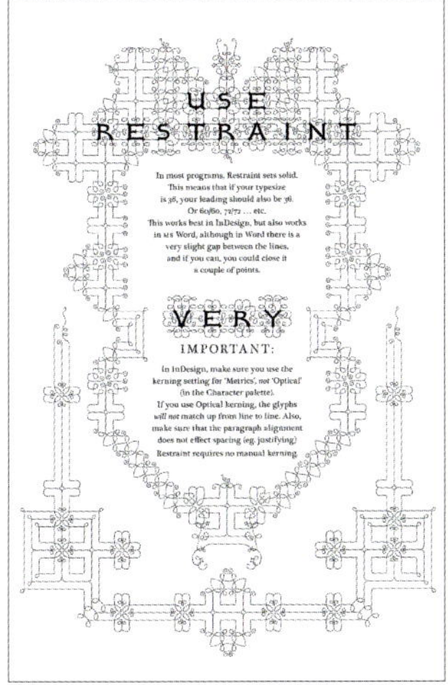

Alternativ dazu kann man die Module nur am Rand ausfüllen und dadurch einen Rahmen erzeugen. Für alle Fälle gilt jedoch: Wer Zurückhaltung übt, beherrscht den feinen Unterschied zwischen Kakofonie und Sinfonie.

In dieser Lizenzvereinbarung für Endnutzer zeigt sich eine wunderbare Typografie und der gekonnte Einsatz des Fonts *Restraint*.

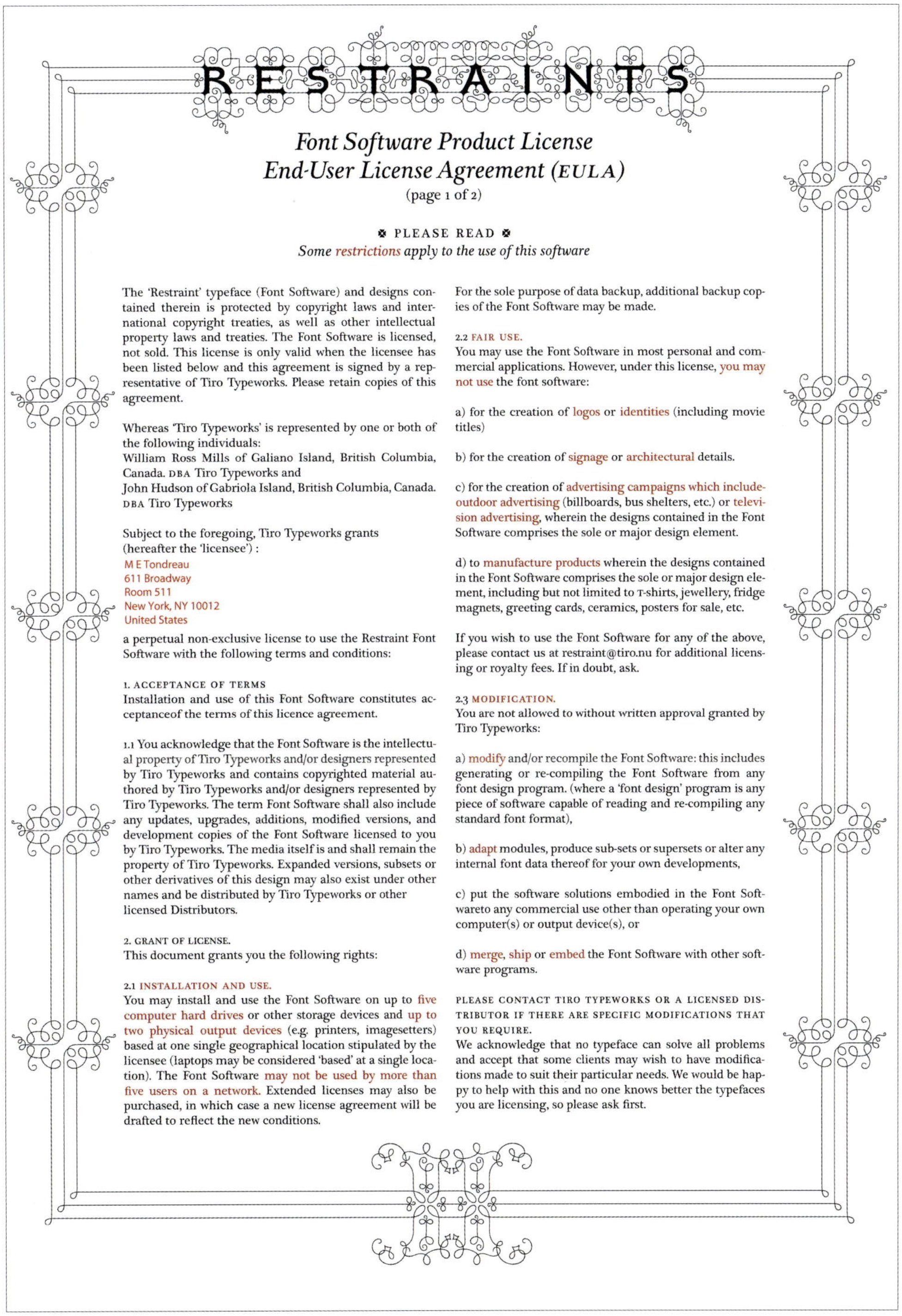

RESTRAINTS

Font Software Product License
End-User License Agreement (EULA)
(page 1 of 2)

PLEASE READ
Some restrictions apply to the use of this software

The 'Restraint' typeface (Font Software) and designs contained therein is protected by copyright laws and international copyright treaties, as well as other intellectual property laws and treaties. The Font Software is licensed, not sold. This license is only valid when the licensee has been listed below and this agreement is signed by a representative of Tiro Typeworks. Please retain copies of this agreement.

Whereas 'Tiro Typeworks' is represented by one or both of the following individuals:
William Ross Mills of Galiano Island, British Columbia, Canada. DBA Tiro Typeworks and
John Hudson of Gabriola Island, British Columbia, Canada. DBA Tiro Typeworks

Subject to the foregoing, Tiro Typeworks grants (hereafter the 'licensee') :
M E Tondreau
611 Broadway
Room 511
New York, NY 10012
United States
a perpetual non-exclusive license to use the Restraint Font Software with the following terms and conditions:

1. ACCEPTANCE OF TERMS
Installation and use of this Font Software constitutes acceptanceof the terms of this licence agreement.

1.1 You acknowledge that the Font Software is the intellectual property of Tiro Typeworks and/or designers represented by Tiro Typeworks and contains copyrighted material authored by Tiro Typeworks and/or designers represented by Tiro Typeworks. The term Font Software shall also include any updates, upgrades, additions, modified versions, and development copies of the Font Software licensed to you by Tiro Typeworks. The media itself is and shall remain the property of Tiro Typeworks. Expanded versions, subsets or other derivatives of this design may also exist under other names and be distributed by Tiro Typeworks or other licensed Distributors.

2. GRANT OF LICENSE.
This document grants you the following rights:

2.1 INSTALLATION AND USE.
You may install and use the Font Software on up to five computer hard drives or other storage devices and up to two physical output devices (e.g. printers, imagesetters) based at one single geographical location stipulated by the licensee (laptops may be considered 'based' at a single location). The Font Software may not be used by more than five users on a network. Extended licenses may also be purchased, in which case a new license agreement will be drafted to reflect the new conditions.

For the sole purpose of data backup, additional backup copies of the Font Software may be made.

2.2 FAIR USE.
You may use the Font Software in most personal and commercial applications. However, under this license, you may not use the font software:

a) for the creation of logos or identities (including movie titles)

b) for the creation of signage or architectural details.

c) for the creation of advertising campaigns which include outdoor advertising (billboards, bus shelters, etc.) or television advertising, wherein the designs contained in the Font Software comprises the sole or major design element.

d) to manufacture products wherein the designs contained in the Font Software comprises the sole or major design element, including but not limited to T-shirts, jewellery, fridge magnets, greeting cards, ceramics, posters for sale, etc.

If you wish to use the Font Software for any of the above, please contact us at restraint@tiro.nu for additional licensing or royalty fees. If in doubt, ask.

2.3 MODIFICATION.
You are not allowed to without written approval granted by Tiro Typeworks:

a) modify and/or recompile the Font Software: this includes generating or re-compiling the Font Software from any font design program. (where a 'font design' program is any piece of software capable of reading and re-compiling any standard font format),

b) adapt modules, produce sub-sets or supersets or alter any internal font data thereof for your own developments,

c) put the software solutions embodied in the Font Software to any commercial use other than operating your own computer(s) or output device(s), or

d) merge, ship or embed the Font Software with other software programs.

PLEASE CONTACT TIRO TYPEWORKS OR A LICENSED DISTRIBUTOR IF THERE ARE SPECIFIC MODIFICATIONS THAT YOU REQUIRE.
We acknowledge that no typeface can solve all problems and accept that some clients may wish to have modifications made to suit their particular needs. We would be happy to help with this and no one knows better the typefaces you are licensing, so please ask first.

33. Module kombinieren

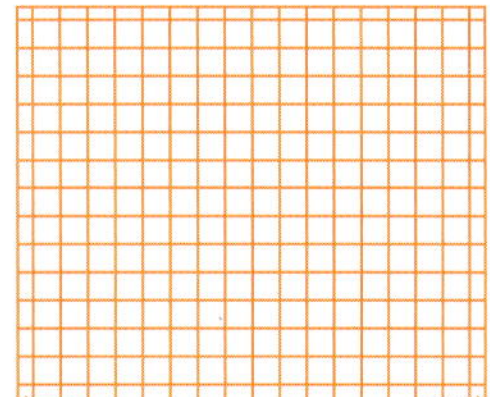

Betrachtet man einen Modulraster als Diagramm, mag er kompliziert erscheinen. Tatsächlich ist das jedoch nicht der Fall und nicht jedes Modul muss auch ausgefüllt werden. Je nach Umfang der Informationen, die auf der Seite Platz finden müssen, kann man ein Modul mit wenigen großen Boxen für Bilder gestalten – und noch wichtiger: für Schlüsselinformationen wie Inhaltsangaben und andere Verzeichnisse.

Die Module sind so angeordnet, dass das Teppichlogo jeweils links unten erscheint.

PROJEKT
Flor Catalog

KUNDE
Flor

DESIGN
The Valentine Group

Das Beispiel dieses Katalogs zum Thema Teppichfliesen zeigt, dass Modulraster ideale Raumteiler sind. Damit lässt sich perfekt eine visuelle Schritt-für-Schritt-Anleitung gestalten.

Farblich abgehobene, leicht les- und erfassbare Einheiten werden in Boxen aufgeteilt und mit einer Inhaltsangabe kombiniert.

Hier werden Farbmuster mit anspruchsvoll-witzigen Fotos und großzügigen Leerräumen kombiniert.

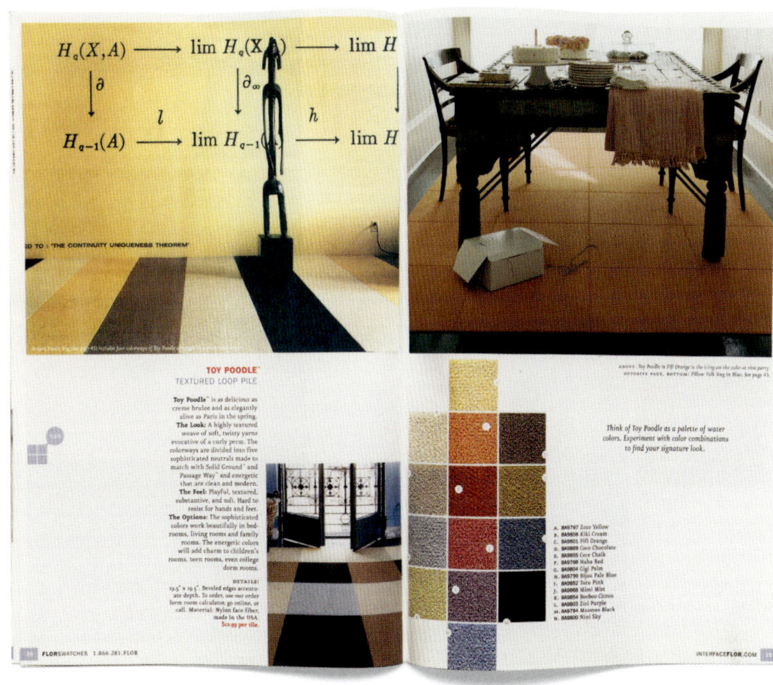

Dieser Bodenbelagsrechner ist selbstverständlich ein starker Modulraster.

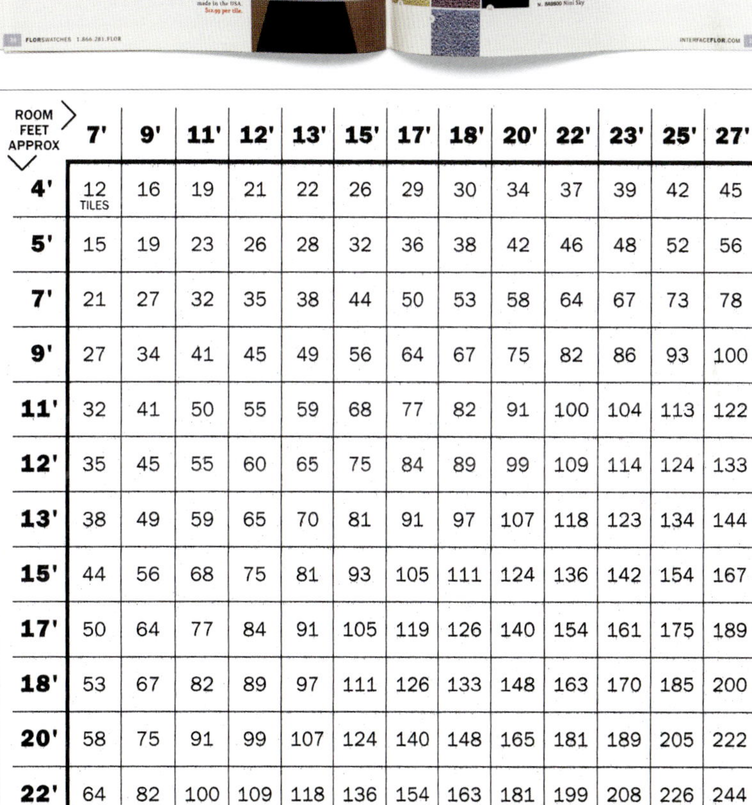

ROOM FEET APPROX	7'	9'	11'	12'	13'	15'	17'	18'	20'	22'	23'	25'	27'
4'	12 TILES	16	19	21	22	26	29	30	34	37	39	42	45
5'	15	19	23	26	28	32	36	38	42	46	48	52	56
7'	21	27	32	35	38	44	50	53	58	64	67	73	78
9'	27	34	41	45	49	56	64	67	75	82	86	93	100
11'	32	41	50	55	59	68	77	82	91	100	104	113	122
12'	35	45	55	60	65	75	84	89	99	109	114	124	133
13'	38	49	59	65	70	81	91	97	107	118	123	134	144
15'	44	56	68	75	81	93	105	111	124	136	142	154	167
17'	50	64	77	84	91	105	119	126	140	154	161	175	189
18'	53	67	82	89	97	111	126	133	148	163	170	185	200
20'	58	75	91	99	107	124	140	148	165	181	189	205	222
22'	64	82	100	109	118	136	154	163	181	199	208	226	244
23'	67	86	104	114	123	142	161	170	189	208	217	236	255
25'	73	93	113	124	134	154	175	185	205	226	236	256	277 TILES

MODULE

34. Der Raum zählt

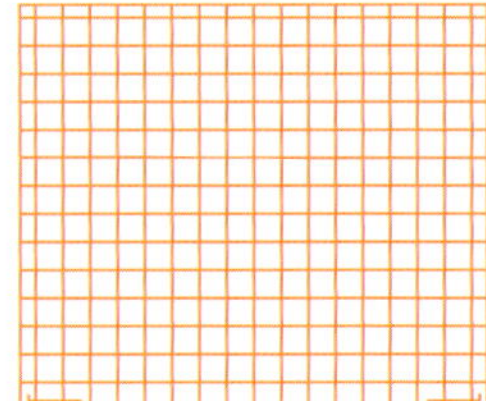

Komplizierte Informationen benötigen einen starken Raster. Planen Sie die Proportionen eines jeden Moduls nach den enthaltenen Informationen, sodass auch eventuell verwirrendes Material klar wird.

Mit ihrem großen Format ziehen Poster besondere Aufmerksamkeit auf sich. Titelzeilen sollten groß sein. So werden Sie schon von Weitem wahrgenommen und fordern zum Lesen des ganzen Textes auf.

PROJEKT
Voting by Design poster

KUNDE
Design Institute,
University of Minnesota

HERAUSGEBER/PROJEKTLEITUNG
Janet Abrams

ART DIRECTION/DESIGN
Sylvia Harris

Dieses Poster stellt eine extrem disziplinierte Aufteilung eines wichtigen Prozesses dar und nutzt jeden Zentimeter Raum. Der Raster steuert die Leseerfahrung.

SEITE GEGENÜBER: Obwohl dieses Poster viele Informationen enthält, lässt es sich dank der Methode, die Erfahrungen in Schritte aufzubrechen, leicht erfassen.

VOTING BY DESIGN

The century began with an electoral bang that opened everyone's eyes to the fragility of the American voting system. But, after two years of legislation, studies and equipment upgrades, major problems still exist. Why?

Voting is not just an event. It's a complex communications process that goes well beyond the casting of a vote. For example, in the 2000 presidential election, 1.5 million votes were missed because of faulty equipment, but a whopping 22 million voters didn't vote at all because of time limitations or registration errors. These and many other voting problems can be traced not just to poor equipment, but also to poor communications.

Communicating with the public is what many designers do for a living. So, seen from a communications perspective, many voting problems are really design problems. That's where you come in.

Take a look at the voting experience map below, and find all the ways you can put design to work for democracy.

A COMMUNICATIONS MAP OF THE AMERICAN VOTER'S EXPERIENCE

EDUCATION	REGISTRATION	PREPARATION	NAVIGATION	VOTING	FEEDBACK
LEARNING ABOUT VOTING RIGHTS AND DEMOCRACY	SIGNING UP TO BECOME A REGISTERED VOTER	BECOMING INFORMED AND PREPARED TO VOTE	FINDING THE WAY TO THE VOTING BOOTH	INDICATING A CHOICE IN AN ELECTION	GIVING FEEDBACK ABOUT THE VOTING EXPERIENCE
WORD-OF-MOUTH	PAPER REGISTRATION FORMS	SAVE-THE-DATE CARD	EXTERIOR STREET SIGNS	HAND-COUNTED PAPER BALLOT	CENSUS SURVEYS
HIGH SCHOOL CIVICS CLASSES	ONLINE REGISTRATION FORMS	VOTER REGISTRATION CARD	PRECINCT SIGNAGE	MACHINE-COUNTED PAPER BALLOT	EXIT POLLS
CITIZENSHIP CLASSES	MOTOR VOTER APPLICATIONS	PUBLIC SERVICE ANNOUNCEMENTS	LINE AND BOOTH IDENTITY	MECHANICAL LEVER	VOTING EXPERIENCE SURVEYS
	VOTER ROLLS	PRE-ELECTION INFO PROGRAMS	PRECINCT WORKERS	PUNCHCARD	
		CAMPAIGN LITERATURE	CAMPAIGN WORKERS	DIRECT RECORD ELECTRONIC	
		SAMPLE BALLOTS		VOTING INSTRUCTIONS	
DISAPPEARING CIVICS CLASSES	FORMS THAT ARE BARRIERS TO PARTICIPATION	TOO MUCH OR TOO LITTLE INFORMATION	GETTING TO THE BOOTH ON TIME	USER-UNFRIENDLY VOTING MACHINES	FUTURE IMPROVEMENTS LACK VOTER INPUT

DESIGN TO THE RESCUE

ALL KINDS OF DESIGNERS CAN PARTICIPATE IN VOTER REFORM. HERE'S WHO SHOULD BE ON ANY VOTING DESIGN DREAM TEAM:

GRAPHIC DESIGNERS

ENVIRONMENTAL GRAPHIC DESIGNERS

INFORMATION DESIGNERS

ARCHITECTS

INDUSTRIAL DESIGNERS

EXPERIENCE DESIGNERS

HOW YOU CAN GET INVOLVED

THERE IS WORK TO BE DONE TO IMPROVE VOTING BY DESIGN, STARTING WITH YOUR OWN COMMUNITY. HERE ARE FIVE THINGS THAT ANY DESIGNER CAN DO, TO MAKE A DIFFERENCE BEFORE THE 2004 ELECTIONS:

1. BECOME A POLLWORKER

2. FORM A VOTING DESIGN COALITION

3. WORK WITH THE POLITICAL PARTY OF YOUR CHOICE

4. CALL YOUR CONGRESSPERSON ABOUT HR 3295

5. FORM A VOTING DESIGN ADVISORY TEAM

University of Minnesota

35. Organische Module

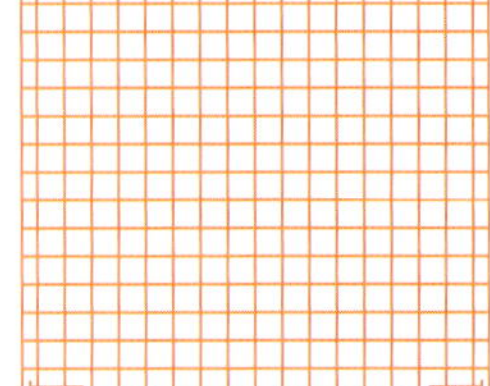

Das Schöne an einem Modulraster ist, dass man nicht quadratisch bleiben muss, denn ein gleichmäßiger Modulraster ermöglicht es, Formen, Größen und Muster zu variieren und dabei immer noch entzückend ordentlich zu wirken.

REGINA, TO THE TRADE; NANCY CORZINE: 212-223-8340.

ELAN, FROM $84; PINE CONE HILL: 413-496-9700.

VINTAGE FABRIC, $95: ABC CARPET & HOME: 212-473-3000.

CHINESE EMPEROR, $165; BEDFORD & CO.: 212-772-7000.

DIYA OATMEAL, $265; JOHN ROBSHAW: 212-594-6006.

APPLE, $48; CARRERAS: 845-758-2200.

ETHNOA KUBA CLOTH, $265; CALYPSO HOME: 631-324-8146.

CAMINO, $195; RALPH LAUREN HOME: 888-475-7674.

19TH C. GOLD METALLIC APPLIQUE, $1350; B. VIZ: 318-766-4950.

TINA, $870; ARMANI CASA: 212-334-1271.

MODERN GOTHIC GRIFFIN, $198; MICHELE VARIAN: 212-343-0033.

DIYA AMBER, $80; JOHN ROBSHAW: 212-594-6006.

SEQUIN COVER, $83; THE CONRAN SHOP: 866-755-9079.

JACK RUSSELL, $98; L'AVENUE DES RÊVES: 212-396-9500.

ZARDOZEE, $90; ALPHA BY MILLI HOME: 212-643-8850.

STRIPES; $215; RANI ARABELLA: 561-802-9900.

112

Die Farbvarianz zu limitieren und sogar für jede Seite eine eigene Farbpalette zu kreieren ist sehr hilfreich. Das erzeugt Harmonie.

PROJEKT
House Beautiful

KUNDE
House Beautiful Magazine

DESIGN
Barbara deWilde

Hier verjüngt sich ein Magazin mit einem knackigen neuen Design.

Die gleichmäßige und klar gegliederte Typografie untermauert jedes Modul, während die ruhige serifenlose Schrift strukturgebende Linien schafft.

36. Diagramme in der Gesamtschau

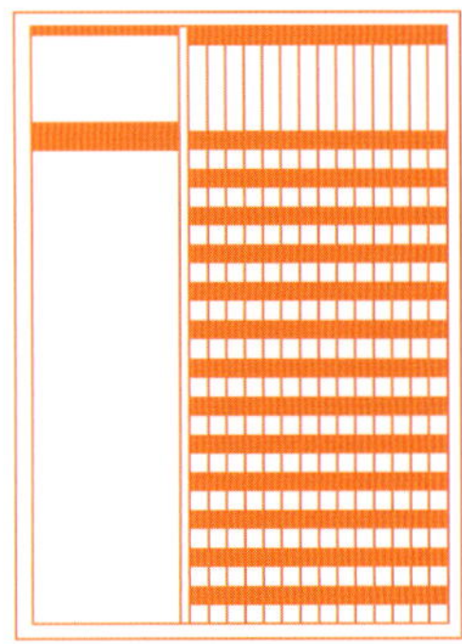

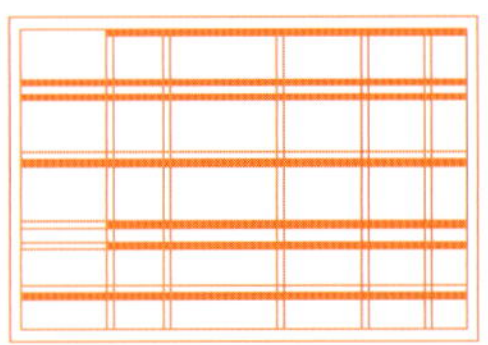

PROJEKT
Fahrpläne für New Jersey Transit

KUNDE
New Jersey Transit

DESIGN
Two Twelve Associates

Diese Fahrpläne des öffentlichen Nahverkehrs in New Jersey zeigen, dass ein klares und einfaches Design auch entstehen kann, wenn man keine Angst vor gerahmten Boxen hat. Pfeile und grafische Symbole verhelfen den Reisenden ebenso zur Übersicht über die komplexen Informationen. Sie mögen abgegriffen sein, aber manchmal ist einfach der kleinste gemeinsame Nenner, das beste Mittel um vielen Lesern gerecht zu werden.

Die Gestaltung von Tabellen, Diagrammen und Fahrplänen gleicht wegen der immensen und starren Zahlenmengen einer Heldentat, die man nicht mutwillig herausfordert. In ihrem Buch *Thinking with* Type rät Ellen Lupton Designern davon ab, das typografische Verbrechen, sie nennt es Datengefängnis, zu begehen, d. h., zu viele Linien und Boxen zu verwenden. Lupton schlägt vielmehr vor, die einzelnen Tabellen, den Raster oder Fahrplan als Ganzes zu betrachten und bei jeder Spalte, Feld oder Reihe zu überlegen, wie sie in das Gesamtbild passt.

Benutzen Sie Farbschattierungen, damit der Leser sich im Dickicht der Informationen besser zurechtfindet. Schattierungen funktionieren bei Schwarz-Weiß und Farbe gleichermaßen. Mit schattierten Balken lassen sich Nummernreihen gut abheben, damit der Leser die Informationen leichter findet. Wenn es um eine übersichtliche Darstellung geht, sind Rahmen und Linien durchaus angebracht. Mit Linien lassen sich spezielle Abschnitte optisch abtrennen und bei Fahrplänen inhaltlich verschiedene Zonen voneinander abheben. Bei noch komplexeren Projekten wie z. B. Zugfahrplänen, die für ein ganzes System erarbeitet werden müssen, haben sich farbliche Unterscheidungen für die einzelnen Bahnlinien bewährt.

Ein Raster sollte den jeweiligen Informationen entsprechen und bei mehrspaltigen Projekten ist eine klare Typografie enorm wichtig. Der Schriftsatz bei Nahverkehrsverbindungen auf Bahnhöfen und Flughäfen kann im Hinblick auf unbeschwertes Reisen und verpasste Anschlüsse ausschlaggebend sein. Lassen Sie ober- und unterhalb der Linien ausreichend Platz, auch wenn die Masse an Informationen dies fast nicht zulässt. Leerraum erhöht die Lesbarkeit, überhaupt das wichtigste Prinzip für Fahrpläne.

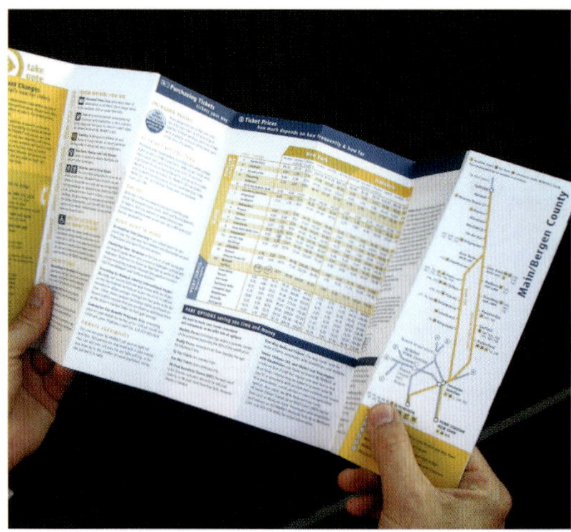

Farblich wechselnde Balken kennzeichnen in diesem Fahrplan die Haltestellen. Linien werden sparsam eingesetzt und trennen Zusatzinformationen ab. Vertikallinien heben Haltestellen von den Endhaltestellen ab, während die Horizontallinien die wichtigsten geografischen Zonen unterteilen.

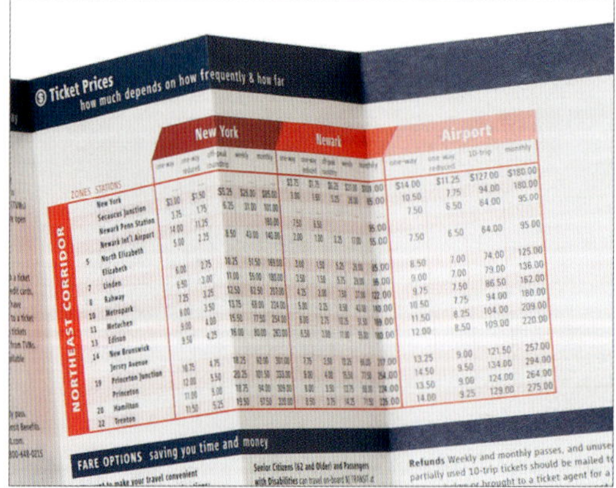

Das System für Fahrpläne lässt sich auch für Gebührentabellen anwenden. Wie hier kennzeichnen farblich abwechselnde Balken Haltestellen und Fahrpreise. Überschriften wie *One Way* und *Off-peak Roundtrip* werden mit Horizontal- und Vertikallinien abgesetzt.

Piktogramme veranschaulichen die Überschriften der Verkaufsbedingungen.

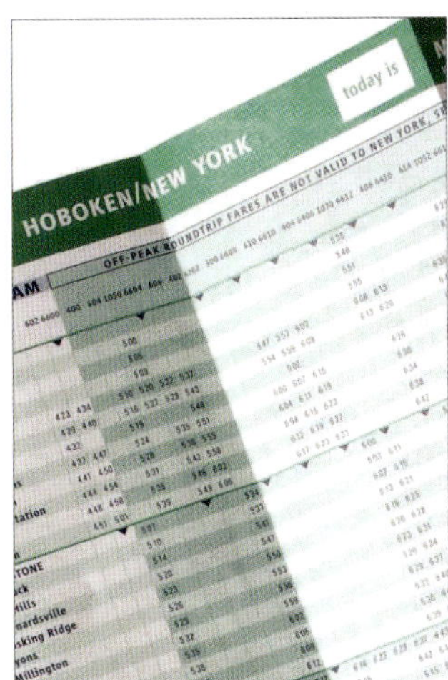

Pfeile kennzeichnen die Expresshaltestellen.

Eine übersichtliche und seriöse Typografie: Bei jeder Zeile und Spalte ist ausreichend Platz gelassen, um die Informationsdichte aufzulockern und gute Lesbarkeit zu erhalten. Bindestriche und Unterringelungen kommen sparsam, aber effektiv zum Einsatz. In den weißen Pfeilen sind die Richtungen untergebracht. Unterbrochene schwarze Balken machen den Fahrplan noch anschaulicher.

37. Diagramme veredeln

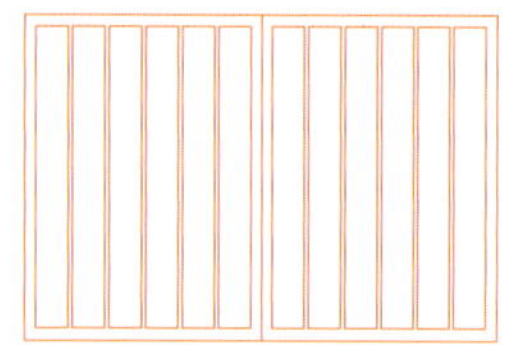

Tabellen und Diagramme enthalten wichtige Informationen, entweder gesetzlich vorgeschrieben oder als Strategie, um Aktionäre oder Investoren zu überzeugen. Gerade in Ge-schäftsberichten können solche Informationen zu monotonen Zahlenreihen werden.

Starke grafische Elemente sorgen hier für Dynamik. Zusammen mit abwechslungsreichen Größen und Strichstärken bringen Sie Wärme und Klarheit in den Bericht. Selbst mit einer begrenzten Palette geben Schriftgrößen und Formen dem Bericht Farbe und Struktur.

Siehe auch Seite
76/77

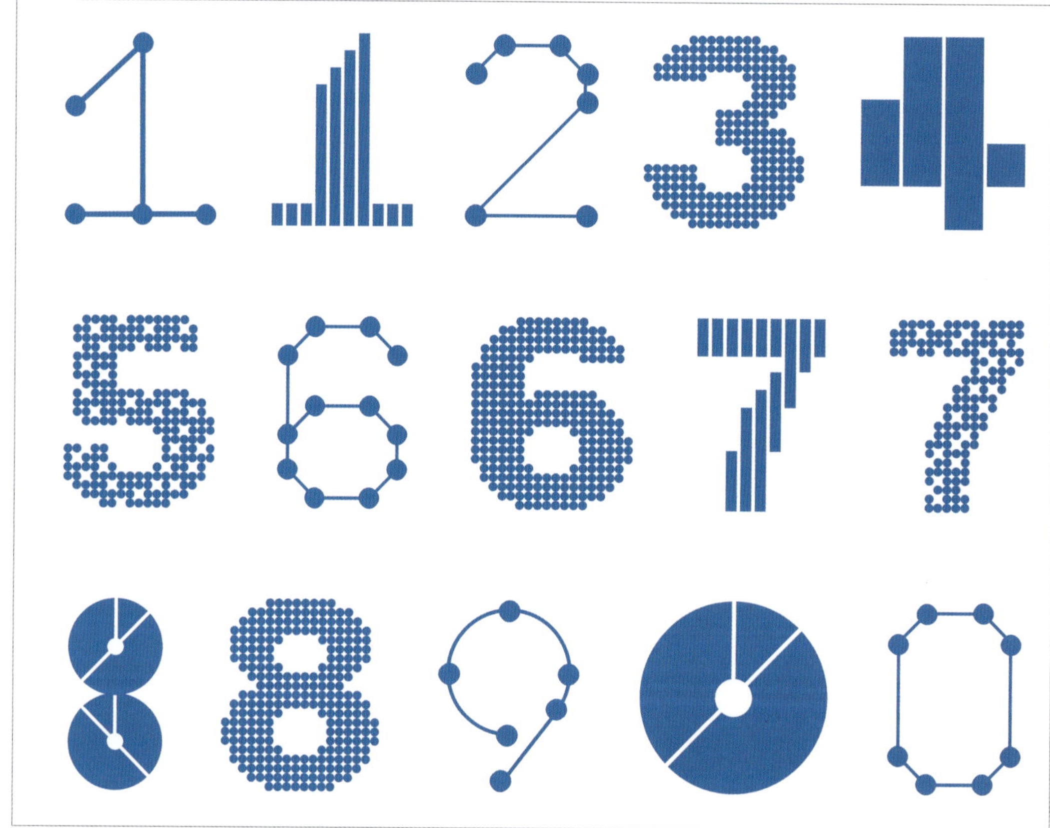

Die grafisch wirkenden Zahlen sind diszipliniert und dennoch reizvoll.

PROJEKT
Zahlen und Geschäftsbericht 2017 der Banc Sabadell (gedruckt)

KUNDE
Banc Sabadell

DESIGN
Mario Eskenazi Studio

DESIGNER
Mario Eskenazi,
Gemma Villegas

Die für den Geschäftsbericht der Banc Sabadell kreierten Zahlen waren so erfolgreich, dass Sie anschließend auch für Werbematerial (Weihnachtsgeschenke usw.) benutzt wurden.

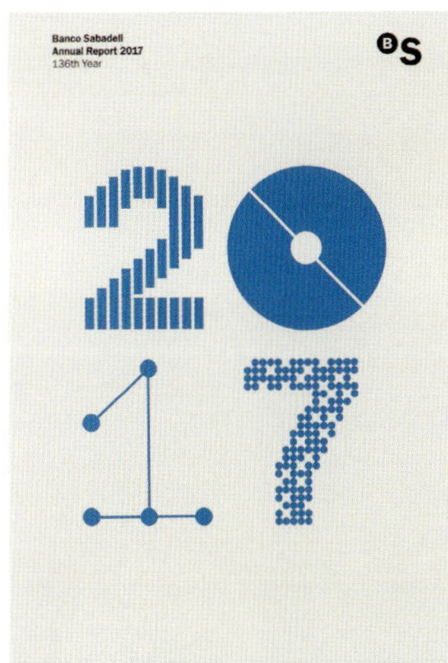

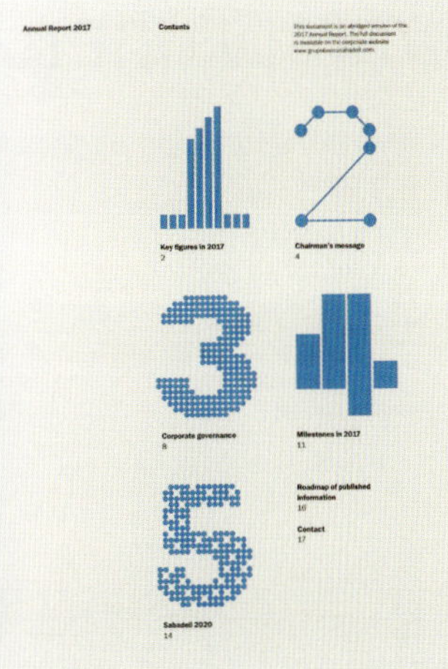

Die clever konzipierten Zahlen finden sich sowohl auf dem Cover als auch im Text und werden als Kapitelnummern wiederholt.

Hervorgehobene Statistiken werden in verschiedenen, aber regelmäßigen Rastern gezeigt, die auf vertikalen und horizontalen Hierarchien und gut organisierten Titeln und Zwischentiteln basieren.

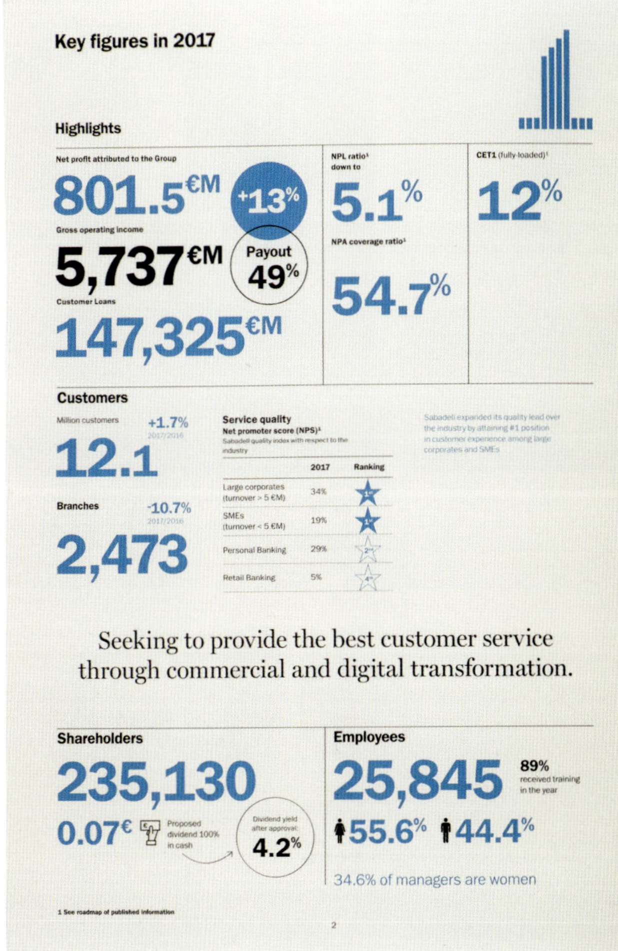

Key figures in 2017

Highlights

Net profit attributed to the Group
801.5€M +13%

Gross operating income
5,737€M Payout 49%

Customer Loans
147,325€M

NPL ratio[1] down to
5.1%

NPA coverage ratio[1]
54.7%

CET1 (fully-loaded)[1]
12%

Customers

Million customers +1.7% 2017/2016
12.1

Branches -10.7% 2017/2016
2,473

Service quality
Net promoter score (NPS)[1]
Sabadell quality index with respect to the industry

	2017	Ranking
Large corporates (turnover > 5 €M)	34%	1st
SMEs (turnover < 5 €M)	19%	1st
Personal Banking	29%	2nd
Retail Banking	5%	4th

Sabadell expanded its quality lead over the industry by attaining #1 position in customer experience among large corporates and SMEs

Seeking to provide the best customer service through commercial and digital transformation.

Shareholders

235,130

0.07€ Proposed dividend 100% in cash

Dividend yield after approval: 4.2%

Employees

25,845 89% received training in the year

55.6% 44.4%

34.6% of managers are women

1 See roadmap of published information

2

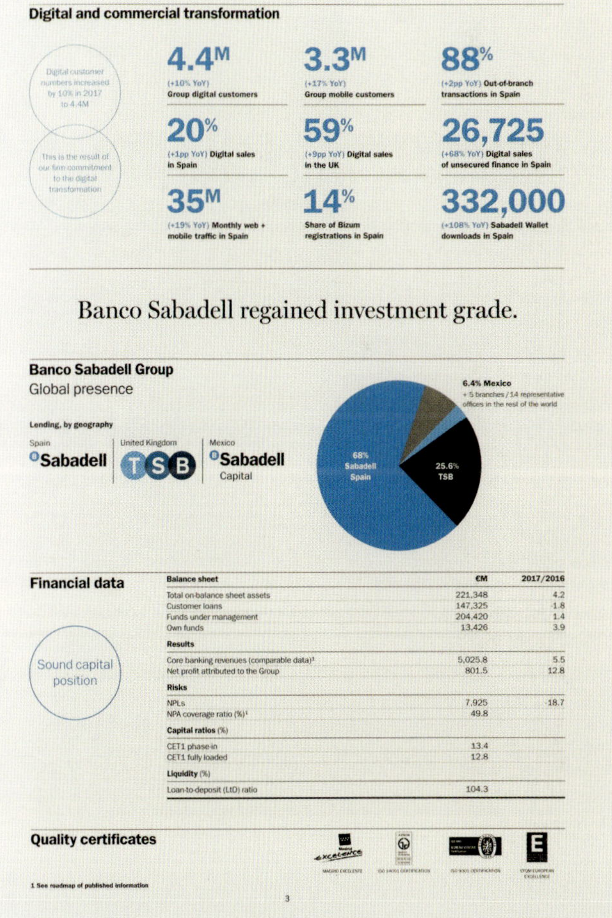

Digital and commercial transformation

Digital customer numbers increased by 10% in 2017 to 4.4M

This is the result of our firm commitment to the digital transformation

4.4M (+10% YoY) Group digital customers

3.3M (+17% YoY) Group mobile customers

88% (+2pp YoY) Out-of-branch transactions in Spain

20% (+1pp YoY) Digital sales in Spain

59% (+9pp YoY) Digital sales in the UK

26,725 (+68% YoY) Digital sales of unsecured finance in Spain

35M (+19% YoY) Monthly web + mobile traffic in Spain

14% Share of Bizum registrations in Spain

332,000 (+108% YoY) Sabadell Wallet downloads in Spain

Banco Sabadell regained investment grade.

Banco Sabadell Group
Global presence

Lending, by geography

Spain: Sabadell | United Kingdom: TSB | Mexico: Sabadell Capital

Financial data

Sound capital position

Balance sheet	€M	2017/2016
Total on-balance sheet assets	221,348	4.2
Customer loans	147,325	-1.8
Funds under management	204,420	1.4
Own funds	13,426	3.9
Results		
Core banking revenues (comparable data)[1]	5,025.8	5.5
Net profit attributed to the Group	801.5	12.8
Risks		
NPLs	7,925	-18.7
NPA coverage ratio (%)[1]	49.8	
Capital ratios (%)		
CET1 phase-in	13.4	
CET1 fully loaded	12.8	
Liquidity (%)		
Loan-to-deposit (LtD) ratio	104.3	

Quality certificates

1 See roadmap of published information

3

38. Freude versprühen

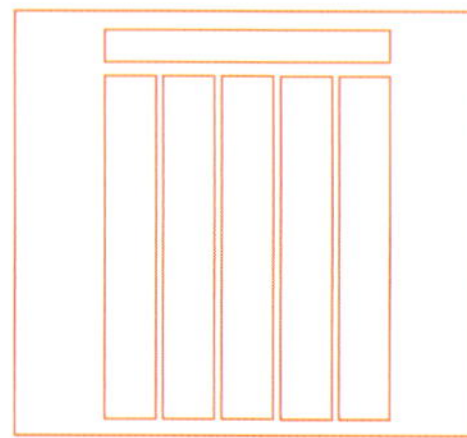

Maßgeschneiderte Zahlen begrüßen den Leser auf der Einstiegsseite und werden in den verschiedenen Medien wiederholt. Kartengrafiken, die auf den gleichen Mustern basieren wie die Zahlen, funktionieren gut bei Balkendiagrammen und Statistiken.

Siehe auch Seite
74/75

DIESE SEITE:

PROJEKT
Geschäftsbericht 2016 der Banc Sabadell (Desktop)

KUNDE
Banc Sabadell

DESIGN
Mario Eskenazi Studio

DESIGNER
Mario Eskenazi,
Gemma Villegas

GEGENÜBERLIEGENDE SEITE

PROJEKT
Geschäftsbericht 2016 der Banc Sabadell (mobile Geräte)

KUNDE:
Banc Sabadell

DESIGN
Mario Eskenazi Studio

DESIGNER
Mario Eskenazi,
Gemma Villegas

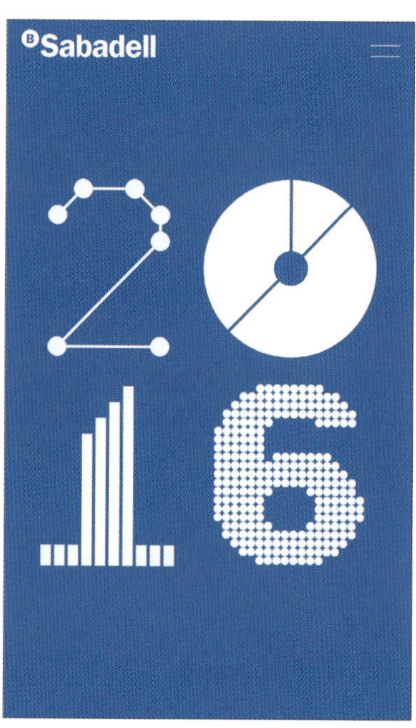

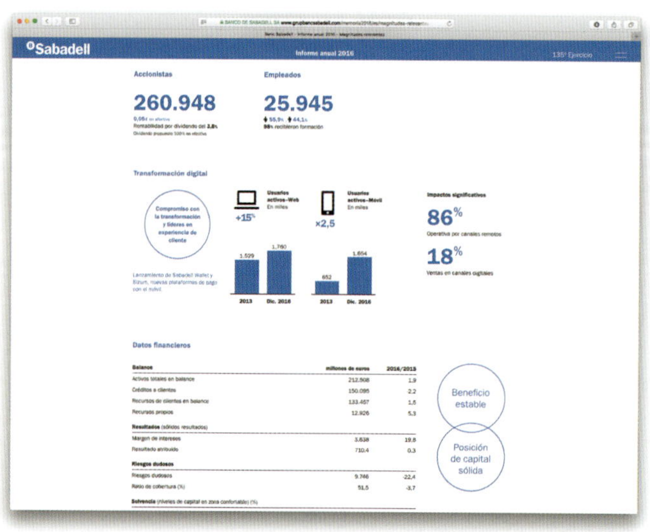

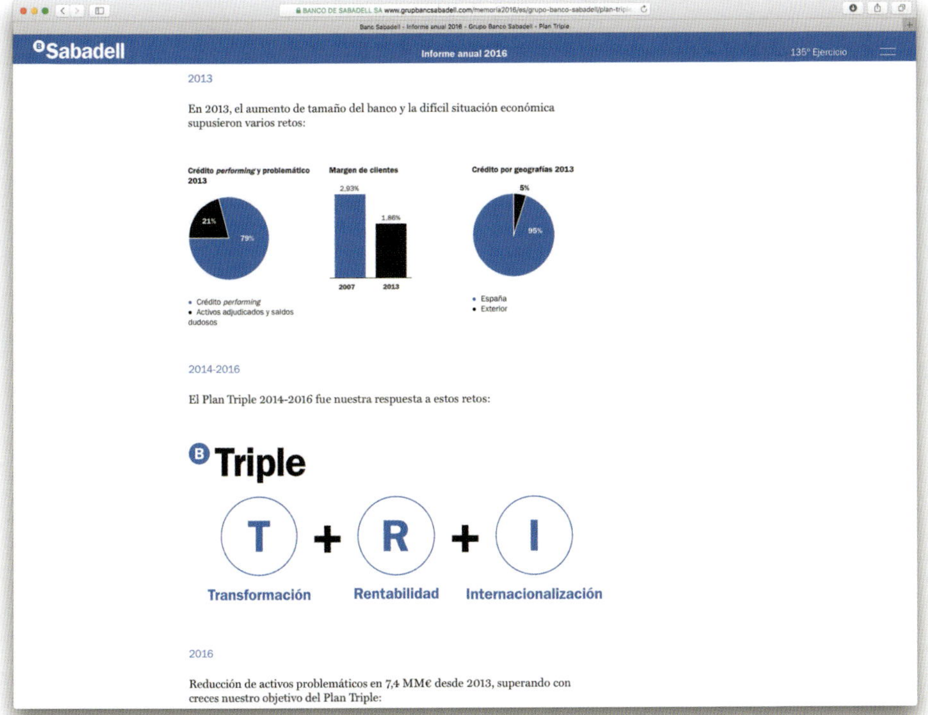

Sabadell
Informe anual 2016
135° Ejercicio
2013
En 2013, el aumento de tamaño del banco y la difícil situación económica supusieron varios retos:
Crédito performing y problemático 2013
Margen de clientes
Crédito por geografías 2013
21%
79%
2,93%
1,86%
2007
2013
5%
95%
Crédito performing
Activos adjudicados y saldos dudosos
España
Exterior
2014-2016
El Plan Triple 2014-2016 fue nuestra respuesta a estos retos:
Triple
T
R
I
Transformación
Rentabilidad
Internacionalización
2016
Reducción de activos problemáticos en 7,4 MM€ desde 2013, superando con creces nuestro objetivo del Plan Triple:

Sabadell
El Plan Triple 2014-2016 fue nuestra respuesta a estos retos:
Triple
T
R
I
Transformación
Rentabilidad
Internacionalización
2016
Reducción de activos problemáticos en 7,4 MM€ desde 2013, superando con creces nuestro objetivo del Plan Triple:
Crédito performing y problemático 2016
12%
88%

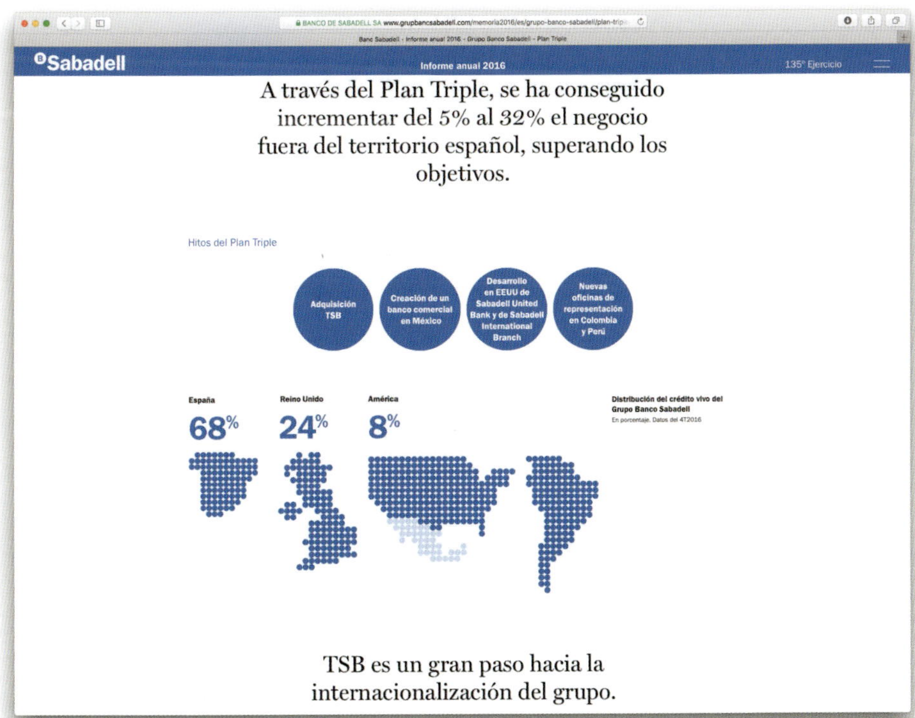

Sabadell
Informe anual 2016
135° Ejercicio
A través del Plan Triple, se ha conseguido incrementar del 5% al 32% el negocio fuera del territorio español, superando los objetivos.
Hitos del Plan Triple
Adquisición TSB
Creación de un banco comercial en México
Desarrollo en EEUU de Sabadell United Bank y de Sabadell International Branch
Nuevas oficinas de representación en Colombia y Perú
España
Reino Unido
América
68%
24%
8%
Distribución del crédito vivo del Grupo Banco Sabadell
TSB es un gran paso hacia la internacionalización del grupo.

Sabadell
68%
24%
8%
Distribución del crédito vivo del Grupo Banco Sabadell
En porcentaje. Datos del 4T2016
TSB es un gran paso hacia la internacionalización del grupo.
TSB
6,4%
Cuota en cuentas corrientes

39. Rahmen clever eingesetzt

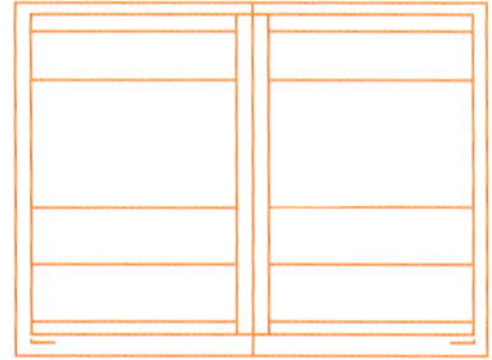

Ideal ist es, wenn tabellarische Informationen ohne eine verwirrende Anzahl von Rahmen und Boxen dargestellt werden. In einigen Fällen aber besteht das Material aus so vielen eigenständigen Elementen, dass Rahmen die beste Lösung sind.

Natürlich ist es möglich, Bestellkarten auch ohne Rahmen und Linien zu gestalten. Verschiedene Unterteilungen und variierende Stärken der Rahmen und Boxen schaffen jedoch nicht nur Überblick, sondern signalisieren auch Struktur und Ordnung.

郵 便 は が き

料金受取人払郵便

新宿北局承認

4121

差出有効期間
平成21年11月
23日まで
★切手不要★

1 6 9 ― 8 7 9 0
1 3 3

東京都新宿区北新宿1-35-20

暮しの手帖社

4世紀31号アンケート係 行

ご住所 〒 ―

電話 ― ―

お名前

メールアドレス @

年齢 [ ] 歳

性別 女 ／ 男

ご職業 [ ]

ご希望のプレゼントに○をつけて下さい。

☐「日東紡のふきん」3枚箱入り

☐「花森安治の表紙絵ポストカード」5枚セット

いただいた個人情報は、誌面作り、当選プレゼントの発送、小社グループの商品案内等の送付に利用させていただき、厳重に管理、保管いたします。

＊ご回答は、184ページの記事一覧をご参照の上、番号でご記入下さい。

A．表紙の印象はいかがですか [ ]
ご意見：

B．面白かった記事を3つ、挙げて下さい [ ] [ ] [ ]

C．役に立った記事を3つ、挙げて下さい [ ] [ ] [ ]

D．興味がなかった、あるいは面白くなかった記事を3つ、挙げて下さい [ ] [ ] [ ]

E．今号を何でお知りになりましたか [ ]
その他：

F．小誌と併読している雑誌を教えて下さい

G．小誌を買った書店を教えて下さい [区市町村]

H．小誌へのご要望、ご意見などございましたらご記入下さい

◎ご協力、ありがとうございました。

Diese und rechte Seite: Die Gestaltung dieser Bestellkarte konzentriert sich auf dem Duktus der Linien. Mit den fetteren wird die Wichtigkeit des Inhalts signalisiert und es werden bestimmte Inhalte betont. Damit wirkt das Layout harmonisch und weniger wichtige Information lässt sich elegant absetzen.

PROJEKT
Kurashi no techo (Everyday Notebook) Magazin

KUNDE
Kurashi no techo (Everyday Notebook) Magazin

DESIGNER
Shuzo Hayashi, Masaaki Kuroyanagi

Eine Bestellkarte, so schön wie funktionell

● 【定期購読】【商品、雑誌・書籍】のお申込みは、こちらの払込取扱票に必要事項を必ず記入の上、最寄りの郵便局に代金を添えてお支払い下さい。
● 169項、183頁の注文方法をご覧下さい。
● 表示金額はすべて税込価格となっております。
● 注文内容を確認させていただく場合がございます。平日の日中に連絡のつく電話番号を、ＦＡＸ番号がございましたら払込取扱票にご記入ください。
● プレゼントの場合はご注文いただいたお客様のご住所、お名前でお送りします。

02 東京

払込取扱票

通常払込料金加入者負担

口座番号 00190=7=45321

金額 6300

加入者名 株式会社 暮しの手帖社

料金　特殊取扱

通信欄
※「暮しの手帖」の定期購読を
20____年____号より1年間（6冊）申し込みます
※プレゼントされる場合、送付先が異なる場合はご送付先を下欄へ記入下さい。
ご住所 〒□□□-□□□□
ご氏名　tel

払込人住所氏名（郵便番号　）※
（電話番号　－　－　）
（FAX　－　－　）

受付局日附印

裏面の注意事項をお読みください。（私製承認東第43990号）
これより下部には何も記入しないでください。

各票の※印欄は、払込人において記載してください。

切り取らないで郵便局にお出しください。

払込金受領証

口座番号 00190=7 45321（通常払込料金加入者負担）

加入者名 株式会社 暮しの手帖社

金額 6300

払込人住所氏名 ※

料金　受付局日附印

特殊取扱

記載事項を訂正した場合は、その箇所に訂正印を押してください。

02 東京

払込取扱票

通常払込料金加入者負担

口座番号 00170=1=59128

金額 ※

加入者名 株式会社 グリーンショップ

料金　特殊取扱

通信欄
※プレゼントされる場合、送付先が異なる場合はご送付先を下欄へ記入下さい。
ご住所 〒□□□-□□□□
ご氏名　tel

払込人住所氏名（郵便番号　）※
（電話番号　－　－　）
（FAX　－　－　）

受付局日附印

裏面の注意事項をお読みください。（私製承認東第44327号）
これより下部には何も記入しないでください。

各票の※印欄は、払込人において記載してください。

切り取らないで郵便局にお出しください。

払込金受領証

口座番号 00170=1 59128（通常払込料金加入者負担）

加入者名 株式会社 グリーンショップ

金額 ※

払込人住所氏名 ※

料金　受付局日附印

特殊取扱

記載事項を訂正した場合は、その箇所に訂正印を押してください。

40. Über das Rechteck hinausdenken

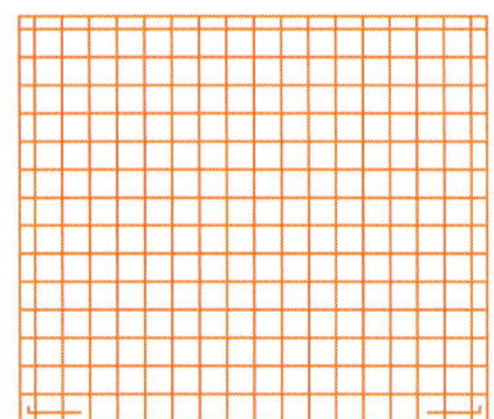

Auf der Grundlage von Rastern lassen sich unkonventionelle Formen gestalten, die den Raum in sinnvolle Abschnitte unterteilen. Ein Kreis kann horizontal oder vertikal in Hälften oder Quadranten aufgeteilt werden oder auch radial in Tortenstücke.

Hier besteht ein starker Kontrast zwischen Bildern und Texteinheiten. Die Typografie ist einfach gehalten. Fette Überschriften erinnern an das Logo und betonen die URL. Die horizontale Schrift auf dem Wagen der Subway korrespondiert mit dem nebenstehenden Text.

NAME:

NEW YORK TRANSIT MUSEUM

Think About It...

When New York City's first subway opened on October 27, 1904, there were about 9 miles of track. Today the subway system has expanded to 26 times that size. About how many miles of track are there in today's system?

Most stations on the first subway line had tiles with a symbol, such as a ferry, lighthouse, or beaver. These tiles were nice decoration, but they also served an important purpose. Why do you think these symbols were helpful to subway passengers?

When subway service began in 1904, the fare was five cents per adult passenger. How much is the fare today? Over time, subway fare and the cost of a slice of pizza have been about the same. Is this true today?

Today's subway system uses a fleet of 6,200 passenger cars. The average length of each car is 62 feet. If all of those subway cars were put together as one super-long train, about how many miles long would that train be? (Hint: There are 5,280 feet in a mile.)

Redbird subway cars, which were first built for the 1964 World's Fair, were used in New York City until 2003. Then many of them were tipped into the Atlantic Ocean to create artificial reefs. A reef makes a good habitat for ocean life—and it is a good way to recycle old subway cars! Can you think of other ways that mass transit helps the environment?

To check your answers and learn more about New York City's subway system, visit our website: **www.transitmuseumeducation.org**. You'll also find special activities, fun games, and more!

The New York Transit Museum's programs are made possible, in part, with public funds from the New York State Council on the Arts, a state agency. All photographs are from the New York Transit Museum Collection.

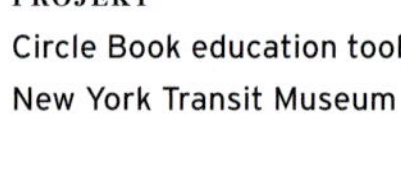

PROJEKT
Circle Book education tool, New York Transit Museum

KUNDE
New York Transit Museum

PROJEKTENTWICKLUNG
Lynette Morse and Virgil Talaid, Education Department

DESIGN
Carapellucci Design

DESIGNER
Janice Carapellucci

In diesem Lernmittel, einer Drehscheibe, wird Didaktik, Information und Aktivität kombiniert – zudem lässt sie sich auch, passend zum Thema, bewegen.

In den dicken Balken sind die Arbeitsanweisungen und Felder zum Ausfüllen untergebracht. Die Farben Blau und Rot entsprechen den Kennfarben der New Yorker Linien A, C und E sowie 1, 2 und 3.

41. Farbe als Blickfang

Jede Publikation, ob einförmig oder abwechselnd im Raster, profitiert von starken Farben, die Kapitel, Geschichten oder Text hervorheben. Farbige Hintergründe im Wechsel mit weißen oder helleren Seiten variieren den Rhythmus und halten die Aufmerksamkeit aufrecht. Seitenleisten oder Untertexte in anderen Farben heben Informationen auch ohne Linien oder Rahmen hervor.

Die erste und die zweite Ausgabe eines Magazins von The Wing, Arbeitszentrum und Treffpunkt für Frauen, verkünden kühn und farbenfroh, dass die Frauen auf dem Vormarsch sind.

PROJEKT
No Man's Land

KUNDE
The Wing

DESIGN
Pentagram

CREATIVE DIRECTION
Emily Oberman

PARTNER
Emily Oberman

SENIOR DESIGNER
Christina Hogan

DESIGNER
Elizabeth Goodspeed

DESIGNER
Joey Petrillo

PROJECT MANAGER
Anna Meixler

Farbige Seiten, Hintergründe und Schriften machen eine starke Publikation noch stärker.

OBEN: Farbige Aufmacher sorgen für Spannung und Abwechslung.

UNTEN LINKS: Doppelseiten mit farbigem Hintergrund mischen alles auf. UNTEN RECHTS: Andersfarbige Zeitleisten und verschiedene Spaltenbreiten heben sich vom Fließtext ab.

McGee had seen and done a lot in her post-skate career. She worked on a fishing boat owned a topless bar got divorced raised her kids ran a trading post with her second husband. At 72, when life tends to slow down for most people McGee is ready for her next adventure

Feature 62 Summer Issue Feature 63 Girls On Wheels

It was a little too chilly for Patti McGee in the air-conditioned skate shop, so she stood by a glass door, soaking in the California sun, presiding over the conversation like a knowing elder.

She munched on a lettuce-wrapped In-N-Out burger (protein style), her blonde hair catching the light, while the shop's owner, Matt Gaudio, told me the story of how McGee, 72, became the brand ambassador for his local skateboard team. Nearly 50 years after winning the first women's national skateboarding championship title in Santa Monica, McGee's likeness is the calling card of Gaudio's Silly Girls Skateboards, a small Fullerton-area girls' skateboarding team with 13 riders.

Behind McGee, displayed in a tall trophy case, was a Barbie doll styled to look just like her or, rather, who she was in 1965: shoeless, hair in a beehive, and doing a handstand on a skateboard. That was the year she became the first woman to win a national skateboarding competition and became a professional skateboarder. McGee made a career out of the sport before Tony Hawk and Rodney Mullen were

WOMEN SKATING THROUGH HISTORY

1963
Wendy Bearer Bull and her brother Danny become the first professionally endorsed skateboarders to be sponsored by Makaha Skateboard Club.

1965
Patti McGee appears on the cover of *LIFE* magazine in May 14, 1965. She goes on to become the first Women's National Skateboard champion.

Laurie Turner becomes the 1965 National Girls' Champion.

1975
Peggy Oki, the only woman on the legendary Zephyr skateboard

1997
The first issue of the *Villa Villa Cola* zine debuts, created by Tiffany and Nicole Morgan, two skateboarding sisters. It uses humor to encourage girls to skateboard and offers advice on how to overcome being intimidated by men in the field. Other zines, *Bruisers* and *50-50: Skateboarding and Gender*, soon follow.

1999
Elissa Steamer is the first woman to appear in Tony Hawk's skateboarding video game series.

2001
Jen O'Brien becomes the first girl to skate at

42. Zurückhaltende Farbpalette

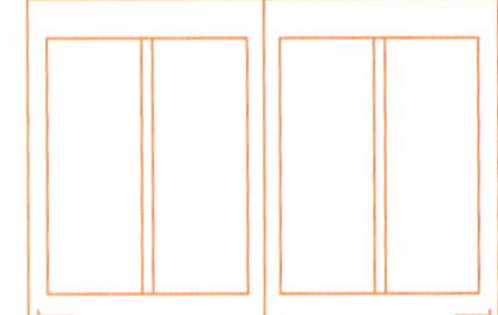

Die zurückhaltende Farbpalette mit nur Gold und Schwarz auf dem Cover sorgt für einen Unterschied zu anderen Publikationen. Der durchdachte Beschnitt wirkt stärker als das volle Bild und deutet auf ein Leben hin, das zu früh endete.

Farben ziehen zwar Blicke auf sich, können die eigentliche Botschaft aber auch erschlagen. Eine zurückhaltende Palette lenkt den Blick auf ein nüchternes oder ernüchterndes Thema.

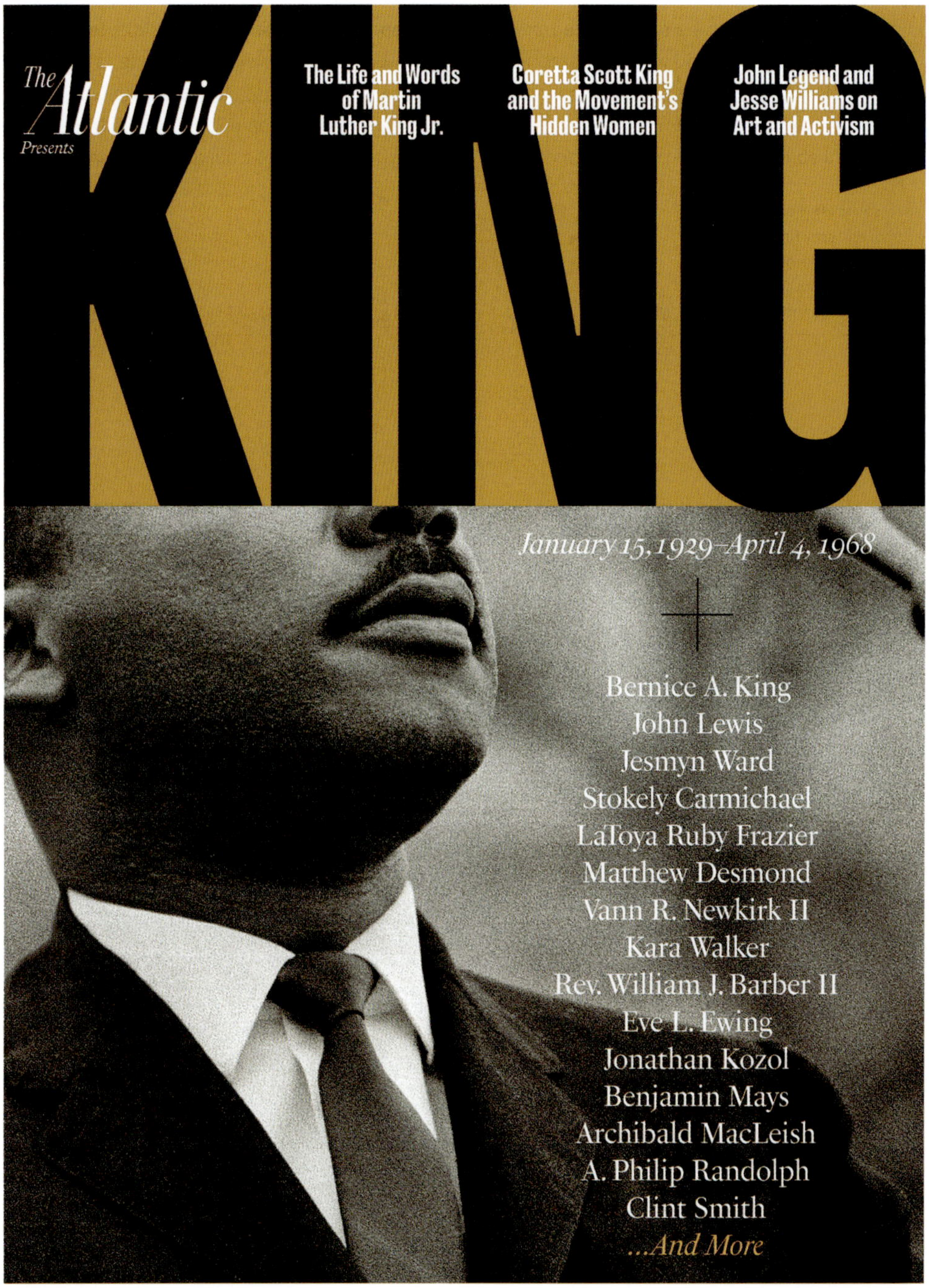

PROJEKT
King, eine Sonderausgabe zum 50. Jahrestag der Ermordung von Martin Luther King Jr.

KUNDE
The Atlantic

CREATIVE DIRECTOR
Paul Spella

ART DIRECTOR
David Somerville

DESIGNAGENTUR
OCD | Original Champions of Design

DESIGNER
Bobby C. Martin Jr.
Jennifer Kinon

Starke Schwarz-Weiß-Seiten, eine kühne Typografie mit geschichtlichen Bezügen und die zurückhaltende Farbpalette mit grünen Akzenten sorgen durchgehend für Dramatik.

Designer sprechen von „Design und Filmmetapher". Übereinander angeordnet zeigen diese Doppelseiten die filmische Verwendung von „The Negro is Your Brother", der Titel der Schlagzeile des Magazins The Atlantic, als dort 1963 der „Letter from Birmingham Jail" von Martin Luther King Jr. veröffentlicht wurde. Die grüne Schlagzeile wirkt wie eine visuelle Stimme aus dem Off.

Freie Räume, die schwarze Schrift und der grüne Akzent lenken die Aufmerksamkeit auf das gesamte Essay und auf besonders wichtige Passagen.

Bei diesem Prinzip geht es zwar um Farbe und ihre Kontrolle, aber diese Doppelseiten zeigen auch andere Aspekte der Arbeit mit Rastern. Bemerkenswert sind die freien Räume, besonders zu Beginn des Essays. Die Bildunterschrift nutzt nicht die gesamte Breite, sondern basiert auf einem separaten Raster. Mit dem freien Raum spiegelt sie das meditative Image von Martin Luther King Jr. wider.

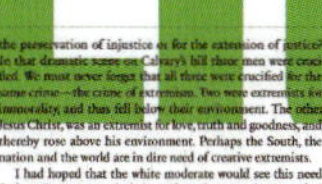

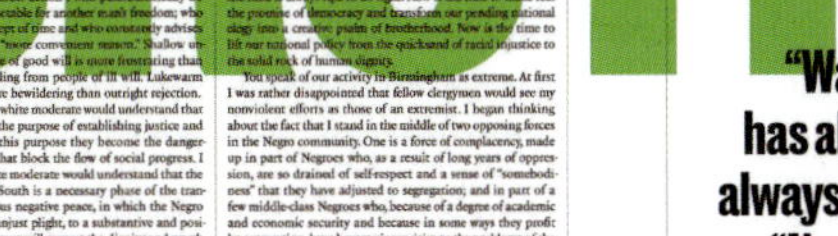

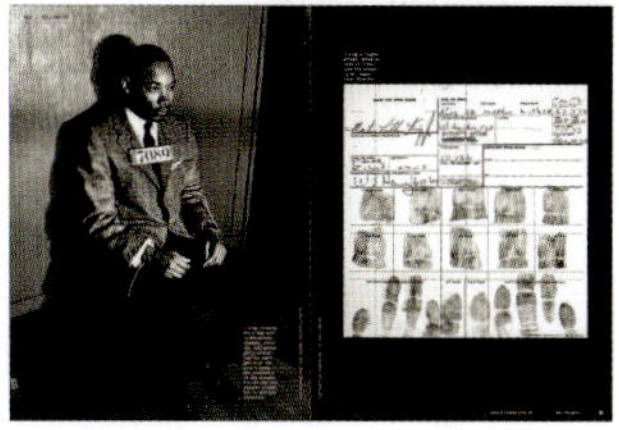

43. Farben sprechen lassen

Wenn eine solide Struktur gegeben ist, und das ist bei Magazinen häufig der Fall, tut es gelegentlich gut, daraus auszubrechen. Neben einer schlichten Typografie werden hier Farben in Gestalt eines großartigen Fotos in den Mittelpunkt (des Layouts) gerückt.

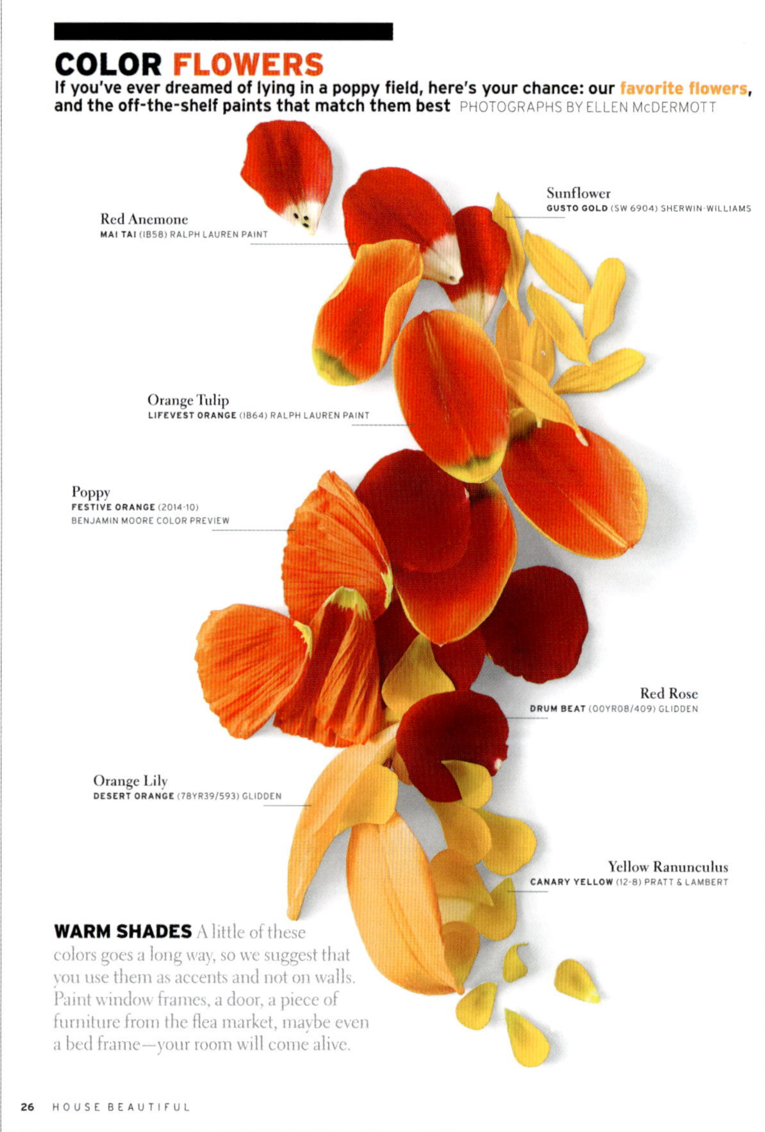

COLOR FLOWERS

If you've ever dreamed of lying in a poppy field, here's your chance: our favorite flowers, and the off-the-shelf paints that match them best PHOTOGRAPHS BY ELLEN McDERMOTT

Red Anemone
MAI TAI (1B58) RALPH LAUREN PAINT

Sunflower
GUSTO GOLD (SW 6904) SHERWIN-WILLIAMS

Orange Tulip
LIFEVEST ORANGE (1B64) RALPH LAUREN PAINT

Poppy
FESTIVE ORANGE (2014-10) BENJAMIN MOORE COLOR PREVIEW

Red Rose
DRUM BEAT (00YR08/409) GLIDDEN

Orange Lily
DESERT ORANGE (78YR39/593) GLIDDEN

Yellow Ranunculus
CANARY YELLOW (12-8) PRATT & LAMBERT

WARM SHADES A little of these colors goes a long way, so we suggest that you use them as accents and not on walls. Paint window frames, a door, a piece of furniture from the flea market, maybe even a bed frame—your room will come alive.

26 HOUSE BEAUTIFUL

DIESE UND RECHTE SEITE: Auch wenn hier die Versuchung groß ist, noch mehr Farbe hereinzubringen, ist eine Beschränkung der Palette, etwa nur Schwarz, klüger. Dadurch wird dem Leser Gelegenheit gegeben, sich auf die satten Farben des Bildes zu konzentrieren. Ein Wettbewerb der Farben wäre hier nur kontraproduktiv.

PROJEKT
House Beautiful

KUNDE
House Beautiful magazine

DESIGN
Barbara deWilde

Ein hervorragendes und kunstvoll gestaltetes Bild hat in diesem Layout keine Konkurrenz zu fürchten.

COLOR

Violet
GENTLE VIOLET (2071-20)
BENJAMIN MOORE COLOR PREVIEW

Peony
SWEET TAFFY (2086-60)
BENJAMIN MOORE COLOR PREVIEW

Magenta Anemone
FORWARD FUCHSIA
(SW 6842) SHERWIN-WILLIAMS

Hydrangea
TROOPER (26-14) PRATT & LAMBERT

Pink Rose
PEACHGLOW (90YR71/144) GLIDDEN

Water Lily
TULIPE VIOLET (30-14) PRATT & LAMBERT

Orchid
VESPER (70RB67/067) GLIDDEN

Hyacinth
ORIENTAL NIGHT (29-14) PRATT & LAMBERT

COOL SHADES A word about finishes: Light colors look darker in a flat finish. Dark colors look brighter in a gloss or semigloss. A flat finish will work well for the lighter shades here, but the deep purples and pinks will definitely look better with a sheen.

Pink Daisy
SMASHING PINK (1303)
BENJAMIN MOORE CLASSIC COLORS

FOR MORE DETAILS, SEE RESOURCES

28 HOUSE BEAUTIFUL

44. Die Verbindung von Farbe und Typografie

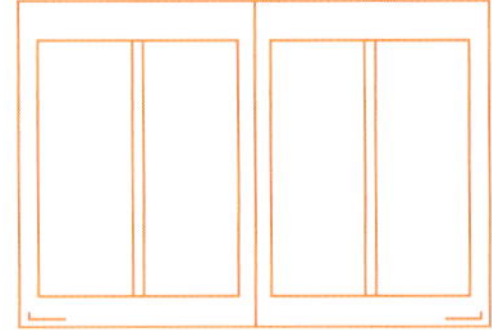

In farbigen Ratgebern oder Büchern mit didaktischem Inhalt ist es oft ratsam, die Farben eher zurückhaltend zu gestalten, damit die Anleitungen nicht von anderen Elementen dominiert werden. Eine geschickte, beschränkte Farbauswahl kann jedoch die Typografie hervorragend zur Geltung bringen.

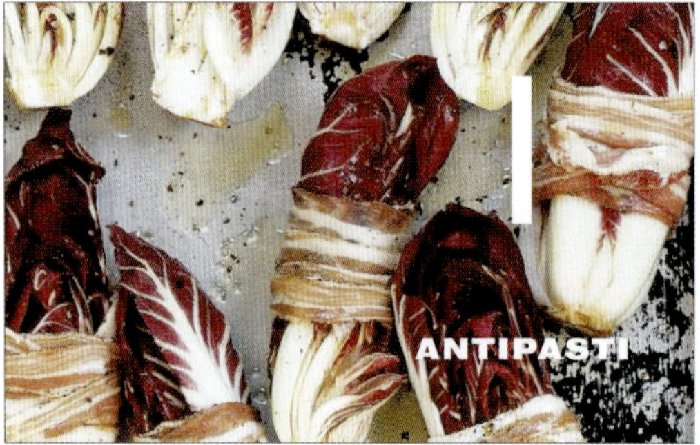

Farbig sehr intensive Motive mit groß gezogenen Bildausschnitten bestreiten die Kapitelanfänge. Die fetten Negativlettern der Headline bieten der Farbintensität die Stirn.

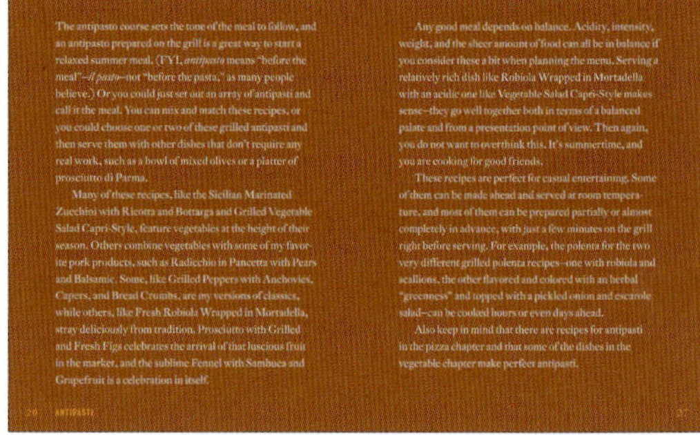

Nach dem grandiosen Kapitelauftakt folgt jeweils eine Doppelseite Einführung. Im Kontrast zur fetten serifenlosen Schrift auf den Aufmacherfotos erscheint der Einführungstext in einer fast grazilen Serifenschrift negativ im Farbfond.

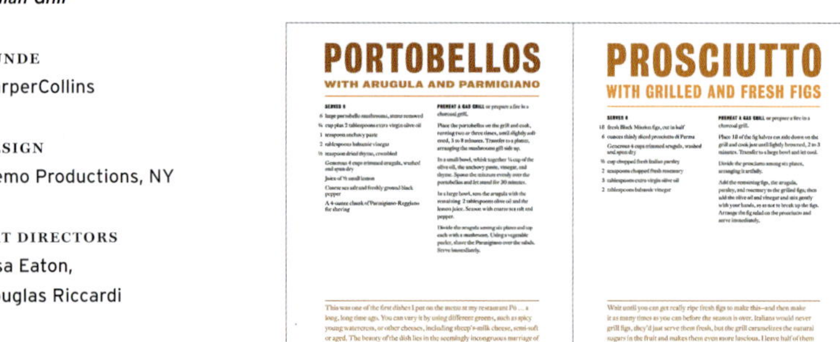

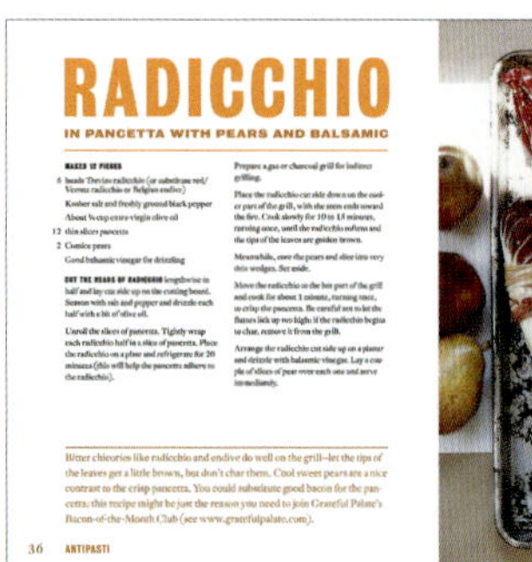

Die drei unteren Bilder auf dieser Seite und rechte Seite: In jedem Kapitel variiert der Farbcode und ergänzt so die Wirkung der Farbfotografien.

PROJEKT
Italian Grill

KUNDE
HarperCollins

DESIGN
Memo Productions, NY

ART DIRECTORS
Lisa Eaton,
Douglas Riccardi

Dieses Kochbuch eines Koches mit markanter Persönlichkeit wird von Rastern untermauert. Hier werden fette farbige Lettern einer nicht minder-frechen Typografie gegenübergestellt. Die Kapitel variieren jeweils nur leicht und alle sind gleich attraktiv.

In Italy, cooking fish is all about freshness and simplicity—as I've said before, the philosophy of Italian fish cookery can be summed up in three words: *Leave it alone.* Complicated sauces and techniques are not part of the repertoire, and, in fact, Italians almost never serve any sauce at all with fish, other than an excellent olive oil. Lemon may sometimes appear, but even that is often considered beside the point. The one exception is *salsa verde*, the fragrant green herb sauce, which may sometimes accompany a fish with character enough to stand up to it, such as a whole grilled branzino (see page 126).

Few Italians would consider cooking anything other than local fish, whether from a mountain stream or the ocean, and I urge you to think in the same way: find a good fish market, and remember that what is freshest is best. If the specific fish called for in your recipe is not available—or doesn't look pristine and glistening—the fishmonger can help you choose another option (I include suggestions for substitutions in many of the recipes). If you are able to get fresh king mackerel for Mackerel "in Scapece" with Amalfi Lemon Salad, you will have the best mackerel dish you've ever tasted; if you can't find it, make the recipe with very fresh bluefish, or move on to another one. Most of the other fish recipes in this chapter, such as Monkfish in Prosciutto with Pesto Fregola and Swordfish Involtini Sicilian-Style, call for widely available varieties. But you'll want to be sure to get the best tuna available—sushi-quality, that is—for Tuna Like Fiorentina, and you really should use wild salmon for the Salmon in Cartoccio with Asparagus, Citrus, and Mint.

Cooking shellfish on the grill is easy, and the recipes in this chapter use several different techniques for achieving simple perfection. Clams in Cartoccio are wrapped in a foil package and allowed to steam in their fragrant juices. The shrimp in Shrimp Rosemary Spiedini alla Romagnola are threaded onto rosemary skewers, which impart their herbal fragrance and look sexy besides. I love cooking shellfish (and cephalopods) on a piastra, a flat griddle or stone placed on the hot grill (see page 000 for more on the subject), because it gives them a great sear and char, as in Sea Scallops alla Caprese or Marinated Calamari with Chickpeas, Olive Pesto, and Oranges.

Thinking globally while buying locally is especially important when you are buying fish. Some "trendy" fish have been overharvested to the point of extinction, and we now know that there can be problems with farmed fish as well, like salmon. The Monterey Bay Aquarium, at www.montereybayaquarium.com, maintains an up-to-date list of species that are being overfished in the United States and in the rest of the world. It's an invaluable resource, and I urge you to consult it when writing your shopping list, as I do both at home and at the restaurants.

86 FISH AND SHELLFISH 87

MARINATED CALAMARI

WITH CHICKPEAS, OLIVE PESTO, AND ORANGES

SERVES 6

CALAMARI

- 3 pounds cleaned calamari (tubes and tentacles)
- ¼ cup extra-virgin olive oil
- Grated zest and juice of 1 lemon
- 4 garlic cloves, thinly sliced
- 2 tablespoons chopped fresh mint
- 2 tablespoons hot red pepper flakes
- 2 tablespoons freshly ground black pepper

CHICKPEAS

- Two 15-ounce cans chickpeas, drained and rinsed, or 3½ cups cooked chickpeas
- ½ cup extra-virgin olive oil
- ¼ cup red wine vinegar
- 4 scallions, thinly sliced
- 4 garlic cloves, thinly sliced
- ¼ cup mustard seeds
- Kosher salt and freshly ground black pepper

OLIVE PESTO

- ¼ cup extra-virgin olive oil
- Grated zest and juice of 1 orange
- ½ cup black olive paste
- 4 jalapeños, finely chopped
- 12 fresh basil leaves, cut into chiffonade (thin slivers)

- 3 oranges
- 2 tablespoons chopped fresh mint

CUT THE CALAMARI BODIES crosswise in half if large. Split the groups of tentacles into 2 pieces each.

Combine the olive oil, lemon zest and juice, garlic, mint, red pepper flakes, and black pepper in a large bowl. Toss in the calamari and stir well to coat. Refrigerate for 30 minutes, or until everything else is ready.

Put the chickpeas in a medium bowl, add the oil, vinegar, scallions, garlic, and mustard seeds, and stir to mix well. Season with salt and pepper and set aside.

93

45. Farben kontrollieren und klären

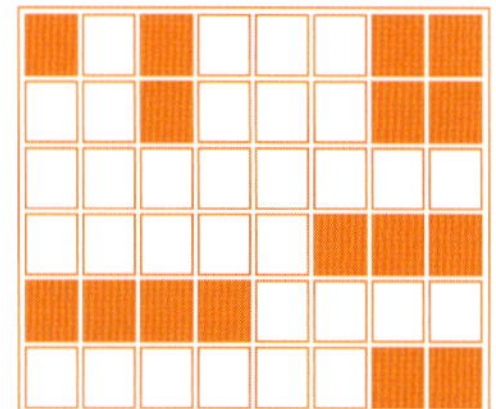

Dieser Ausstellungskalender ist ein Beispiel für einheitliche Größen innerhalb eines Rasters, in dem spielerisch Farbmodule sowie Text- und Bildmodule auf engem Raum variiert werden. Boxen und Farben bilden System und Struktur, wodurch die Informationen übersichtlich gestaltet werden können. Wenn viele verschiedene Informationen in einem Projekt untergebracht werden müssen, können in einem Raster Farbmodule mit Zeitangaben, Texteinheiten, URLs, Aufrufe oder Banner mit dem Thema des Projekts hervorgehoben werden.

PROJEKT
Program calendar

KUNDE
Smithsonian, Cooper-Hewitt, National Design Museum

DESIGN
Tsang Seymour Design, Inc.

DESIGN DIRECTOR
Patrick Seymour

ART DIRECTOR
Dieses System für einen Ausstellungskalender bietet eine gleichförmige Kommunikation. Dynamische Farb- und Bildvariationen sind jedoch ebenso möglich.

Auf der Rückseite des Ausstellungskalenders werden die Hauptausstellungen mit Ausstellungsdauer noch einmal kurz beschrieben. Ergänzend dazu üppige und überbordende Bilder, die optisch Spannung und Verdichtung erzeugen.

Unterschiedliche Bildgrößen und vereinzelte Freisteller passen zu den farbigen Boxen und springen teilweise ins Auge.

Stellen Sie fest, wie viel Fläche Sie zur Verfügung haben. Die teilen Sie in Quadrate gleicher Größe auf. Bestimmen Sie dann einen einheitlichen Rand für das Projekt. Benutzen Sie die Quadrate wie einzelne Boxen. Die können Sie dann in Zweier- oder Dreierpacks horizontal oder vertikal anordnen bzw. stapeln. Je nach Inhalten wie Datum, Monat, Preis, Veranstaltung etc. können Sie den Boxen bestimmte Farben zuweisen. Wenn die Informationen schwer verdaulich sind, sollten die Farben den Informationsfluss erleichtern und erhellen.

In den Modulen können auch Fotos und Illustrationen untergebracht werden. Wie beim Text kann ein Bild in einem Modul, in zwei vertikal angeordneten Modulen, in vier horizontal verlaufenden Modulen oder vier übereinandergestapelten Modulen Platz finden. Kurz gesagt erlauben die Farbmodule eine ganze Reihe von Variationsmöglichkeiten, ohne dadurch die Struktur und den Gesamteindruck zu gefährden. Unterbrechen Sie den Raster hie und da mit Freistellern, um das Interesse wachzuhalten und um einem lebendigen Programm noch mehr Rhythmus und optische Weitläufigkeit zu verleihen.

Durch die starke Struktur des Farbreigens können die Informationseinheiten für sich bestehen. Die Farbmodule begründen eine gut lesbare Hierarchie der Informationen von kleinen Schriftgraden bis wuchtigen Überschriften und fetten Lettern. Unterschiedliche Schriftgrößen sowie Groß- und Kleinbuchstaben ermöglichen dem Leser schnelles Erfassen von Zeitpunkt, Ort und Beschreibung der Veranstaltung. Große Überschriften in den zusammengefassten Modulen sorgen für Rhythmus und Überraschung. Daneben gibt es korrespondierend die Gestaltung ähnlicher Textinhalte wie Marketing, Auftraggeber oder Museum, Handlungsaufrufe und Kontaktdaten.

Für doppelseitige Projekte kann ein Modulraster mit klar definierten Flächen, die aber auch durchbrochen werden, sehr von Vorteil sein.

Linke Seite: Jeder Monat hat eine eigene Kennfarbe und ist so chronologisch klar zugeordnet.

FARBE ALS ORGANISIERENDES ELEMENT

46. Farbige Schrift als Blickfang

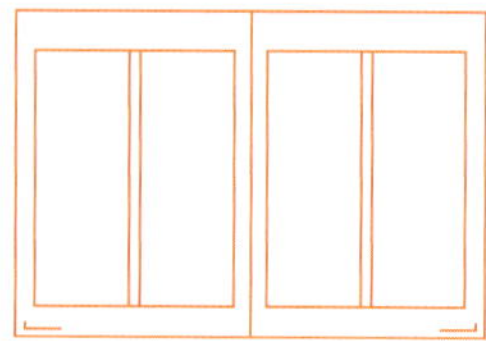

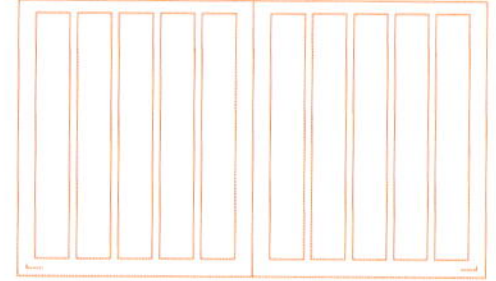

Zu viele Farben wirken auch wieder unruhig und verwirrend. Das richtige Maß jedoch versetzt den Leser in die Lage, die Inhaltsschwerpunkte besser zu erkennen. Eine klare Hierarchie bei den Überschriften wird noch leichter erfassbar, wenn damit farbliche Akzente einhergehen.

白玉すいとん

あり合わせの根菜と一緒に白玉団子を煮込んだ手軽な汁料理。「主食もおかずも一度に食べられる。撮影など仕事の合間の昼食としても活躍した汁ものです。祖母もよく準備の手を休めて食べていました」。すいとんと言えばうどん粉が王道だが、阿部さんはより手軽な白玉粉を好んで使った。豆腐を練り込んだ白玉は、もちもちと柔らかな食感。

材料 4人分

大根6cm　人参1/3本　ごぼう10cm　しめじ1/3房　まいたけ1/3房　油揚げ1枚　三つ葉8本　煮干し10本　薄口醤油大さじ2　豆腐約1/6丁　白玉粉約1/3カップ

作り方

1 煮干しは頭と内臓を取り、鍋などで乾煎りしてから、水につけておく。(約6カップ、分量外)

2 人参、大根は皮をむいて薄めのイチョウ切り、ごぼうは皮をたわしなどでよくこそいで洗い、薄く斜めに切って水にさらしておく。

3 油揚げは熱湯をかけ、油抜きして食べやすい大きさに切り、しめじ、まいたけは小房に分けておく。

4 1に2を加え火にかけ、ひと煮立ちしたら3も加え、薄口醤油を半量入れてしばらく煮る。

5 白玉粉に豆腐を混ぜ(写真)、みみたぶくらいの柔らかさにして形を整え、熱湯に入れて浮き上がってくるまで茹でたのち冷水に取る。

6 4に5を加え、ひと煮立ちしたら、残りの薄口醤油を加え味を調えて、ざく切りにした三つ葉を加える。

PROJEKT
Croissant Magazin

KUNDE
Croissant Magazin

DESIGNER
Seiko Baba

ILLUSTRATION
Yohimi Obata

Mit Farben lässt sich Text subtil betonen. Sie bringen Klarheit und Pep in Magazindoppelseiten. Das hier abgebildete heißt MOOK, eine Spezialausgabe der *Croissant*-Herausgeber. Der Titel lautet *Mukashi nagara no kurashi no chie*; frei übersetzt „Altehrwürdige Lebensweisheiten".

人参ご飯

根菜の甘みがやさしい炊き込みご飯2種。

大根飯

いつものフライもチーズを使い、ハイカラに。

丸干しのチーズフライ

白玉すいとん

豆腐入り白玉の食感が楽しい、素朴な汁もの。

Wenn ein Buchstabe größer und farbig gesetzt wird, lenkt das natürlich die Aufmerksamkeit auf diese Überschrift.

じゃが芋団子

薩摩芋もち

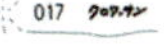

茄子の胡麻煮

煮干しとごぼうの立田揚げ

だしに使う煮干しも、立派なメインに。

Hier werden die Informationseinheiten farblich voneinander abgesetzt. Besonders bei Anleitungen ist eine klare Differenzierung nützlich und wichtig. Auf dieser Doppelseite eines Kochbuches sind die Zwischentitel farbig. Die Zahlen sind auch rot, um sie vom Text abzuheben.

お付き合い編

人付き合いを潤滑にする言葉づかい。

Die unterschiedlichen Ausgestaltungen des „Q" (für „Questions") in Strichstärke, Größe, Schattierung oder Farbe schaffen Struktur und erzeugen Aufmerksamkeit.

47. Kalendermodule mit Farbe aufwerten

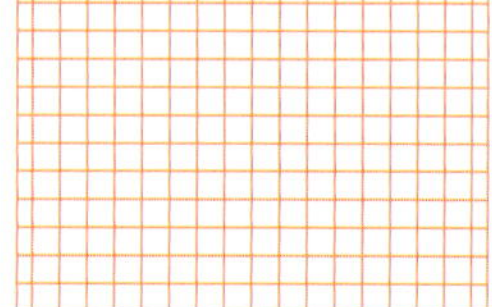

Die Farbgebung bei der Gestaltung eines Veranstaltungsprogramms erleichtert die Differenzierung verschiedenster Elemente wie Wochentage. Die Information sticht ins Auge und arbeitet mit der ganzen Doppelseite. Die Farben können auch so gewählt werden, dass sie die Farbpalette auf dem Foto ergänzen.

Wenn es bei einem Projekt besonders wichtig ist, dass die Zeitangaben klar und deutlich hervortreten, sind gedämpfte, sich nicht in den Vordergrund drängende Farben sehr günstig. Ungesättigte Farben (also solche mit einem hohen Grauanteil) sind ideal für das Übereinanderdrucken.

ALVIN AILEY AMERICAN DANCE THEATER

Nov. 28 – Dec. 31

Tickets: Start at $25

The dancers of Alvin Ailey American Dance Theater have it all – exquisite technique, breathtaking artistry and passionate spirit. The Company's 2007 season offers electrifying world premieres by Camille A. Brown and Frederick Earl Mosley; sumptuous new productions of classic works by Alvin Ailey and Talley Beatty and a Company premiere of Robert Battle's *Unfold*. AAADT becomes the first American company to perform a complete work by Maurice Béjart when they take on the radiant grandeur of his famous *Firebird*. Come to experience something brand new, see a forgotten favorite or revisit the beloved masterpiece *Revelations* once again.

Ein reicher Hintergrund und erstaunliche Dynamik stehen dem strukturierten Kalender in Komplementärfarben gegenüber.

PROJEKT
Veranstaltungskalender

KUNDE
New York City Center

DESIGN
Andrew Jerabek

Die Farbschattierungen der Kalenderboxen korrespondieren im Ton mit der Farbgebung der Fotografie

Die Farben für die Boxen sind so einfühlsam und treffend ausgewählt, dass sie die großartige Fotografie ideal ergänzen und keinesfalls mit ihr im Wettbewerb stehen.

Hier verstärken Herbstfarben ein beeindruckendes Foto mit safranfarbenen Akzenten.

48. Mit Farben Ordnung schaffen

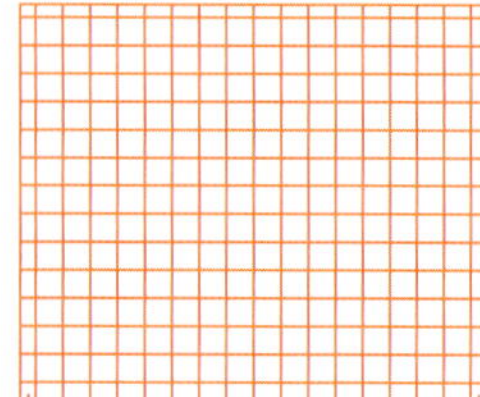

Wenn Informationen farblich gekennzeichnet werden, kann der Leser schnell die benötigte Auskunft finden. Farben, kombiniert mit grafischen Symbolen, sind bei weitem leichter und schneller zu erfassen als nur Text oder Farben.

Abhängig vom Kunden oder von Material und Projekt können gedämpfte oder auch kräftige Farben gewählt werden. Gesättigte Farben (also solche mit geringerem Grauanteil) stechen allerdings weit mehr ins Auge.

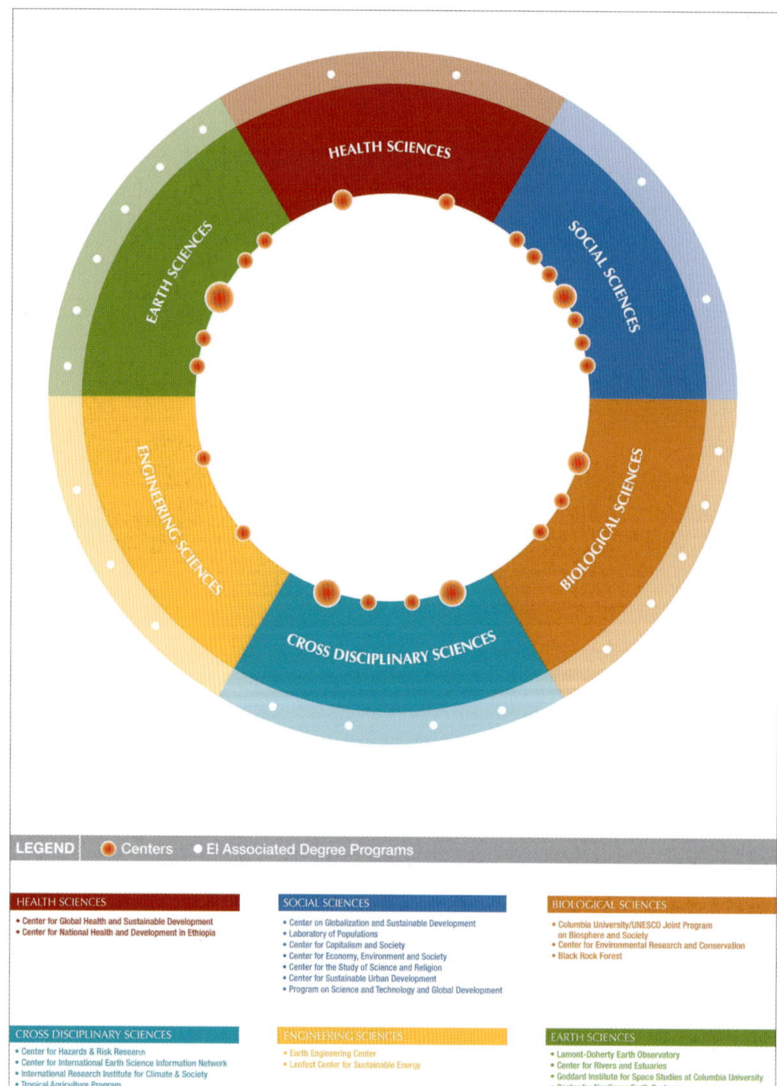

Jede Fachrichtung verfügt über eine Reihe von Forschungszentren und Studienabschlüssen und ist mit einem eigenen Farbsystem gekennzeichnet.

PROJEKT
Identity-Programm

KUNDE
Earth Institute at Columbia University

DESIGN
Mark Inglis

CREATIVE DIRECTOR
Mark Inglis

Hier ist eine Gruppe von Fachrichtungen des Earth Institute an der Columbia University farbcodiert.

Grafische Symbole lassen sich ebenfalls perfekt in ein Farbsystem integrieren.

Der Einsatz von Farben funktioniert auch bei Symbolen, Balken und Text.

49. Farbe als Mittel zur Trennung

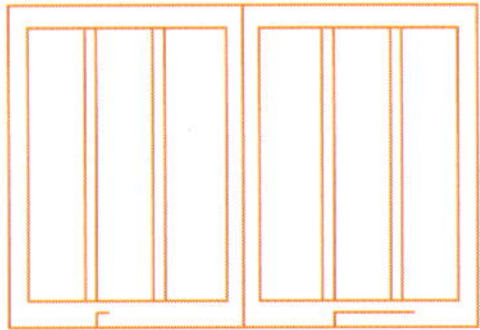

Natürlich lassen sich einzelne Elemente durch vertikal und horizontal klar definierte Spalten trennen, aber verschiedene Farbschattierungen sorgen für mehr Pfiff und grenzen die einzelnen Beiträge voneinander ab. Weiße Schrift vor einem farbigen Hintergrund sorgt für noch stärkere Effekte.

Siehe auch Seite 100/101

FARBUMKEHR

Umgekehrte Farben wirken dramatisch, können aber bei großen Mengen am kleinem Text die Lesbarkeit beeinträchtigen, besonders bei Serifenschriften. Serifenlose Schriften eignen sich hier besser. Auf Bildschirmen kann die Lesbarkeit noch schlechter sein, sodass man auf größere, fettere Schriften zurückgreifen muss.

AT SPOTCO, WE BEGIN WORK-ing on a show by understanding its Event. I didn't invent this phrase—it was loaned to me by producer Margo Lion. But what I came to understand it as is quite simply the reason you see a show. Or even more simply, the reason you tell someone else to see a show. It can be straightforward,

Das Beispiel oben zeigt die Abbildungsgröße.

PROJEKT
Broadway; From Rent to Revolution

KUNDE
Drew Hodges, Autor; Rizzoli, Verleger

CREATIVE DIRECTOR
Drew Hodges

DESIGN
Naomi Mizusaki

Erinnerungen verschiedener Mitarbeiter an Spotco, die berühmteste Broadway-Agentur, sind in verschiedenen Farben herausgestellt.

Jede Farbe steht für einen Beitrag eines anderen Mitarbeiters. Der Text für den Abschluss, der schwarze Kasten, dreht sich nicht um eine Person, sondern um „das Event", ein wichtiges Element der Philosophie der Agentur.

1987 | ACT ONE

DREW HODGES
FOUNDER AND CHIEF EXECUTIVE OFFICER

WE OPEN ON A YOUNG design firm called Spot Design. It was named for a dog the landlord said we couldn't have. So I named the office as my pet.

After attending art school at School of Visual Arts in New York City, I had left working for my college mentor Paula Scher and began freelancing solo out of my apartment. I was working in the kitchen of my loft, across from the now-defunct flea markets on 26th Street and Sixth Avenue. This is the same kitchen where producers Barry and Fran Weissler came to see the early designs for *Chicago*—but I get ahead of myself. Ultimately, we were five designers and one part-time bookkeeper doing entertainment and rock 'n' roll work. We were young and laughed a lot. Ten years later, we had been privileged to work with Swatch Watch, MTV, Nickelodeon, the launch of Comedy Central and *The Daily Show*, as well as record work for Sony Music, Atlantic Records, and Geffen/DreamWorks records, where our most notable projects were album packages for downtown diva Lisa Loeb and iron-lunged Aerosmith. We grew adept at strategy, design, and collaboration with many downtown artists, illustrators, and photographers,—all people we would come to take full advantage of as we began our theater work. I went to the theater—it was a New York City joy for me. I had gone since early high school, riding the train to the city. But I never dreamed I would get any nearer than second-acting *Dreamgirls* from the mezzanine.

Two bold incidents happened to change that. First, Tom Viola and Rodger McFarlane were heroes of mine for the work they did through what was to become Broadway Cares/Equity Fights AIDS. A friend and art director from Sony Music named Mark Burdett was assigned to work with Spot on a ad for the Grammy Awards program the prime position of the back cover was Martin Luther King Jr. Day, an clients were all away. So we moaned doing yet another ad filled with alb minis of the labels and latest Streisa release with hollow congratulations a waste of a great opportunity. We posited that Sony could be the first record company to take a stand aga AIDS by making a donation and att ing a red ribbon to the back of each issue of the program. And to pitch t idea, we called Rodger and Tom at t offices to help us fulfill it. The ribbo was theirs after all. Remarkably, the were working on the holiday and answered the phone. They agreed to and the rest is history. I believe it w the first awards ceremony to public script the concern over AIDS, and d friendships were formed. Later that year, Tom and Rodger called. They h an ad due in three days for their new show, *The Destiny of Me*, Larry Kra sequel to *The Normal Heart*. We sen them a design based on a photo of m right hand—I guess we felt it seeme personal—and they loved it. This w our first theater poster.

But it would have been a short-liv career without the second event. Tw years later, we had just finished doi the Aerosmith album for Geffen. Ro Sloane, David Geffen's star creative director, called and asked us to mee with the producers of *Rent*—Geffen would be releasing the album. I too meeting with the ad agency in charg and got the assignment and a ticket the hottest show in town a week aft had opened Off-Broadway. Within a year, we would have designed *Rent* *Chicago*, and Jeffrey Seller sat me d in a mall in Miami to ask if we had thought about starting an ad agency seemed a big risk—but it also seeme like a world where you could actual meet the people doing the creative you were assigned to promote. And began to try and figure out just how ad agency worked anyway.

8

BRIAN BERK

CO-FOUNDER AND CHIEF OPERATING OFFICER / CHIEF FINANCIAL OFFICER

IN THE SPRING OF 1997, after designing the successful ad campaigns for *Rent* and *Chicago*, we decided to attempt to open a theatrical ad agency. The first question was: What would we need to be able to pull this off? For starters, we would need equipment, office space, a staff, and most importantly, clients.

The equipment was easy. In order to keep upfront costs down, we could lease—a few computers and a fax machine. From there, we could scrape by until we had some clients.

Office space: The design studio was currently housed in Drew's apartment. We knew that for potential clients to consider hiring us, we would need to be in the theater district, and we would need to have a large conference room for the weekly ad meetings. I set out to look at space. One space was located in 1600 Broadway. The building was fairly run-down (and we would later learn it had a mouse problem), but it did have a long and interesting history. It was built in the very early 1900s as a Studebaker factory and showroom. In the 1920s, it was converted to offices and at one point housed the original offices of Columbia Pictures, Universal Pictures, and Max Fleischer Studios, creator of Betty Boop. This seemed a fitting place for a theatrical ad agency. By 1997, the building held a combination of offices and screening rooms. (It has since been torn down.) The space we looked at consisted of two small offices, a big bullpen area for our designers, and one large conference room with the most amazing view of Times Square. We actually found a photograph of a movie executive sitting at his desk in the room that would eventually become our conference room. It has the same wood paneling and window with the view. However, the bearskin rug, which is seen on the floor in the photo, is long gone. The space had character. We had to furnish it on the cheap. We hired a set decorator friend to style the office circa 1940s, so all the used office furniture we purchased would look like a very conscious design choice. We moved to the space in June 1997.

Staff: We already had a creative director (Drew), four graphic designers, an office manager, and me. I handled finance, administration, and facilities. We needed someone to head account services, an assistant account executive and a graphic production artist to produce all the ads. We'd hire a writer once we had some clients. For the production artist, we knew just who to hire: Mary Littell. She had worked for us before and was great. The person to head account services was harder to find. We needed someone who had worked at an ad agency before and understood media. From what Drew learned, one of the most respected account managers in the industry was Jim Edwards, or as was said by several producers, "He is the least hated." He had worked at two of the existing theatrical ad agencies. But would Jim join a startup? He was game and joined our team. Jim walked in the door on July 21, 1997. Mary was at her desk working on dot gain so she would be ready if we were ever hired to place an ad. Now, all we needed were some clients.

JIM EDWARDS

CO-FOUNDER AND FORMER CHIEF OPERATING OFFICER

OF DREW, BRIAN, MARY, BOB Guglielmo, Karen Hermelin, and Jesse Wann, I was the only one who had worked at an ad agency before. Little things like a copy machine that can make more than one copy every thirty seconds was not part of our infrastructure. I started on a Monday, and the pitch for *The Diary of Anne Frank* was that Friday. We didn't have any clients so that entire week was all about the pitch. Thursday night we were there late and inadvertently got locked in the office (how that is even possible still strikes me as odd). We couldn't reach anyone who had a key so we had to call the fire department to let us out. They did—and were adorable too.

Once we had a show, we became a legitimate advertising agency, which led to David Mamet's *The Old Neighborhood*, John Leguizamo's *Freak*, and Joanna Murray-Smith's *Honour* within months of being open for business.

Since SpotCo was a brand-new company, we had no credit with any of our vendors. The *New York Times* made us jump through so many hoops about establishing a relationship with them. I think we had to have a letter from the producers of *Anne Frank* saying they were hiring us as their ad agency. When it came time to reserve our first *Times* ad, about a month had passed since all those rules were handed down. Back then, you simply called the *Times* reservation desk and reserved the space. That's what I had been doing for years so I did it again, inadvertently forgetting that the ad needed to be paid for well in advance. I knew everyone there so they accepted my reservation without question. The ad ran. No one said anything. The bill came about a month later, and we paid it. About two months in, the *Times* noticed that we were sliding under their rules but since we were paying our bills and were current, they granted us credit. By that first Christmas, we had established credit everywhere, which was and is a big deal. Not many new companies can make that kind of claim.

In the first eighteen months of SpotCo, I never worked harder in my life. The hours were long (I gained a lot of weight during that time—do not put this in the book), and it wasn't easy, but we also saw direct results from all the hard work. The work was good, and people noticed what we were doing—and we were making money! The Christmas party of 1998 was particularly memorable. That day, we had just released the full-page ad for *The Blue Room*, which was pretty provocative because all we ran was the photo and a quote about how hot the ticket was. It was a big deal for us and kind of heady. The party was at some restaurant, and Brian had secured a private room. There were only three tables of ten, and we shared our Secret Santa gifts. Everyone was really into it and every time someone opened a gift, Amelia Heape would shout, "Feel the *love*, people! Feel the *love*!" Indeed.

TOM GREENWALD

CO-FOUNDER AND CHIEF STRATEGY OFFICER

WHAT AM I, NUTS? IT WAS early 1998. I had a good job, an amazing wife, three adorable kids under five years old, and a modest but nice house in Connecticut. In other words, I was settling in nicely to the 9 to 7, suburban commuting life. But I kept hearing about this guy Drew Hodges. First I heard about him through my friend Jeffrey Seller, who had worked with Drew on the designs for *Rent*. Then I heard about him through my friend David Stone, who was just about to start working with him on *The Diary of Anne Frank*. Then, I realized, they're talking about the tall guy who talked a lot and had barreled his way through meetings at Grey Advertising (where I worked at the time) while designing the artwork for *Chicago*. So when Jim Edwards called me and said, "Hey, I'm joining up with this guy Drew and we're turning his design shop into an ad agency and did you want to meet him." I had to think about it. No way was I going to give up my job security, right? The odds of any theatrical ad agency surviving at all were miniscule, much less one with . . . wait, let me add them up . . . one client. And besides, I'd probably have to take a cut in pay. Only an insane person would consider it. "Sure," I told Jim. "Set it up."

So I went in to talk to Drew, and after about eighteen seconds, I'd made up my mind. When Drew asked if I had any questions, I had only one. "Where do I sign?" I joined SpotCo as the head (only) copywriter, head (only) broadcast director, and head (only) proofreader. On the downside, we were a very lean department. On the upside, I never had any problems with my staff.

Now here we are, almost eighteen years later. My wife is still amazing, my three kids are still adorable (although no longer under five), and my house is still modest but nice (although we redid the TV room). I never did settle in to that 9 to 7 job, though. Instead, I decided to take a chance, a flyer, and a crazy ride—and it's worked out pretty well. So yeah, I guess I was a little nuts. But it turns out craziness has its perks.

THE EVENT

*AT SPOTCO, WE BEGIN WORK*ing on a show by understanding its Event. I didn't invent this phrase—it was loaned to me by producer Margo Lion. But what I came to understand it as is quite simply the reason you see a show. Or even more simply, the reason you tell someone else to see a show. It can be straightforward, or subtle. As much as I love advertising, branding and design and what it can do for a show, word of mouth is the number one reason a show becomes a hit. So the most effective branding hopes to invite, encourage, and curate that word of mouth. It has to be true to the experience of that show—otherwise it will fail miserably. If every great musical begins with an "I want" song, then every show's voice begins with "I want you to think of this show this way" tune.

What we as an agency have also learned is you choose your Event or it will be chosen for you, and you (and ticket buyers) may not like the sound of that choice. Lets call that the non-event.

At the beginning of each show presented here, we have listed the non-event first—the thing we feared the Broadway customer would default to when describing a show, knowing little else. This is followed by the Event (capital E!) that we wanted to communicate. Hopefully, this informs our successes and disappointments and, more to the point, why a certain idea was ultimately chosen. —D.H.

50. Farbigkeit durch Strichstärke, Größe und Formen

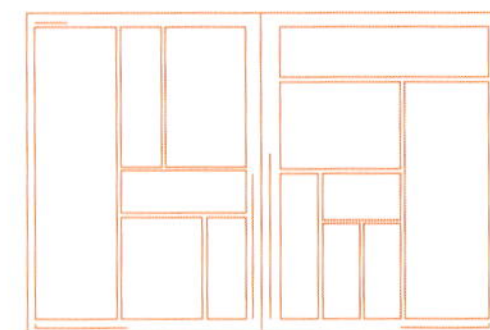

Bei einem Schwall an Informationen setzen Designer manchmal auf Schwarzweiß oder möglichst wenige Farben. Oder das Budget gibt nicht mehr als eine Farbe her. Auch wer nur wenige oder gar nur eine Farbe zur Verfügung hat, kann durch Fonts, Schriftfamilien und Strichstärken sowie durch den Anteil an Bildelementen Farbigkeit und Struktur erzeugen. Ein vielseitiger Raster unterstützt viele Breiten und Größen. Wer alle diese Optionen nutzt, schafft Variantenreichtum und Faszination. Freie Räume, die ja auch grafische Elemente sind, bringen Energie und Kontrast.

SCHATTIERUNGEN

Wenn die Vorgaben oder der Designer auf eine Farbe oder wenige Farben setzen, kann man mit Schattierungen, d. h. Abstufungen einer Farbe, arbeiten. Auf einem Hintergrund mit 10 % Schwarz ist Text in 100 % Farbe noch gut zu erkennen. Je dunkler der Hintergrund, desto schlechter lesbar sind 100 % Farbe (je nach Schriftgröße und Papier). Eine Möglichkeit ist die Nutzung einer hellen oder weißen Schrift. Im gesamten Buch sind die Seitenleisten wie diese mit 70 % Schwarz auf weißem Hintergrund gedruckt.

Siehe auch Seite
98/99, 112/113

PROJEKT
King, eine Sonderausgabe zum 50. Jahrestag der Ermordung von Martin Luther King Jr.

KUNDE
The Atlantic

CREATIVE DIRECTOR
Paul Spella

ART DIRECTOR
David Somerville

DESIGNAGENTUR
OCD | Original Champions of Design

DESIGNER
Bobby C. Martin Jr., Jennifer Kinon

Unterschiedliche Schriften in verschiedenen Größen und der dramatische Einsatz von Größen und freien Räumen schaffen eine höchst erfolgreiche Sonderausgabe.

MLK | THE MAN

Eli Lee is an editorial fellow at *The Atlantic.*

The Arc of a Life

Martin Luther King Jr. was just 26 when he came to prominence, by leading a bus boycott in Montgomery, Alabama. He was only 39 when he was killed.

By Eli Lee

January 15, 1929: Michael King Jr. is born in Atlanta to Michael King Sr., a prominent local preacher and civil-rights leader, and Alberta King, a former schoolteacher. In 1934, his father changes both of their first names to Martin-by various accounts, to correct a birth-certificate mistake or to honor the theologian Martin Luther. His childhood is comfortable; he excels in school, skipping the ninth and 12th grades. He first becomes conscious of racism at age 6, when a white friend's father prohibits his son from playing with Martin.

June 8, 1948: King graduates from Morehouse College, in Atlanta, with a bachelor's degree in sociology. Later, he earns a bachelor-of-divinity degree from Crozer Theological Seminary, in Chester, Pennsylvania, and a doctorate in systematic theology from Boston University. (In 1991, a BU committee would determine that King had plagiarized passages of his dissertation, "A Comparison of the Conceptions of God in the Thinking of Paul Tillich and Henry Nelson Wieman," from other scholars' work.)

"I would turn to the Almighty, and say, 'If you allow me to live just a few years in the second half of the 20th century, I will be happy.'"

From his "I've Been to the Mountaintop" sermon, delivered on April 3, 1968, at the Mason Temple in Memphis, Tennessee

June 18, 1953:
King marries Coretta Scott, an activist and aspiring singer from Alabama studying at the New England Conservatory of Music. After his death, she would advance her husband's legacy by founding the Martin Luther King Jr. Center for Nonviolent Social Change.

1955–56: Activists organize a boycott of the bus system in Montgomery, Alabama, after Rosa Parks, a black woman, is arrested for refusing to give up her seat to a white man. A former head of the state's NAACP calls on King, now the 26-year-old pastor of a local black church, to lead the boycott-because he's "young and intelligent with leadership ability" and has a "wonderful speaking voice." The protest lasts 381 days, ending in victory after the U.S. Supreme Court rules that segregation on public buses is unconstitutional.

January 10–11, 1957: Notables in the civil-rights movement form the Southern Christian Leadership Conference to coordinate nonviolent protest actions in the South. They soon elect King as the organization's president.

COURTESY OF MOREHOUSE COLLEGE; CHARLES MOORE; DON CRAVENS/THE LIFE IMAGES COLLECTION/GETTY

12 THE ATLANTIC MARTIN LUTHER KING JR.

R. SATAKOPAN; FRANCIS MILLER/THE LIFE PICTURE COLLECTION/GETTY; PARAMONOV ALEXANDER; CHARLES KELLY/AP

February 1959: At the invitation of Prime Minister Jawaharlal Nehru, King visits India for a month, meeting with social reformers, government officials, and associates of the late Mahatma Gandhi, whose acts of civil disobedience to free the country from British rule inspired King's own approach to bringing about change. "I left India more convinced than ever before," King writes at the time, "that non-violent resistance is the most potent weapon available to oppressed people in their struggle for freedom."

August 28, 1963: At the March on Washington for Jobs and Freedom, then the largest protest in U.S. history, King addresses an estimated 250,000 people on the National Mall. His soaring call for racial justice comes to be known as the "I Have a Dream" speech, after his ad-libbed ending. Today, the speech is seen as both a rhetorical masterpiece and a defining moment of the civil-rights movement.

December 10, 1964:
King receives the Nobel Peace Prize. At age 35, he is its youngest recipient so far. He promises to donate the prize's $54,123 award to the civil-rights movement.

March 1965: In response to the continued disenfranchisement of millions of black people across the South, the SCLC and other civil-rights groups demand voting rights in a 54-mile march from Selma, Alabama, to the state capital of Montgomery. Local police and white mobs react brutally to the nonviolent protest, beating many participants; two white demonstrators are murdered. The march hastens passage of the federal Voting Rights Act later in the year.

April 4, 1967: *In a speech titled "Beyond Vietnam," delivered at Riverside Church in Manhattan, King declares his opposition to the Vietnam War. He publicly criticized the war two months earlier, in a speech to the Nation Institute (see page 92). But the widely publicized Riverside Church speech upsets many of King's usual allies, who accuse him of hurting the cause of civil rights by alienating the American government and public. With this stand, King expands his calls for social justice at home into a broader, pacifist message.*

April 4, 1968:
While in Memphis to protest black sanitation workers' poor treatment by the city, King is shot as he stands on a balcony at the Lorraine Motel. He is declared dead about an hour later. The murderer, James Earl Ray, flees the country and is arrested two months later at Heathrow Airport, in London; he is known to be a racist, but his exact motive is never made clear. (Sentenced to 99 years in prison, he dies of natural causes in 1998.) The assassination sparks riots in more than 100 U.S. cities. President Lyndon B. Johnson declares a national day of mourning.

November 2, 1983: After years of advocacy from labor unions and civil-rights groups, President Ronald Reagan signs legislation to commemorate the slain leader's birthday with a federal holiday on the third Monday in January. Martin Luther King Jr. Day is celebrated nationwide for the first time in 1986.

MARTIN LUTHER KING JR. THE ATLANTIC 13

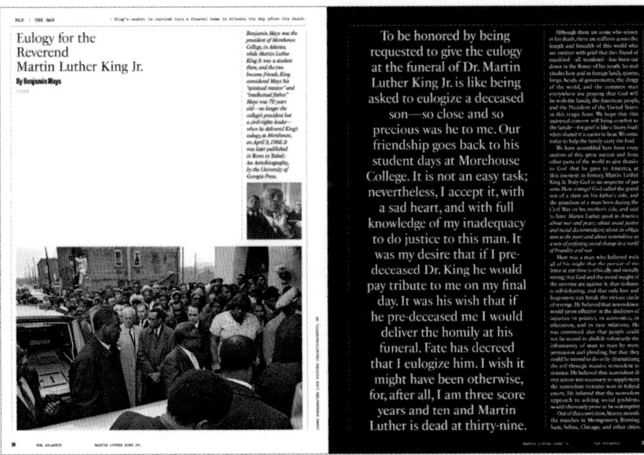

Eulogy for the Reverend Martin Luther King Jr.

By Benjamin Mays

To be honored by being requested to give the eulogy at the funeral of Dr. Martin Luther King Jr. is like being asked to eulogize a deceased son—so close and so precious was he to me. Our friendship goes back to his student days at Morehouse College. It is not an easy task; nevertheless, I accept it, with a sad heart, and with full knowledge of my inadequacy to do justice to this man. It was my desire that if I predeceased Dr. King he would pay tribute to me on my final day. It was his wish that if he pre-deceased me I would deliver the homily at his funeral. Fate has decreed that I eulogize him. I wish it might have been otherwise, for, after all, I am three score years and ten and Martin Luther is dead at thirty-nine.

"I make bold to assert that it took more courage for King to practice nonviolence than it took for his assassin to fire the fatal shot."

Seite gegenüber: Ein Meisterwerk mit historischen Fotos, Zitaten und Berichten. Diese Doppelseite schöpft viele Möglichkeiten des Rasters aus, um eine Struktur zu erreichen. Zusammen beschwören die drei verschiedenen Schriften die 60er Jahre herauf und bringen den klar organisierten Kapiteln viel Frische mit verschiedenen Größen, aber einheitlichen Zeitleisten.

Diese Seite oben: Diese vier Aufmacher wirken ohne Farbe allein durch die Größen dramatisch.

Diese Seite unten: Der Nachruf auf Reverend Martin Luther King Jr. ist respektvoll und dennoch dynamisch. Wenige Farben, Hervorhebungen durch weiße Schrift auf dunklem Hintergrund, durch mehr oder weniger Breite oder durch freie Stellen im Raster lassen die Räume genauso sprachgewaltig erscheinen wie die Essays und die Bilder.

51. Aufteilungen für Schilder und Plakate

Hinweisschilder müssen logisch, klar strukturiert und konsequent gestaltet sein und stellen daher an Grafikdesigner besondere Anforderungen. Die Raster für die auf einem Schild enthaltenen grafischen Darstellungen – insbesondere wenn sie um Verkaufsstände herum angebracht sind – können folgende Elemente enthalten:

- Informationen mit unterschiedlichen Wahlmöglichkeiten, die nacheinander gesucht werden: 1, 2 etc.
- sekundäre, aber dennoch wichtige Wahlmöglichkeiten wie die Sprache
- tertiäre Informationen, die allgemeine Fragen beantworten, wie die Nummer eines Gates am Flughafen, Wegweiser zu Toiletten und Imbissständen oder Restaurants
- verschiedene andere Optionen, die sich erst dann ergeben, wenn man den Hinweisschildern folgt: Ein Besucher stellt etwa fest, dass er denselben Weg zurückgehen muss

Damit der Besucher die Schilder ohne Schwierigkeiten auch im Vorbeigehen sehen und lesen kann, sollte der Text leserlich und klar strukturiert sein. Die Farben sollten ins Auge springen, ohne die Informationen zu verwässern.

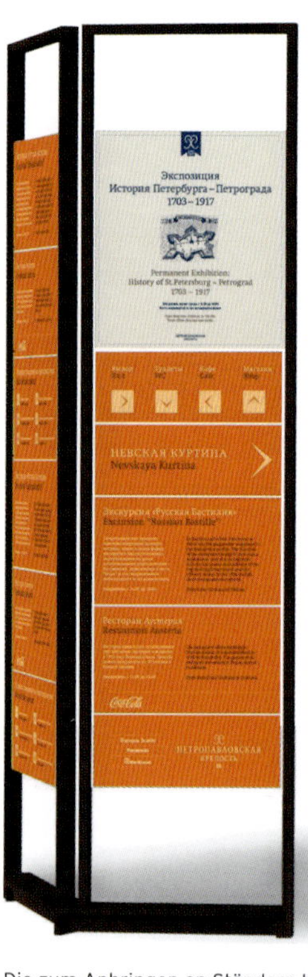

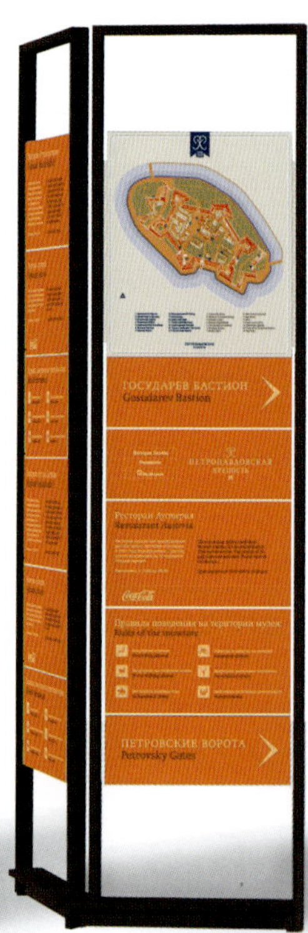

Die zum Anbringen an Ständern bestimmten Hinweisschilder und Schautafeln sind als Informationsbänder gestaltet.

PROJEKT
Identity und Information

KUNDE
The Peter and Paul Fortress, St. Petersburg, Russia

ART DIRECTION
Anton Ginzburg

DESIGN
Studio RADIA

Die Tafeln mit Informationen über die Peter-und-Paul-Festung in St. Petersburg, Russland, weisen den Besuchern auf Englisch und Russisch den Weg. Das Projekt wurde nur teilweise fertiggestellt.

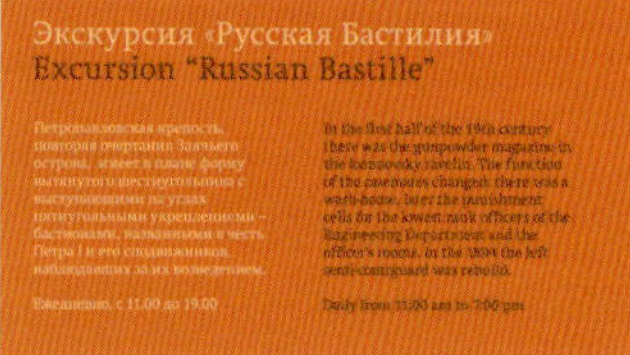

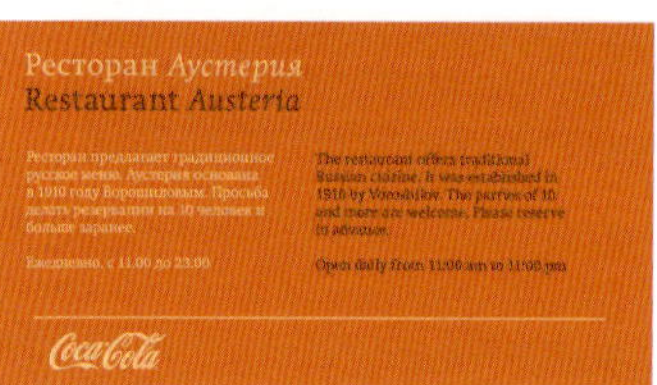

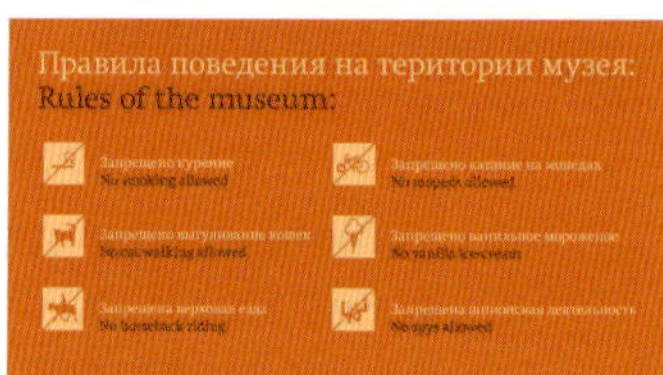

Ausschnitte der Schautafeln zeigen die Informationen, die die Grafikdesigner angeben und darstellen mussten.

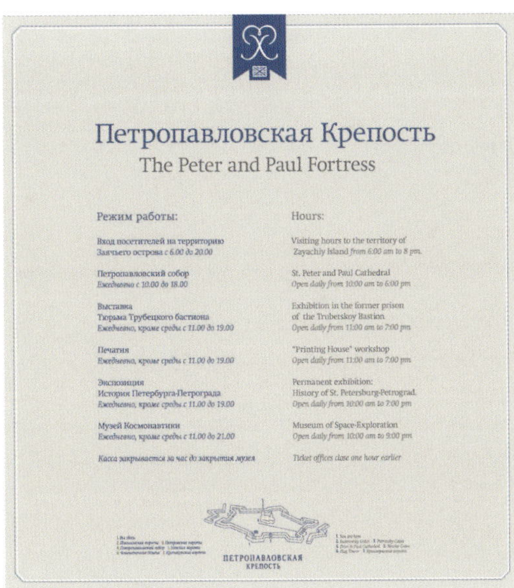

Der Text eines Schildes, das an einem Ständer angebracht wird, ist in einer klaren, klassischen Typografie geschrieben und verweist auf die Geschichte der Stadt.

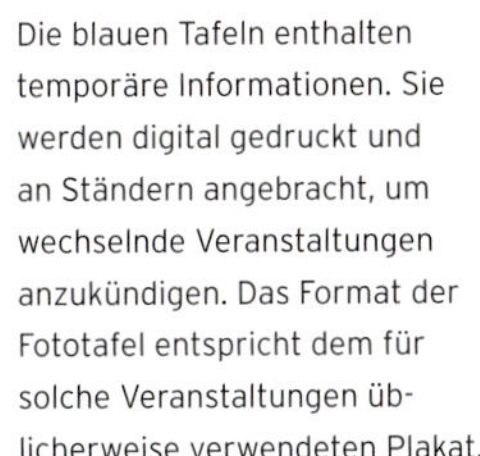
Die blauen Tafeln enthalten temporäre Informationen. Sie werden digital gedruckt und an Ständern angebracht, um wechselnde Veranstaltungen anzukündigen. Das Format der Fototafel entspricht dem für solche Veranstaltungen üblicherweise verwendeten Plakat.

52. System mit Bändern

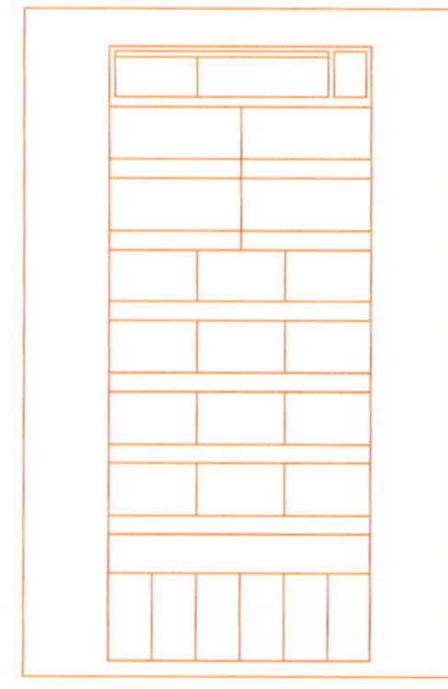

Horizontale Hierarchien eignen sich zur Trennung von Informationen. Informationen können als Teil des Navigationssystems in Bändern angeordnet sein. Diese Hierarchie gilt für mobile Geräte ebenso wie für die Webseite.

OBEN UND GEGENÜBER: Horizontale Bänder bilden einen Rahmen, die Navigationsleiste ist oben. Eine Auswahl in der Navigationsleiste führt zu einer Kaskade horizontal angeordneter Informationen.

PROJEKT
Jewish Online Museum
Webseite

KUNDE
Jewish Online Museum

BRANDING/FRONT-END DEVELOPMENT
Threaded

DRAHTGITTERMODELLE
Lushai

WEB-ENTWICKLUNG
Ghost Street
Reactive

Die Webseite für das Jewish Online Museum - die erste ihrer Art nicht nur in Neuseeland, sondern auch international - soll eine einnehmende und lehrreiche Quelle für zahlreiche Besucher sein, aber in erster Linie als Sammlung und zugängliche Quelle für die jüdische Gemeinschaft in Neuseeland dienen.

DIE BEDEUTUNG DES PROGRAMMIERERS

Der Frontend-Designer programmiert nicht das Backend. Auch wenn das eigentlich logisch ist, weiß der Kunde das oft nicht. Gute Programmierer sind schwer zu finden. Für die Webseite des Jewish Online Museum arbeiteten die Designer zusammen mit Lushai, dem Entwickler der Drahtgittermodelle, an der Entwicklung dynamischer Drahtgittermodelle für den Bedarf der Nutzer des JOM. Die Workshops befassten sich mit der Nutzerinteraktion und der ansprechenden Übertragung auf Mobilgeräte.

GANZ RECHTS: Die mobilen Displays entsprechen denen der Webseite, auch wenn die Bänder durchbrochen werden, um Erinnerungen hervorzuheben.

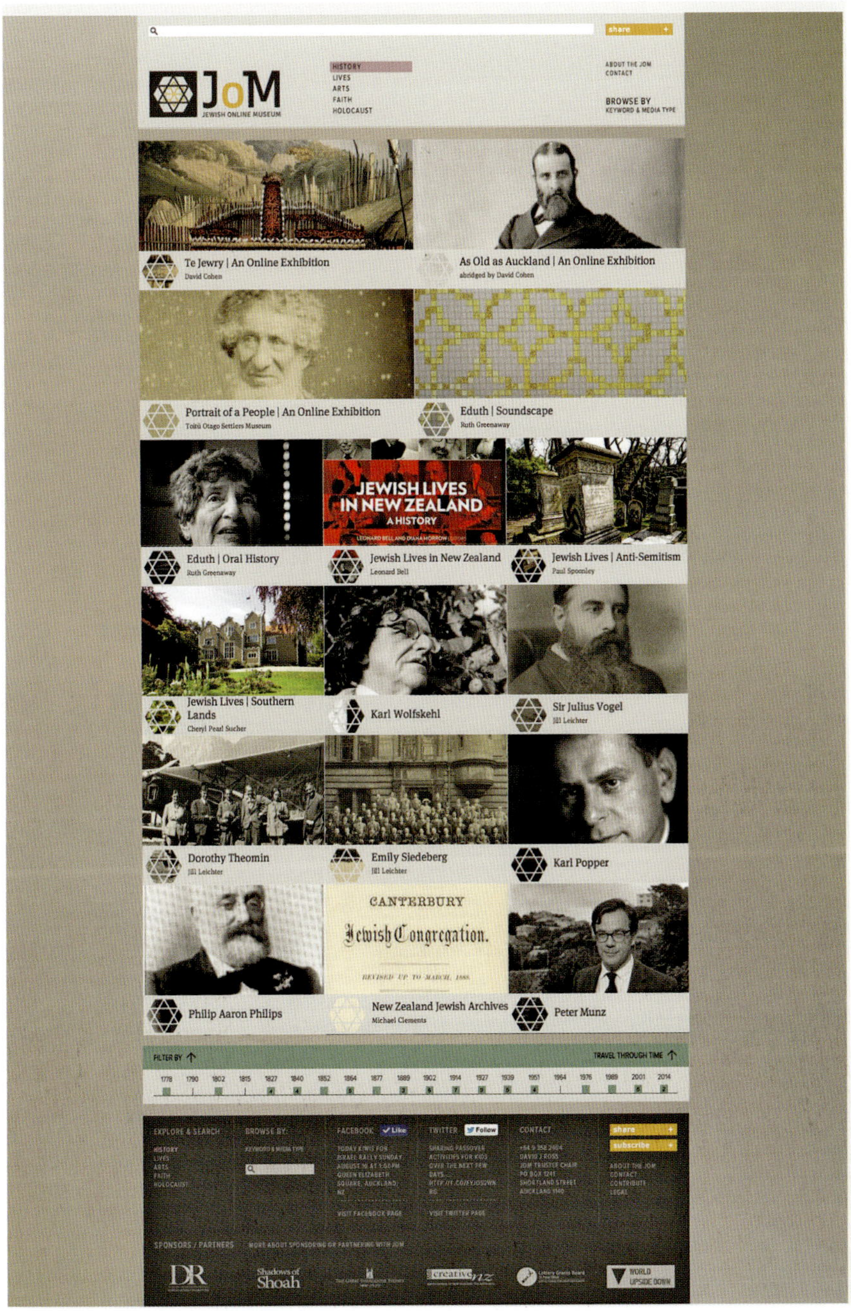

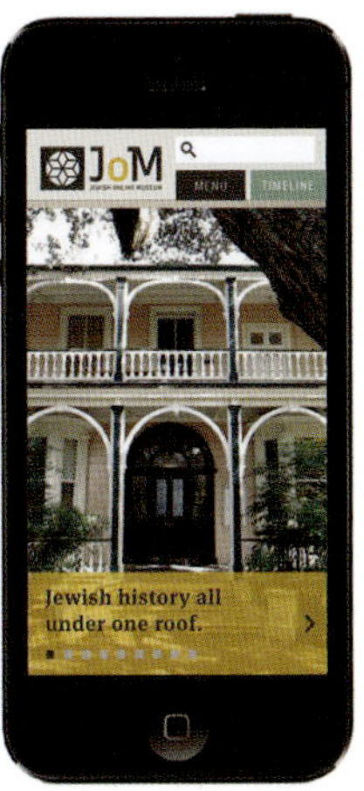

share
HISTORY
LIVES
ARTS
FAITH
HOLOCAUST
ABOUT THE JOM
CONTACT
BROWSE BY
KEYWORD & MEDIA TYPE
JoM
JEWISH ONLINE MUSEUM
Te Jewry | An Online Exhibition
As Old as Auckland | An Online Exhibition
Portrait of a People | An Online Exhibition
Eduth | Soundscape
JEWISH LIVES IN NEW ZEALAND
A HISTORY
Eduth | Oral History
Jewish Lives in New Zealand
Jewish Lives | Anti-Semitism
Jewish Lives | Southern Lands
Karl Wolfskehl
Sir Julius Vogel
Dorothy Theomin
Emily Siedeberg
Karl Popper
CANTERBURY
Jewish Congregation.
Philip Aaron Philips
New Zealand Jewish Archives
Peter Munz
FILTER BY
TRAVEL THROUGH TIME
share
subscribe
Shadows of Shoah
WORLD UPSIDE DOWN
JoM
MENU
TIMELINE
Jewish history all under one roof.
JoM
MENU
TIMELINE
HISTORY
BIOGRAPHY
ARTS
FAITH
HOLOCAUST
ABOUT
CONTACT
CLOSE X
JoM
MENU
TIMELINE
My birth coincided with the passing of the Nuremberg Laws

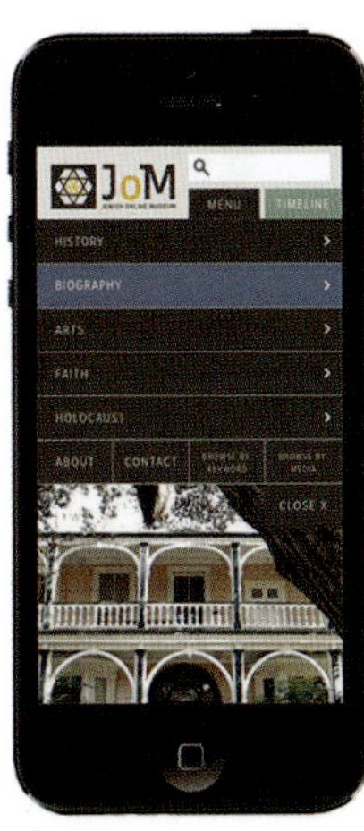

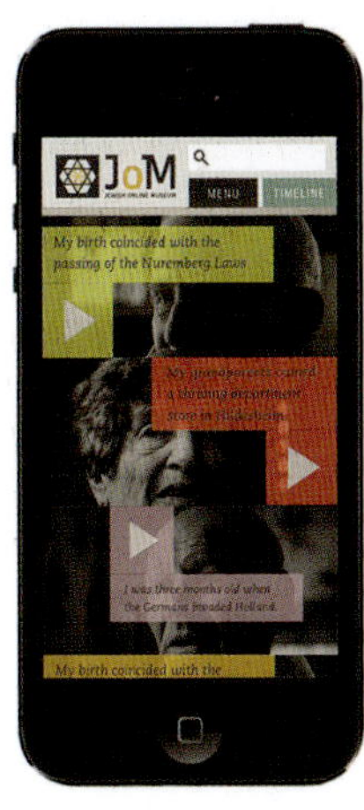

53. Horizontale über freie Flächen definieren

Ausreichend freie Flächen auf einer Textseite sorgen für Ordnung und Ausgeglichenheit. Ist noch mehr freie Fläche vorhanden, lassen sich einleitende Elemente wie Titel und Texte von erläuternden Teilen wie Bildunterschriften oder Schritt-für-Schritt-Anleitungen trennen. Auf diese Weise abgetrennte Bereiche erleichtern dem Leser dann später das Navigieren enorm.

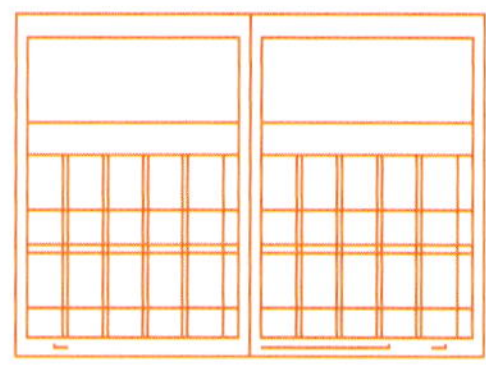

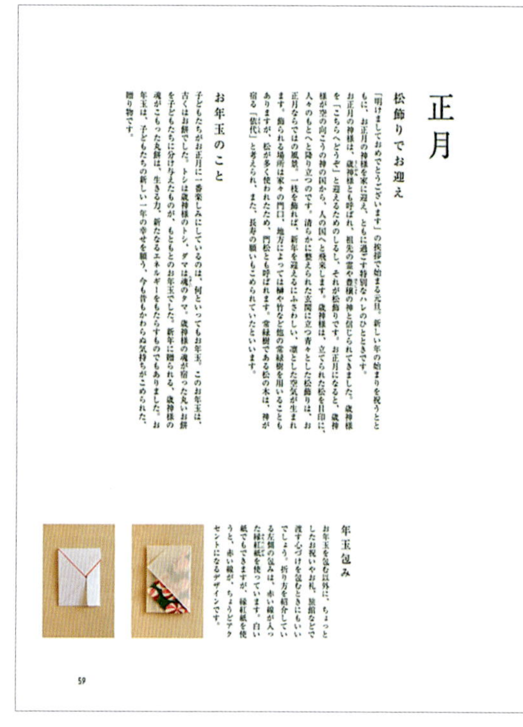

正月

松飾りでお迎え

「明けましておめでとうございます」の挨拶で始まる元旦。新しい年の始まりを祝うとともに、お正月の神様を家に迎え、ともに過ごす特別なハレのひとときです。

お正月の神様は、歳神様とも呼ばれ、祖先の霊や豊穣の神と信じられてきました。歳神様を「こちらへどうぞ」と迎えるためのしるし、それが松飾りです。お正月になると、歳神様が空の向こうの神の国から、人の国へと飛来します。歳神様は、立てられた松を目印に、人々のもとへと降り立つのです。清らかに整えられた玄関に立つ青々とした松飾りは、お正月ならではの風景。一枝を飾れば、新年を迎えるにふさわしい、凛とした空気が生まれます。飾られる場所は家々の門口。地方によっては榊や竹など他の常緑樹を用いることもありますが、松が多く使われたため、門松とも呼ばれます。常緑樹である松の木は、神が宿る「依代」と考えられ、また、長寿の願いもこめられていたといいます。

お年玉のこと

子どもたちがお正月に一番楽しみにしているのは、何といってもお年玉。このお年玉は、古くはお餅でした。トシは歳神様のトシ、ダマは魂のタマ。歳神様の魂が宿った丸いお餅を子どもたちに分け与えたものが、もともとのお年玉でした。新年に贈られる、歳神様の魂がこもった丸餅は、生きる力、新たなるエネルギーをもたらすものでもありました。お年玉は、子どもたちの新しい一年の幸せを願う、今も昔もかわらぬ気持ちがこめられた、贈り物です。

年玉包み

お年玉を包む以外に、ちょっとしたお祝いやお礼、旅館などで渡す心づけを包むときにもいいでしょう。折り方を紹介している左側の包みは、赤い線が入った縁紅紙を使っています。白い紙でもできますが、縁紅紙を使うと、赤い線が、ちょうどアクセントになるデザインです。

59

新年の折形

指導　折形デザイン研究所

折形は、
和紙を使った包み方の作法。
日本のしきたりから生まれた、
幸せを祈る心のかたちです。
門松を立て、お雑煮を囲んで
新しい一年を迎える
晴れやかなひとときに、
折形を添えてみませんか。
静かな気持ちで
白い紙を広げ、
清々しい一年を
迎える準備を始めましょう。
お正月はもうすぐです。

松飾り

大きな門松になると、家によっては、立てる場所などが難しいもの。一枝の松飾りなら、マンションなどの玄関にも自然になじみます。「木の花包み」といわれる折形を、松飾りに仕立てています。裏側にこよりなどで紐をつけ、ドアや壁に貼ったり、引っ掛けるとよいでしょう。

58

PROJEKT
Kurashi no techo (Everyday Notebook) Magazin

KUNDE
Kurashi no techo (Everyday Notebook) Magazin

DESIGNER
Shuzo Hayashi, Masaaki Kuroyanagi

Auf Seiten oder Doppelseiten mit vielen Bildern und Informationen werden durch einen horizontalen Aufbau Überschriften und sich daran anschließende einzelne Schritte abgegrenzt. So entsteht ein Gefühl von Ruhe und Ordnung und die optische Gliederung der Informationen wird erleichtert.

お膳

Durch freie Flächen werden Text und Bilder klar getrennt und Informationsfelder festgelegt.

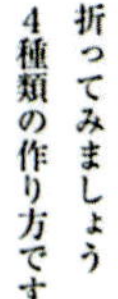

実際に和紙を折ってみましょう 4種類の作り方です

折形と日本のしきたり

折形は、室町時代に始まった、武家に伝わる礼法と伝えられます。折形をはじめとする日本のしきたりの数々は、本来の意味や由来は忘れられながらも、暮らしの中に生き続け、今に伝わったもの、と折形デザイン研究所の山口信博さんはおっしゃいます。

お正月にお雑煮をいただき、結婚のお祝いには水引をかけたご祝儀を贈る。生活に根付いたしきたりを、民俗学者の折口信夫は、「生活の古典」と呼びました。

「……私どもの生活は、功利の目的のついて廻らぬ、いわばむだとも思われる様式の、由来不明なる「為来り」によって、純粋にせられることが多い。その多くは、家庭生活を優雅にし、しなやかな力を与える。門松を樹てた後の心持ちのやすらいを考えてみればよい。……」（『古代研究Ⅰ――祭りの発生』〈古代生活の研究〉中公クラシックスより）

松飾り

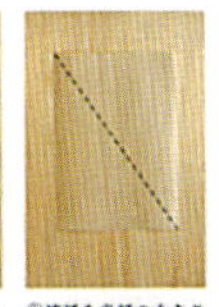

①檀紙を半紙の大きさに切って、左下の角を対角線で折り上げます。

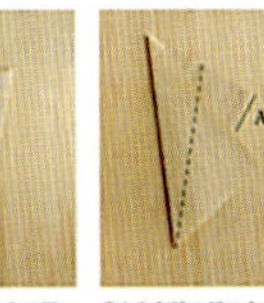

②Aを左端の辺に合わせて折ります。

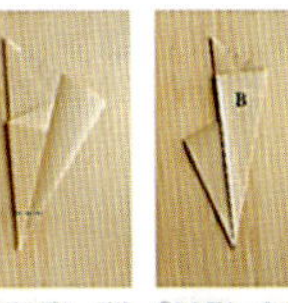

③Bを開き、その開いた折り目に合わせて、左端の辺を折ります。

④Bを元に戻し、下端を裏側に折り上げます。

⑤いったん開いて、同寸の赤い和紙を重ねて折り直し、麻苧を片輪結びにします。赤い紙を最初から重ねて折ってもいいでしょう。

年玉包み

①18㎝四方の縁紅紙を対角線で折り、左辺を三等分して、上の角、下の角の順に折ります。

②右端の角を左に折ります。

③上の角を上に引き上げて、右辺を中央に合わせて折ります。

④上の角の2枚の紙の間から三つ折りのお札や硬貨を入れ、上の角を、下の角の2枚の紙の間に差し込みます。

⑤出来上がりです。

------は手前に折る折り目、------は裏に折り返す折り目、——は辺を示します。

62

お正月は、しきたりが特に身近になる時期。まずは小さな折形から、日本の豊かな心を感じてみてはいかがでしょう。

贈り物を包むことは紙を選ぶときから始まっています

折形には、和紙で出来た半紙を使います。和文具店などで手に入りますが、手漉きの和紙を使うと、やはり一味違うもの。今回は、折形デザイン研究所の美濃和紙、「折形半紙1/2」を使いました。半紙は多少サイズに幅がありますが、ここでは折形半紙の243×343㎜を目安にしています。今回の折形は、全て折形デザイン研究所のオリジナルです。

折形には真・行・草の格があり、紙と包み方の組み合わせを、贈り物や相手に合わせて選びます。包み方は同じでも、紙を変えれば格が異なってきます。松飾りと67頁右上の祝儀包みに使用した檀紙は、皺という凸凹のある格の高い紙。年玉包みに使用した縁紅紙は、縁に赤い線が入った正方形の和紙です。赤は、彩りのアクセントに使われ、少しのぞかせたり、内側に重ねた色を「におい」といいます。赤い「におい」は、今回は民芸紙を使いましたが、他の赤い和紙でも。67頁の青の紙幣包みは、片端に赤い線が入った折形半紙を使っています。絵の具で端に線を描いてもいいでしょう。

屠蘇散包み

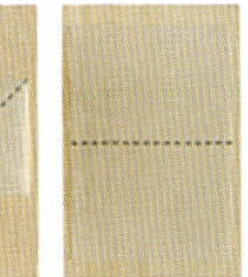

①横半分に切った半紙を縦に置き、下辺を上辺に合わせて折り上げます。

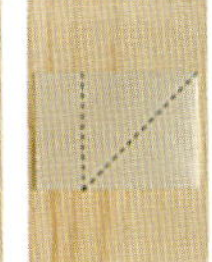

②右下の角を上辺に合わせて折り、左辺を右に折り返します。

③下辺を左辺に合わせて折り上げます。

④上端の4枚の紙を2枚ずつに開いて屠蘇散を入れ、右上の角を、下の角に合わせて差し込みます。

⑤赤い紙を、下の三角の上端から少し出る大きさに切って、差し込みます。

箸包み

①半紙を横半分に切って図のように置き、左上の辺を、右上の辺と平行で、右に正方形ができる位置に折ります。

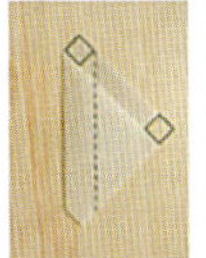

②上に、右と同じ正方形ができるように、右にある2つの角を、2枚一緒に左に折ります。

③左にある2つの角が右辺に接するように折ります。

④下端を裏側に折り上げます。

⑤いったん開いて、同寸の赤い和紙を重ねて折り直します。赤い紙を最初から重ねて折ってもいいでしょう。

63

Eine gut durchdachte horizontale Anordnung unterteilt einleitende Elemente in Bereiche. Durch Bilder und Bildunterschriften, die sich über die Doppelseite erstrecken, entsteht ein horizontaler Fluss und gleichzeitig wird jede Kombination aus Bild und Bildunterschrift zu einem klar und leicht lesbaren Schritt in den Anweisungen des Artikels.

54. Zeitleisten erhellen

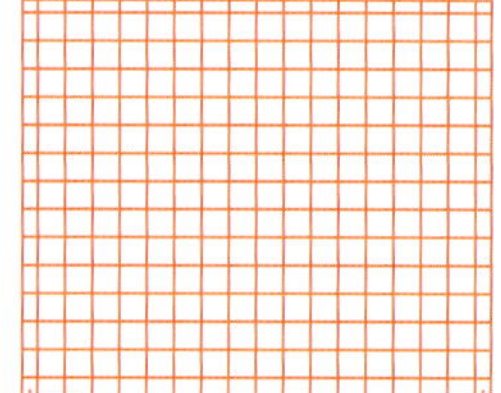

Eine Zeittafel sollte niemals nur als rein sachliche Information betrachtet werden. Da sie auch das Leben einer Person oder einer Epoche darstellen kann, muss das Design den Inhalt wiedergeben.

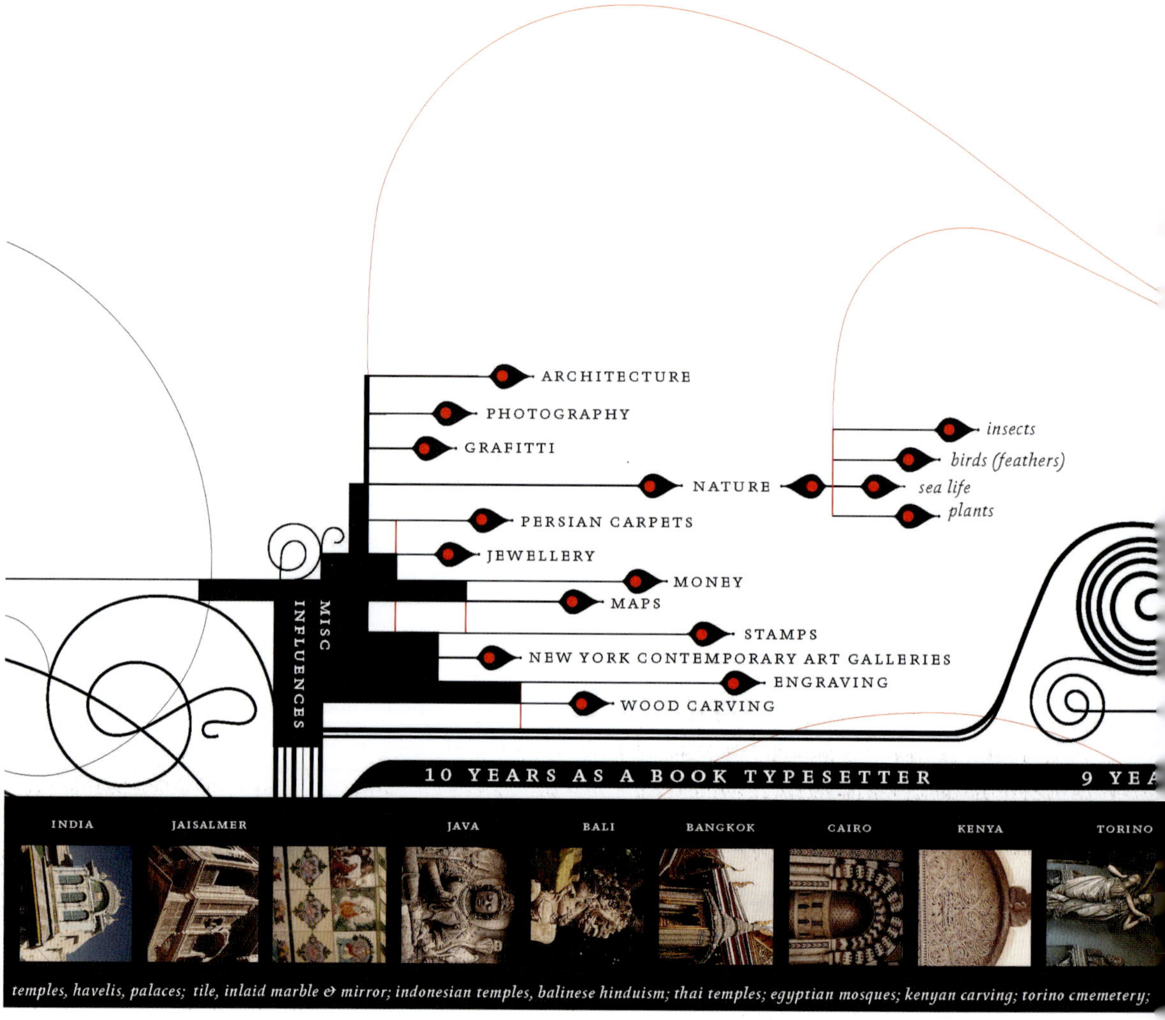

PROJEKT
Influence map

KUNDE
Marian Bantjes

DESIGNER AND ILLUSTRATOR
Marian Bantjes

In Marian Bantjes Illustration ***Influences & Artistic Vocabulary*** stehen künstlerisches Können und Detail im Vordergrund. Ein Beweis dafür sind die aus Einflüssen wie Bewegung, fließender Darstellung und Ornamentierung gewonnenen Erfahrungen. In ihrer zehnjährigen Tätigkeit als Buchillustratorin konnte Bantjes ihr großartiges typografisches Talent entwickeln.

ARCHITECTURE
PHOTOGRAPHY
GRAFITTI
NATURE
insects
birds (feathers)
sea life
plants
PERSIAN CARPETS
JEWELLERY
MONEY
MAPS
MISC INFLUENCES
STAMPS
NEW YORK CONTEMPORARY ART GALLERIES
ENGRAVING
WOOD CARVING
10 YEARS AS A BOOK TYPESETTER
INDIA
JAISALMER
JAVA
BALI
BANGKOK
CAIRO
KENYA
TORINO
temples, havelis, palaces; tile, inlaid marble & mirror; indonesian temples, balinese hinduism; thai temples; egyptian mosques; kenyan carving; torino cmemetery;

Der lyrische Charakter beruht auf den gebogenen Linien, aber auch auf ihrer Stärke. Die Spationierung der Buchstaben in den Bildunterschriften schafft Textur und Leichtigkeit. Das &-Zeichen ist wunderschön. Obwohl die Darstellung voller Bewegung ist, steht dem eine geradlinige Ausrichtung gegenüber.

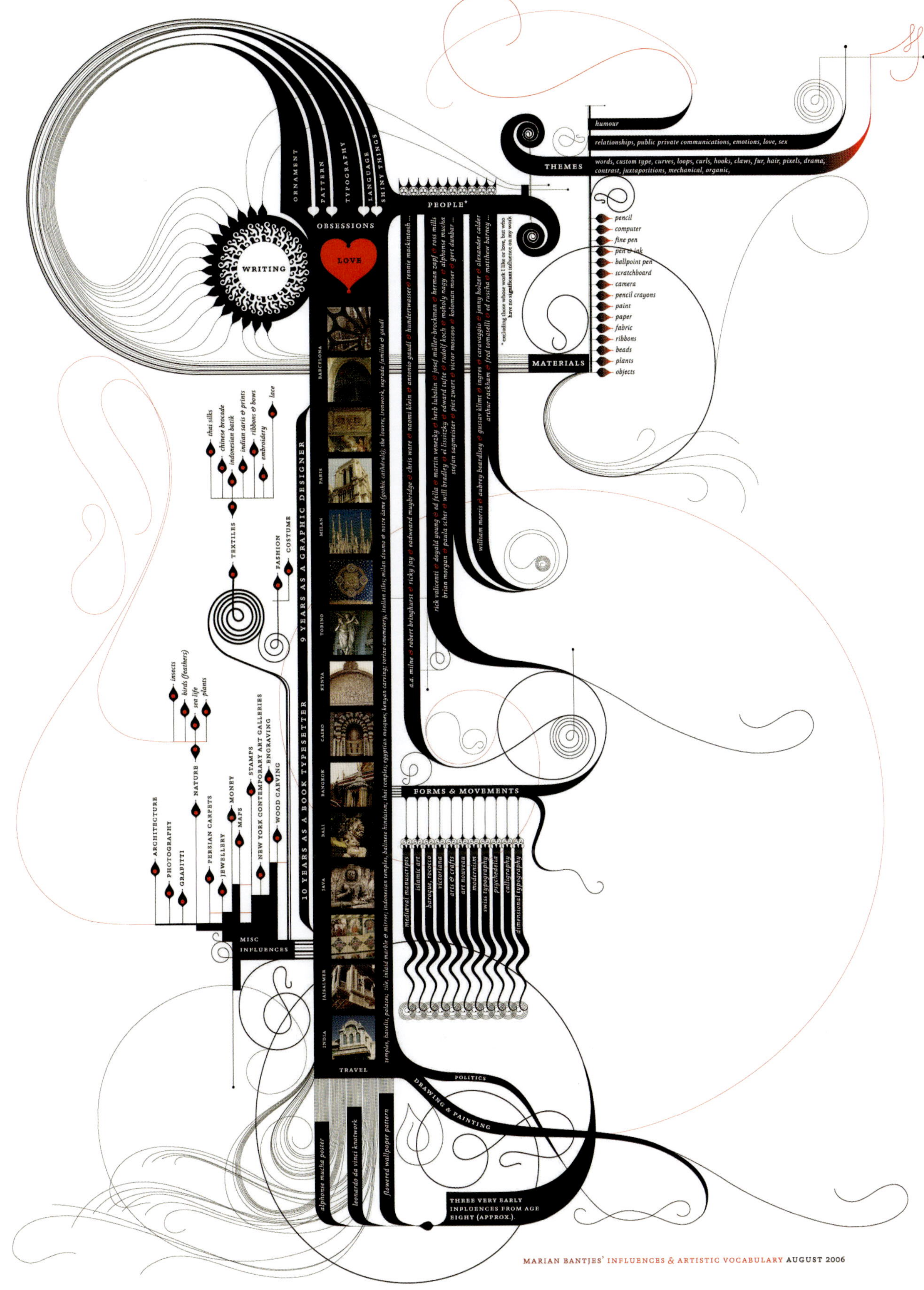

MARIAN BANTJES' INFLUENCES & ARTISTIC VOCABULARY AUGUST 2006

55. Die Navigationsleiste als Banner

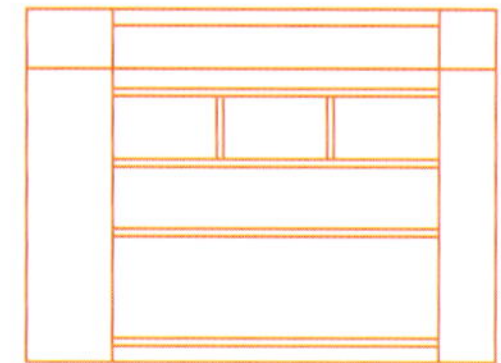

Elemente lassen sich gut durch die Aufteilung der vorhandenen Räume trennen. Ein klarer horizontaler Balken kann als Banner dienen, um das Interesse auf die Titelgeschichte oder die wichtigsten Informationen zu lenken. Außerdem bietet der Einsatz von Farbe am oberen Rand des Balkens die Möglichkeit, die Informationen aus der Headline quasi herausfließen zu lassen. Dadurch wird ganz natürlich eine gelungene Spannung zwischen den Polen Negativ und Positiv, Hell und Dunkel sowie Beherrschend und Untergeordnet erzeugt.

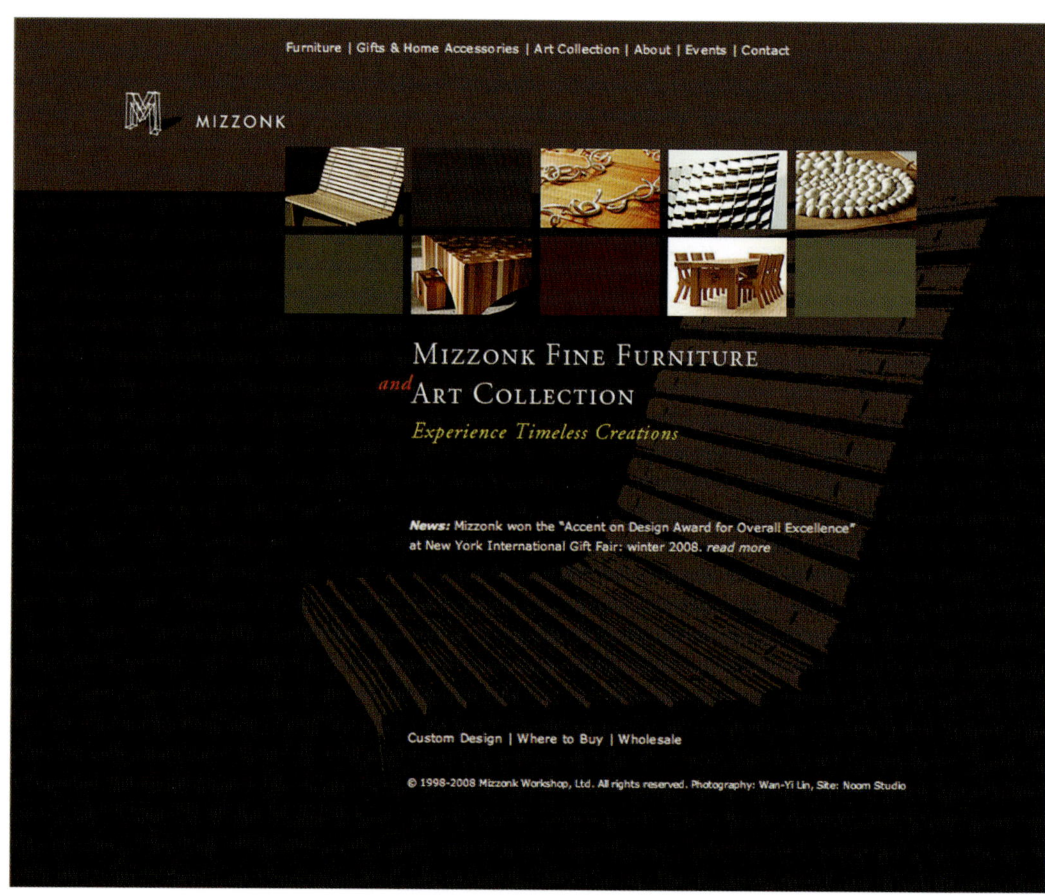

Innerhalb der horizontalen Anordnung kann man die Startseite schnell von oben nach unten überfliegen.

PROJEKT
www.mizzonk.com

KUNDE
Mizzonk Workshop

DESIGN
Punyapol "Noom" Kittayarak

Schmale Zeilen kennzeichnen die Webseite eines Unternehmens für maßgefertigte Möbel in Vancouver, British Columbia.

Die Navigationsleiste bleibt auch auf Subscreens als klarer horizontaler Wegweiser erhalten.

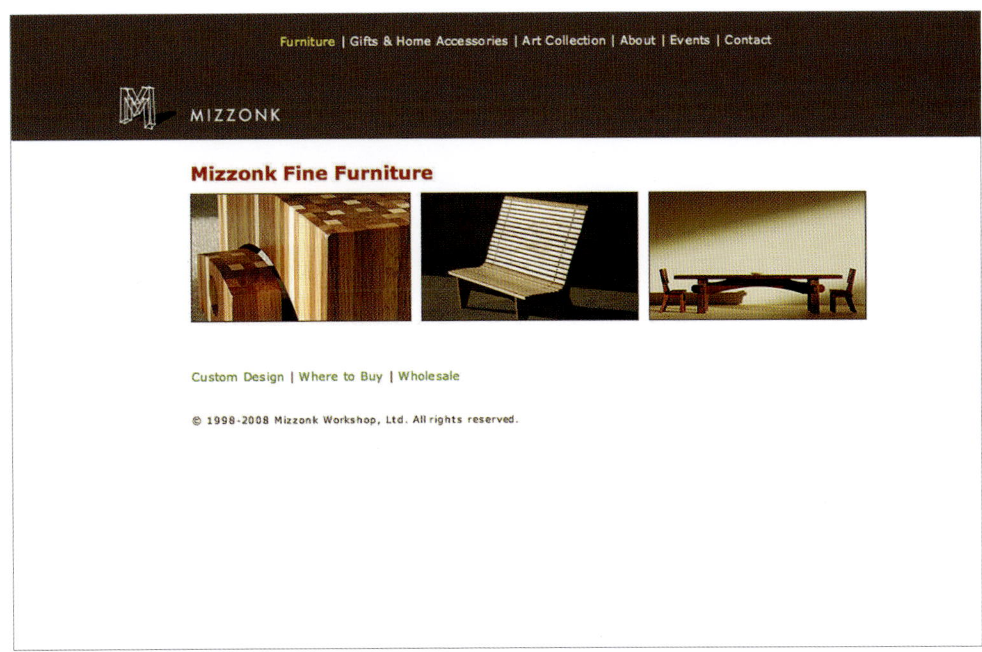

Nicht alle Elemente sind gleich groß und gleich tief. Wenn der Text unter die Grundlinie des Bildes fällt, entsteht eine lyrische Fließbewegung.

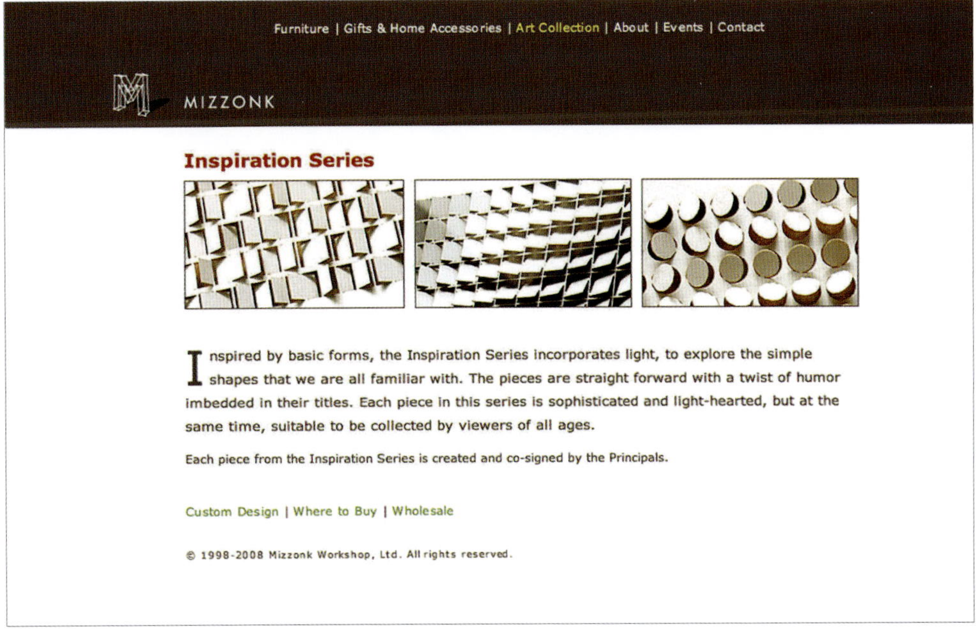

56. Klar und doch verspielt

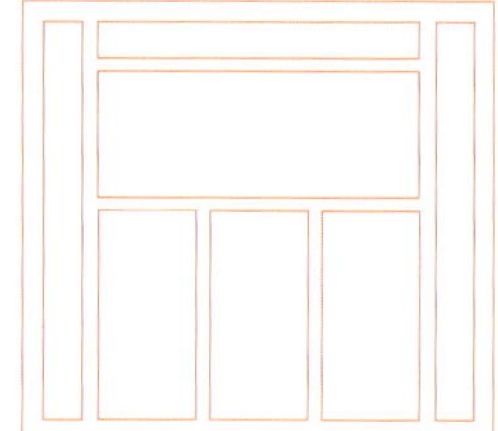

Bei diesem Etikett gibt es kein Vertun. Zwei Farben und zwei Sprachen machen alles klar und alles, was Sie wissen müssen (0 %!), ist übersichtlich dargestellt. Der clevere Einsatz eines Rasters ist unbeschwert und verspielt (und hält den Benutzer nüchtern).

Siehe auch Seite
100/101

PROJEKT
Identität und Verpackung für das alkoholfreie Bier Free Damm.

KUNDE
Cervezas Damm, Barcelona

DESIGN
Mario Eskenazi Studio

DESIGNER
Mario Eskenazi,
Marc Ferrer Vives

Eine herrliche Verpackung mit Organisation und viel Freigeist.

Seite gegenüber, beide Fotos:
Das Etikett würde ohne Flasche auch als Poster funktionieren. Der Druckbogen (oben) bildet ein eigenes modulares Raster. Das Etikett hat nur zwei Farben, aber Schriftgröße, Schriftart und die weißen Buchstaben haben enorme Durchschlagskraft.

Diese Seite, beide Fotos:
Die Identität funktioniert als Design und als Verpackung. Das ist ein erstklassiges Branding.

57. Einmal kippen

Schrift kann auf der horizontalen und der vertikalen Achse gleichzeitig wirken. Große Schrift dient als Behältnis für die übrigen Informationen. Die Breite der jeweiligen Namen kann durch unterschiedliche Schriftgrößen, -breiten und -stärken beeinflusst werden.

PROJEKT
Theater ad for *Cyrano de Bergerac*

KUNDE
Susan Bristow, Lead Producer

DESIGN
SpotCo

CREATIVE DIRECTOR
Gail Anderson

DESIGNER
Frank Gargiulo

ILLUSTRATOR
Edel Rodriguez

Dieses Plakat hebt den einprägsamsten Teil des Titels hervor und vermeidet viel Text. Der würde aufgrund des ausdrucksstarken Vor- und des kleiner gedruckten Nachnamens kaum beachtet werden.

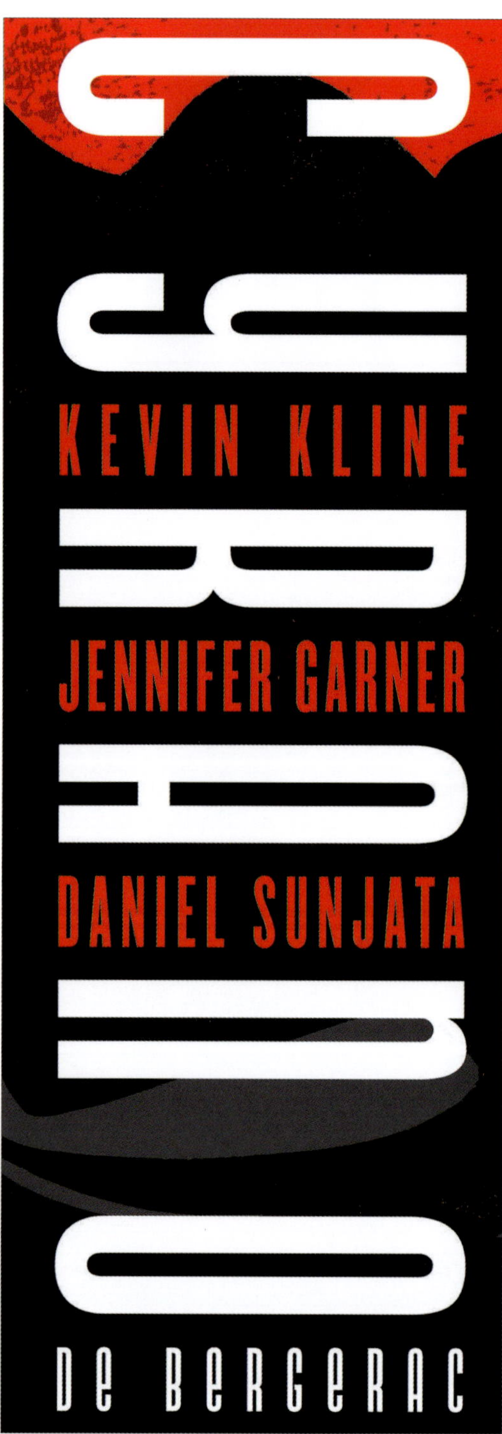

LINKS UND RECHTE SEITE: Saubere Anordnung und beschränkte Farbpalette müssen nicht zwangsläufig statisch wirken. Vielmehr stellt die fette Schrift die Informationsspalte in den Mittelpunkt. Um auf die Hauptfigur des Theaterstückes hinzuweisen, haben die Designer ein faszinierendes Profil mit einer außergewöhnlichen Typografie kombiniert.

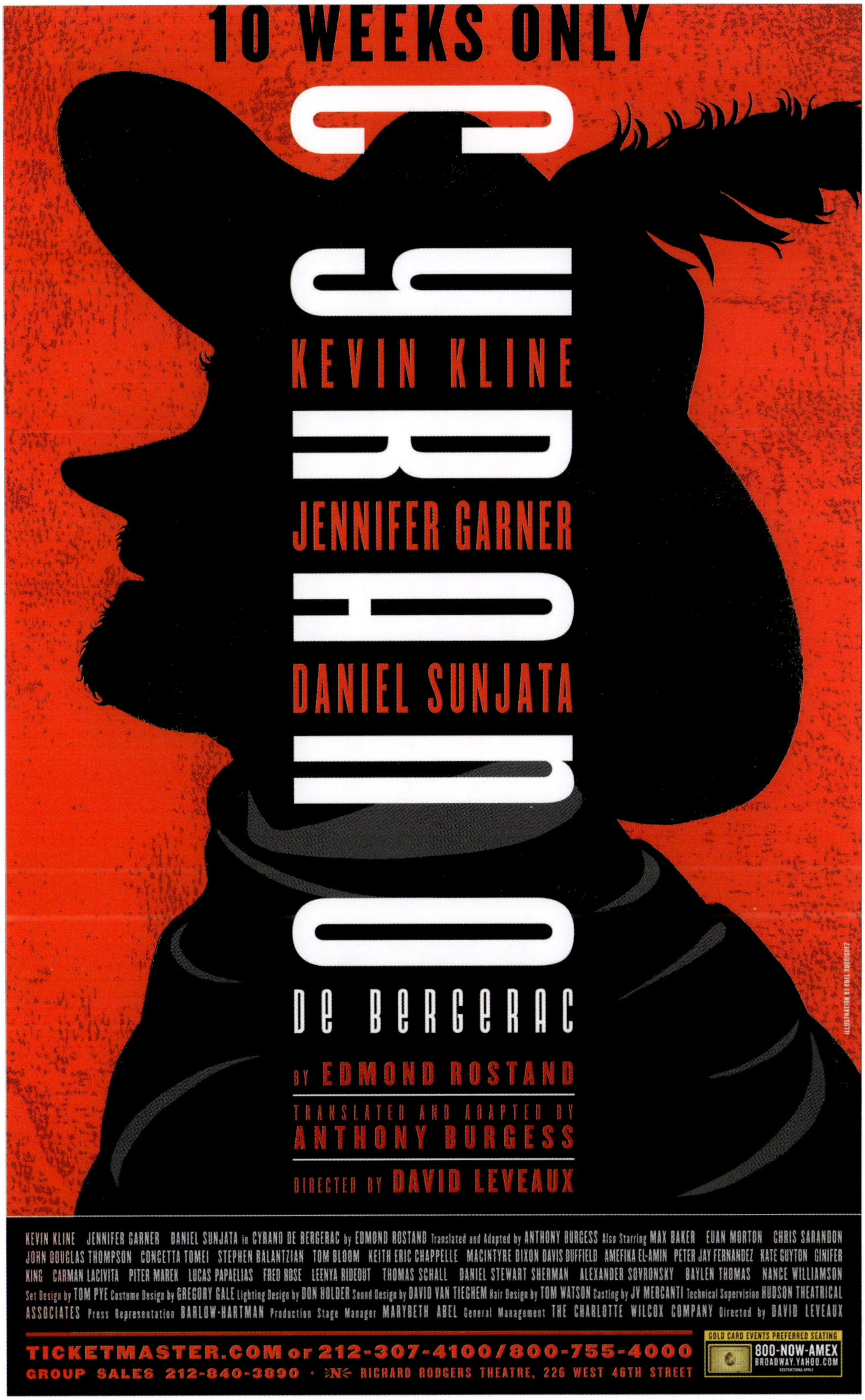

10 WEEKS ONLY

CYRANO

KEVIN KLINE

JENNIFER GARNER

DANIEL SUNJATA

DE BERGERAC

BY EDMOND ROSTAND

TRANSLATED AND ADAPTED BY ANTHONY BURGESS

DIRECTED BY DAVID LEVEAUX

KEVIN KLINE JENNIFER GARNER DANIEL SUNJATA in CYRANO DE BERGERAC by EDMOND ROSTAND Translated and Adapted by ANTHONY BURGESS Also Starring MAX BAKER EUAN MORTON CHRIS SARANDON JOHN DOUGLAS THOMPSON CONCETTA TOMEI STEPHEN BALANTZIAN TOM BLOOM KEITH ERIC CHAPPELLE MACINTYRE DIXON DAVIS DUFFIELD AMEFIKA EL-AMIN PETER JAY FERNANDEZ KATE GUYTON GINIFER KING CARMAN LACIVITA PITER MAREK LUCAS PAPAELIAS FRED ROSE LEENYA RIDEOUT THOMAS SCHALL DANIEL STEWART SHERMAN ALEXANDER SOVRONSKY BAYLEN THOMAS NANCE WILLIAMSON Set Design by TOM PYE Costume Design by GREGORY GALE Lighting Design by DON HOLDER Sound Design by DAVID VAN TIEGHEM Hair Design by TOM WATSON Casting by JV MERCANTI Technical Supervision HUDSON THEATRICAL ASSOCIATES Press Representation BARLOW-HARTMAN Production Stage Manager MARYBETH ABEL General Management THE CHARLOTTE WILCOX COMPANY Directed by DAVID LEVEAUX

TICKETMASTER.COM or 212-307-4100/800-755-4000

GROUP SALES 212-840-3890 · RICHARD RODGERS THEATRE, 226 WEST 46TH STREET

GOLD CARD EVENTS PREFERRED SEATING

800-NOW-AMEX

BROADWAY.YAHOO.COM

58. Sauber bleiben

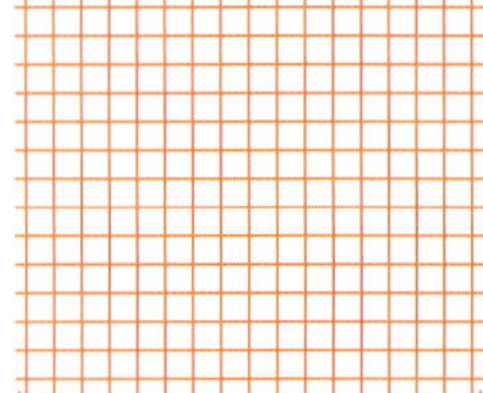

Ein Schriftbild auf einem Raster und für ein Raster wirkt frisch und sauber. Das schließt lustige Variationen der Bögen und andere Abwandlungen keineswegs aus.

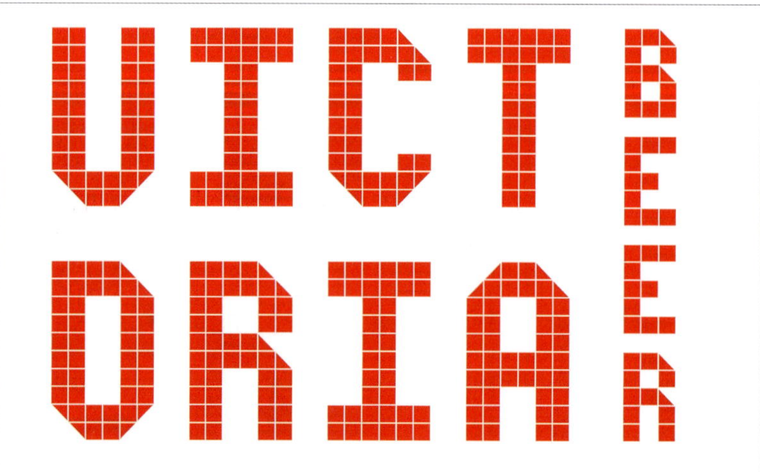

Dieses Alphabet für das Victoria-Bier feiert nicht nur das 90-jährige Jubiläum, sondern auch die Rückkehr der Brauerei nach Malaga nach 20 Jahren im Exil. „Malaguena y exquisita" bedeutet „aus Malaga und exquisit".

PROJEKT
Cervezas Victoria
Typograpie

KUNDE
Cervezas Victoria (Damm)

DESIGN
Mario Eskenazi Studio

DESIGNER
Mario Eskenazi,
Dani Rubio,
Marc Ferrer Vives

Dieses Alphabet, gestaltet 2017/2018 auf Keramikfliesen für eine Ausstellung zum 90-jährigen Jubiläum der Brauerei Victoria in Malaga, Spanien, wird auch für Werbegeschenke und Kampagnen benutzt.

MALAGUEÑA
Y EXQUISITA
MAL
AGA

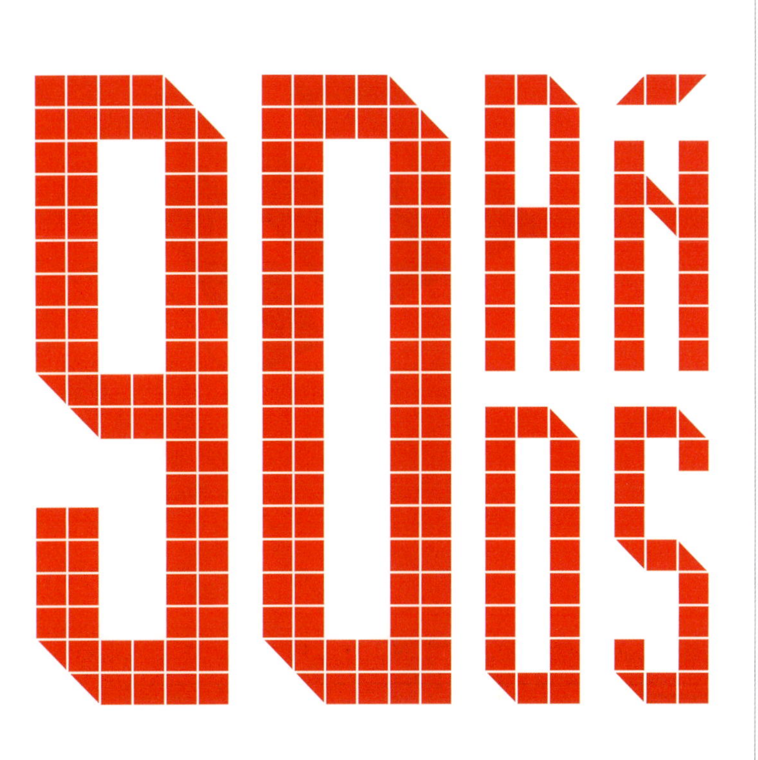
90 AÑOS

MALAGUEÑA
Y EXQUISITA

59. Rasterspiele

Jazz und Typografie haben eines gemeinsam: Sie lassen sich synkopieren. Selbst in einem engen und gut durchdachten Raster lässt sich mit unterschiedlichen Breiten, Stärken und Platzierungen eine typografische Jam-Session veranstalten. Was geschieht dann, wenn man alles wieder auf die Seite legt?

Dank der Dynamik der kleinen serifenlosen Schrift und einer größeren Zeile gewinnt die Schrift an Fahrt. Wird sie dann noch schräg gestellt und über zwei übereinanderliegende Freisteller gedruckt, scheint sie wirklich zu swingen.

PROJEKT
Plakate für Werbung und Promotion

KUNDE
Jazz at Lincoln Center

DESIGN
JALC Design Department

DESIGNER
Bobby C. Martin Jr.

Jazz im Lincoln Center ist exzellent, diszipliniert und voller Energie. Das Design gleicht der Schweizer Grafik, ist aber synkopiert – und einfach cool.

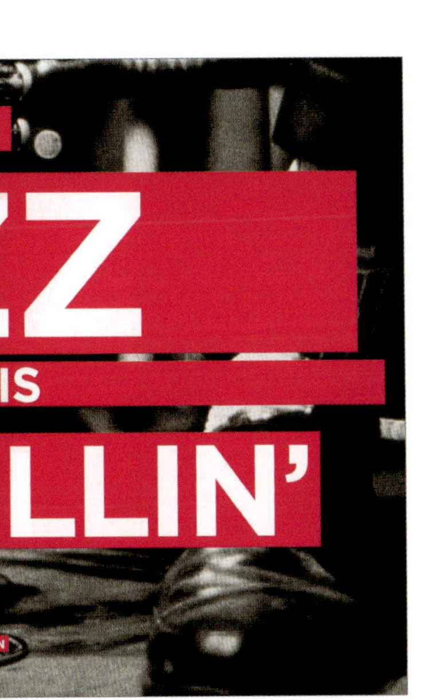

Negativschrift in Boxen unterschiedlicher Größen bildet gegenüber geschickten Bildaussschnitten einen scharfen, rhythmischen Kontrapunkt.

WILLIE NELSON

LINCOLN CENTER JAZZ ORCHESTRA
WYNTON MARSALIS

JAZZ AT LINCOLN CENTER’S 06-07 SEASON

THE MUSIC OF
MILES DAVIS

WHEN
JAZZ
IS
KILLIN’

JOHN ZORN
DAVE DOUGLAS

DIANNE REEVES

NOTES EXPLODE.
CROWDS PRESS.
TICKETS EVAPORATE.

06-07 SUBSCRIPTIONS
WHILE THEY LAST!

FOR MORE INFORMATION
Call 212-258-9999 or visit www.jalc.org/subs

JOE ZAWINUL

AFRO-LATIN JAZZ ORCHESTRA
ARTURO O’FARRILL

THE MUSIC OF
GEORGE GERSHWIN

& MANY MORE

THE MUSIC OF
JOHN COLTRANE

Cadillac
Lead New York Sponsor

jazz at lincoln center

60. Den Leser einbinden

Manchmal muss sich ein Raster vom Raster lösen. Lokale oder globale Kulturinformationen, die zwar in einer Schrift, aber in unterschiedlichen Größen, Formen und Stärken übermittelt werden, machen den Leser neugierig und sind ein Aufruf zum Handeln.

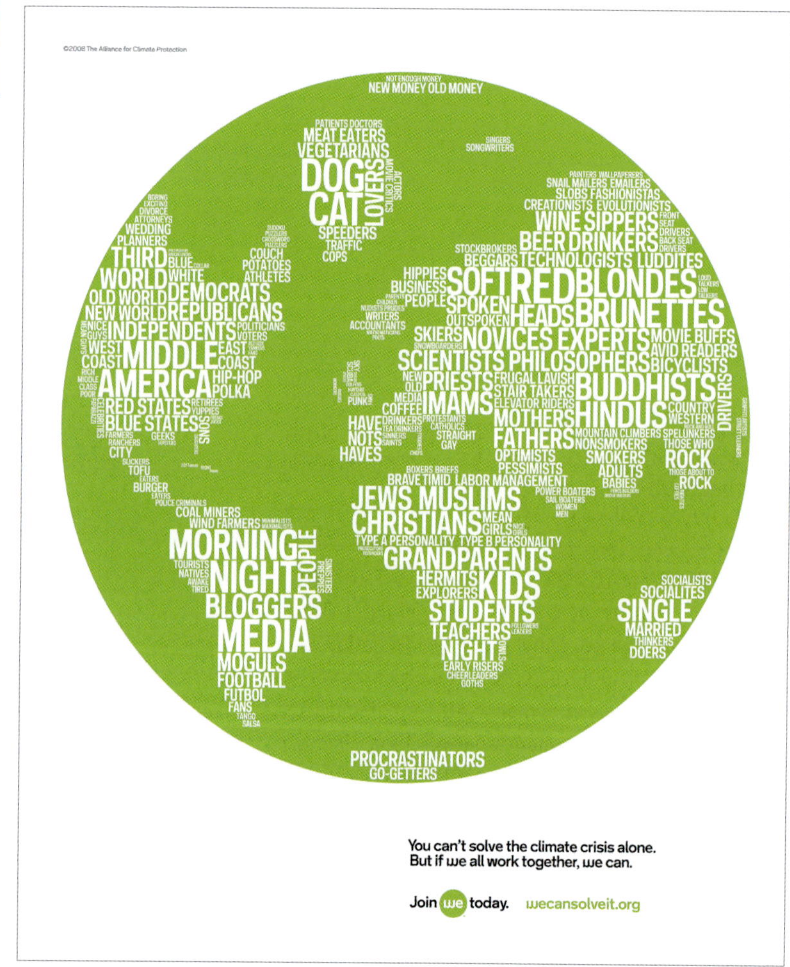

Die Wahl der Wörter und Schriftgrößen könnte nach speziellen Gesichtspunkten erfolgt sein – oder auch nicht. Große Schriften schreien buchstäblich nach Beachtung, kleinere dienen eher als visuelles Bindeglied. Das Plakat wirbt in einem sinnigen, grellen Grün für den Klimaschutz.

PROJEKT
Werbung der Alliance for Climate Protection

KUNDE
WeCanDoSolveIt.org

DESIGN
The Martin Agency; Collins

DESIGNER
The Martin Agency: Mike Hughes, Sean Riley, Raymond McKinney, Ty Harper; Collins: Brian Collins, John Moon, Michael Pangilinan

Diese Werbung für eine Umweltschutzorganisation nutzt den Vorteil fetter Typografie, um sich Geltung zu verschaffen.

NOT ENOUGH MONEY
NEW MONEY OLD MONEY
PATIENTS DOCTORS
MEAT EATERS
VEGETARIANS
DOG
CAT
LOVERS
MOVIE CRITICS
ACTORS
SPEEDERS
TRAFFIC
COPS
SINGERS
SONGWRITERS
BORING
EXCITING
DIVORCE
ATTORNEYS
WEDDING
PLANNERS
SUDOKU
PUZZLERS
CROSSWORD
PUZZLERS
COUCH
POTATOES
ATHLETES
THIRD
WORLD
BLUE COLLAR
WHITE
DEMOCRATS
OLD WORLD
NEW WORLD
REPUBLICANS
NICE GUYS
MEAN GUYS
INDEPENDENTS
POLITICIANS
VOTERS
WEST COAST
MIDDLE
EAST COAST
RED SOX
YANKEES
FANS
RICH
MIDDLE CLASS
POOR
AMERICA
HIP-HOP
POLKA
PAPARAZZI
CELEBRITIES
RED STATES
RETIREES
YUPPIES
BLUE STATES
SONS
FREAKS
JOCKS
DAUGHTERS
FARMERS
RANCHERS
GEEKS
HIPSTERS
CITY
SLICKERS
TOFU
EATERS
BURGER
EATERS
LEFT
RIGHT
POLICE CRIMINALS
COAL MINERS
WIND FARMERS
MINIMALISTS
MAXIMALISTS
MORNING
NIGHT
PEOPLE
TOURISTS
NATIVES
AWAKE
TIRED
SINISTERS
PREPPIES
BLOGGERS
MEDIA
MOGULS
FOOTBALL
FUTBOL
FANS
TANGO
SALSA
PAINTERS WALLPAPERERS
SNAIL MAILERS EMAILERS
SLOBS FASHIONISTAS
CREATIONISTS EVOLUTIONISTS
WINE SIPPERS
FRONT SEAT DRIVERS
BEER DRINKERS
BACK SEAT DRIVERS
STOCKBROKERS
BEGGARS
TECHNOLOGISTS
LUDDITES
HIPPIES
BUSINESS
PEOPLE
PARENTS
CHILDREN
NUDISTS PRUDES
WRITERS
ACCOUNTANTS
MATHEMATICIANS
POETS
SOFT
SPOKEN
OUTSPOKEN
RED
HEADS
BLONDES
BRUNETTES
LOUD TALKERS
LOW TALKERS
SKIERS
SNOWBOARDERS
NOVICES EXPERTS
MOVIE BUFFS
AVID READERS
SCIENTISTS
PHILOSOPHERS
BICYCLISTS
SKY
SCUBA
DIVERS
GOLFERS
HUNTERS
CLASSICAL
PUNK
FANS
WORKERS
BOSSES
NEW
OLD
MEDIA
PRIESTS
FRUGAL LAVISH
STAIR TAKERS
ELEVATOR RIDERS
BUDDHISTS
DRIVERS
IMAMS
COFFEE
DRINKERS
TEA DRINKERS
PROTESTANTS
CATHOLICS
HAVE
NOTS
HAVES
SINNERS
SAINTS
FOODIES
CHEFS
STRAIGHT
GAY
MOTHERS
FATHERS
HINDUS
COUNTRY
WESTERN
ROCK AND ROLL
GRAFFITI ARTISTS
STREET CLEANERS
MOUNTAIN CLIMBERS
SPELUNKERS
NONSMOKERS
SMOKERS
THOSE WHO
ROCK
OPTIMISTS
PESSIMISTS
ADULTS
BABIES
THOSE ABOUT TO
ROCK
FENCE BUILDERS
BRIDGE BUILDERS
LEFTIES
RIGHTIES
BOXERS BRIEFS
BRAVE TIMID
LABOR MANAGEMENT
POWER BOATERS
SAIL BOATERS
WOMEN
MEN
JEWS MUSLIMS
CHRISTIANS
MEAN
GIRLS
NICE
GIRLS
TYPE A PERSONALITY TYPE B PERSONALITY
PROSECUTORS
DEFENDERS
GRANDPARENTS
HERMITS
EXPLORERS
KIDS
STUDENTS
TEACHERS
FOLLOWERS
LEADERS
NIGHT
OWLS
EARLY RISERS
CHEERLEADERS
GOTHS
SOCIALISTS
SOCIALITES
SINGLE
MARRIED
THINKERS
DOERS
PROCRASTINATORS
GO-GETTERS

61. Ordnung trotz schmaler Stege

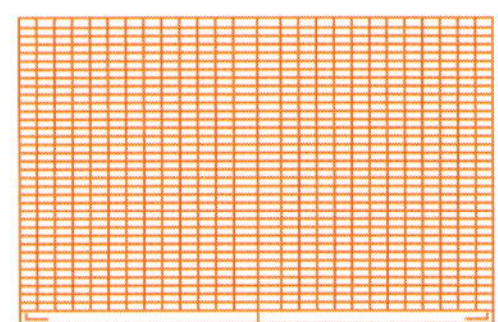

Bei einem gut konzipierten Raster macht es nichts aus, wenn die Ränder schmal sind. Sind die Bilder sauber auf klaren Rasterlinien angeordnet sowie Fläche und Typografie genau geregelt, können schmale Außenstege durchaus Teil eines sorgfältig angelegten Konzeptes sein. Die Ordnung auf einer ausgeglichenen Seite kann als Unterlage für schmale Stege dienen und ein geregeltes Layout begrenzen.

Wenn man das weiß, sollte man am Anfang einen Rand für Fehler frei lassen. Sowohl für Anfänger als auch für erfahrene Praktiker sind Ränder eine knifflige Sache. Das Anlegen eines Rasters mit Variablen verlangt Können und Erfahrung, aber auch die Bereitschaft, etwas auszuprobieren und Fehler zu machen. Die meisten herkömmlichen Offsetdrucker und Verleger zucken bei zu schmalen Stegen. Winzige Außenstege lassen wenig Spielraum für leichte Schwingungen der Papierrolle, wenn diese durch die Druckmaschine läuft. Aus diesem Grund achten Grafikdesigner oft auf großzügige Außenstege.

Typographie et Architecture

«que se passe-t-il quand le **texte sort de l'échelle habituelle** de l'imprimé ou de l'écran et va se confronter à notre corps, prendre **une importance physique**, quand il veut prendre une autre place et se développe dans la profondeur **comme une sculpture**?» **par presse papier** ■ ■ ■

Avec la "façade aux mille lettres" (aussi appelée moucharabieh typographique polyglotte) du musée des écritures du monde, à Figeac, et le "T" du tramway de Nice, Pierre di Sciullo vient de livrer deux nouveaux travaux qui font la part belle à la typographie dans l'espace public… L'occasion de revenir sur quelques-uns des travaux de ce graphiste, typographe, dessinateur de caractères, et de questionner avec lui les rapports entre typographie et architecture.Nous voici donc à la gare de Gretz-Armainvilliers, par un froid dimanche de décembre. Pierre di Sciullo, un béret noir – plutôt inattendu sous ces latitudes – vissé sur la tête, nous attend au bas de la passerelle métallique, pour nous conduire, tout près d'ici, à l'atelier qu'il anime avec Catherine di Sciullo et Juliette Cheval. Une discussion s'installe

Marie Bruneau (MB): Pierre, finalement, quel est ton métier? graphiste? typographe?
Pierre di Sciullo (PdS): J'ai reçu, à ce propos, une leçon dont je me souviendrai toute ma vie. Assez intimidé, j'avais pris mon élan pour appeler Adrian Frutiger: «Bonjour, je suis typographe, et j'aimerais bien venir vous voir.» Il me répond aussitôt, presque cassant: «Dans ce cas, jeune homme, vous n'êtes peut-être pas avec la bonne personne: vous auriez dû rencontrer Emil Ruder; malheureusement, il n'est plus de ce monde. En ce qui me concerne, je ne suis que dessinateur de lettres.» J'ai vraiment compris ce jour-là la différence qui existe entre les dessinateurs de caractères qui créent des alphabets, les typographes qui mettent en forme le texte dans la page (avec généralement des caractères créés par d'autres), et les graphistes, qui sont souvent – mais pas toujours – typographes. En ce qui me concerne, je coche les trois cases… Ce qui explique qu'en tant que dessinateur de caractères, j'aie parfois tendance à être un peu expéditif! C'est un travail extrêmement minutieux, et long. Dans ce que j'appelle ma "boîte à outils", il y a de nombreux caractères qui ne sont pas encore complètement développés…
MB: Comment t'es-tu intéressé à la typographie?
PdS: Je situe le point de non-retour au numéro 5 de Qui? Résiste, le "Manuel de la femme", publié en mai 1985. C'est à ce moment-là que je me suis posé des questions du type: «C'est quoi, finalement, la vraie forme d'un "M"?». Rien de ce que je voyais ne me satisfaisait vraiment, et j'en suis venu à dessiner les titres à la main, avec une règle et un compas… et à composer le mot "femme" avec deux "M" (deux "aime"?) dissymétriques, et des "E" capitales à quatre barres horizontales, juste avant que The Face ne publie ceux de Neuvile Broody… (J'en suis assez fier même si je suis à peu près sûr qu'on trouve ça cinquante ans plus tôt, chez Kurt Schwitters!). Plus tard Adrian Frutiger m'a expliqué qu'il ne fallait surtout pas dessiner des lettres avec une règle et un compas; il m'a remis un petit livre bourré

«c'est quoi, finalement, la vraie forme d'un "M"?»

48 : 2.2008

2.2008 : 49

Eine ausgeglichene Seite mit vollkommen geradlinigen Anordnungen zeigt die Flexibilität des Rasters. Alle Elemente liegen auf einer Linie, dennoch entsteht durch die große Schrift Dynamik. Freier Fläche stehen schmale Außenstege gegenüber. Durch das harmonische Wirken mehrerer Schriften in verschiedenen Stärken und Größen ist die Typografie ebenfalls ausgeglichen.

PROJEKT
étapes: magazine

KUNDE
Pyramyd/*étapes:* magazine

DESIGN
Anna Tunick

Der klare Raster dieser französischen Zeitschrift über Grafikdesign stellt einen so wirkungsvollen Eindruck von Ordnung dar, dass schmale Ränder Teil eines Planes sind, möglichst viele Informationen unterzubringen.

Alle Elemente auf dieser Doppelseite sind so konsequent geradlinig angeordnet, dass selbst die freie Fläche zwischen den Bildern den schmalen Rändern Rechnung trägt.

Der Unterbau des (zwölfspaltigen) Rasters ermöglicht es, bestimmte Spalten nicht zu füllen. Dadurch werden Ränder ausgeglichen und es wird einer Doppelseite mit viel Inhalt Raum zum Atmen gegeben.

j'ai vu le moment où l'on allait inaugurer le bâtiment sans mon travail. pourquoi? parce que l'on ne parvenait pas à s'accorder sur sa dénomination exacte: "sculpture typographique" ou "enseigne"?

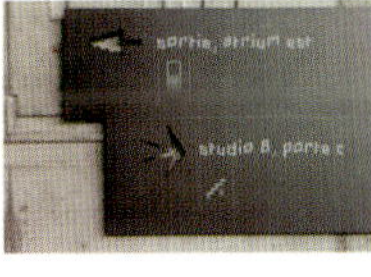

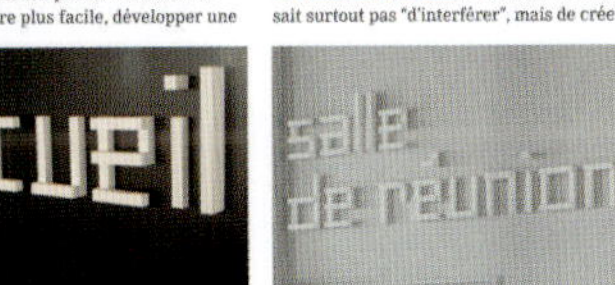

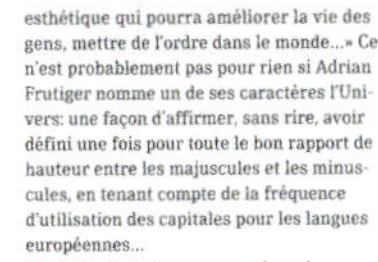

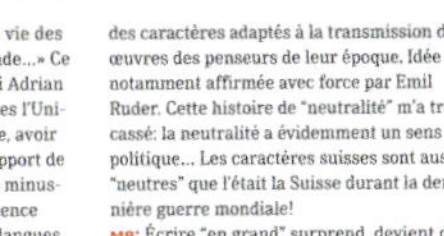

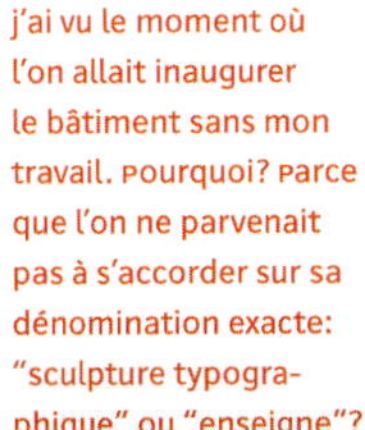

62. Auf den Punkt gebracht

Manche Themen enthalten extrem viele ausführliche, tief gehende und komplexe Informationen. Sollen sie alle auf begrenztem Raum untergebracht werden, dann ist es nötig, einzelne Punkte hervorzuheben.

Das kann etwa wie folgt geschehen: die Nutzung freier Flächen für ein Impressum, farbige oder farbcodierte Seitenleisten, Aufzählungen, auf bestimmte Rubriken hinweisende Icons und farbig gestaltete Titel und wichtige Textpassagen.

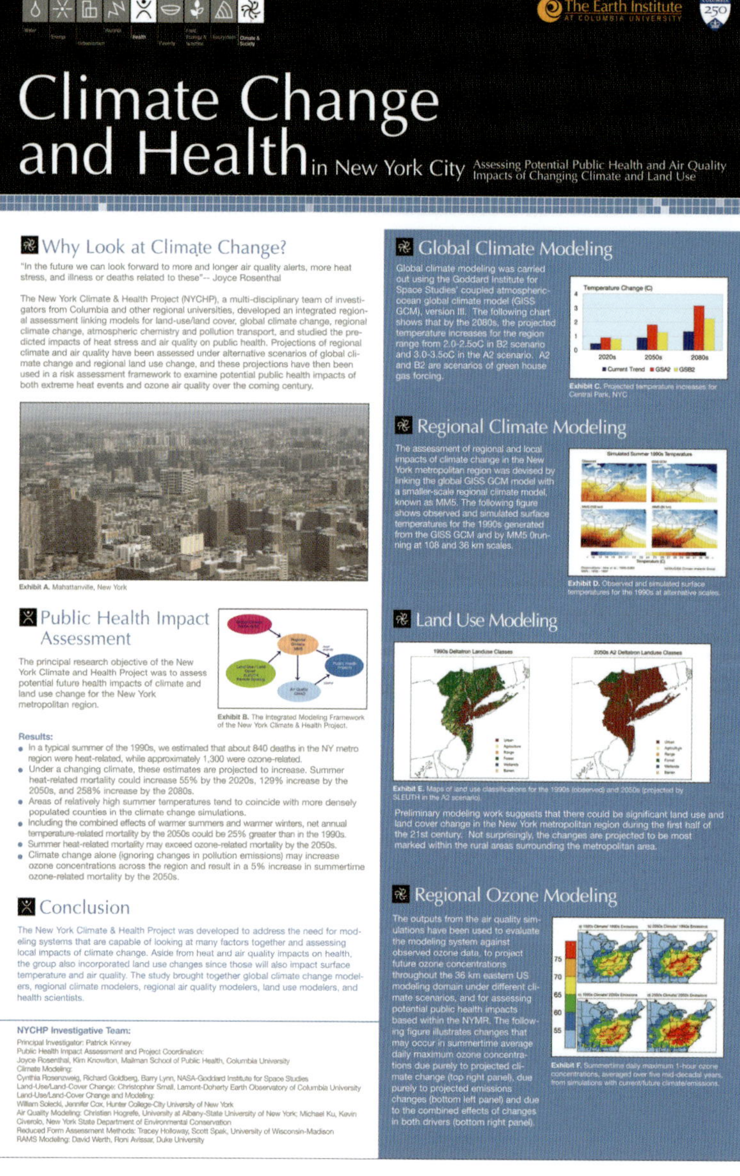

Sämtliche Icons sind in der Kopfzeile eines jeden Displays zu sehen. Die Icons der jeweiligen Themen sind hervorgehoben und stehen sozusagen als Bake am Beginn eines Absatzes.

PROJEKT
Materials and Displays for a Public Event

KUNDE
Earth Institute at Columbia University

CREATIVE DIRECTOR
Mark Inglis

DESIGN
Sunghee Kim

Bei diesen komplexen, detaillierten Lehrdisplays wird eine Kombination aus Icons und Farben eingesetzt, die auf die in einem Abschnitt oder Absatz behandelten Themen hinweisen. Verschiedene grafische Mittel wie Icons, Überschriften, Titel, Text, Bilder und Diagramme heben die Abschnitte hervor und erleichtern die Navigation durch die Informationen. Dabei wird an den idealen Zielen Fläche, Textur, Farbe, Struktur, weiße gegen schwarze Fläche und leserliche Schrift festgehalten. Wo unterschiedliche Lehrmittel eingesetzt werden, da können saubere, geradlinige Anordnungen über Interesse oder Desinteresse des Lesers entscheiden.

Mit einem schwarzen Band, das auf allen Displays als Impressum erscheint, wird die allgemeine Aufmachung der Displays vereinheitlicht. Das stets gleich große schwarze Band enthält Angaben wie die Icons, die Logos der Columbia University und das Institut der Universität, die Überschrift und die Untertitel.

Der Abschnitt unter dem schwarzen Band enthält jeweils nicht nur das Icon, sondern auch Überschriften, die in verschiedenen, für jedes Display codierten Schriften und Farben gedruckt sind.

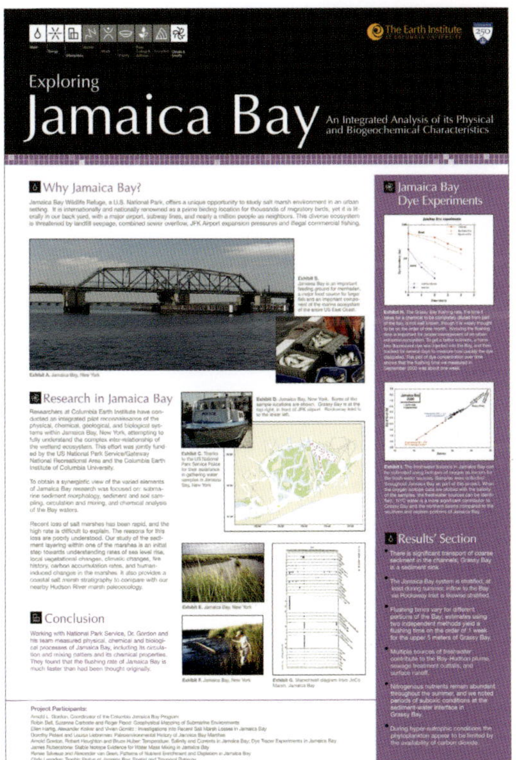

Die Typografie ist klar und deutlich, Aufzählungszeichen gliedern die Informationen. Ergebnisse und Schlussfolgerungen werden in der Kennfarbe der wissenschaftlichen Disziplin, die das Display betreibt, hervorgehoben.

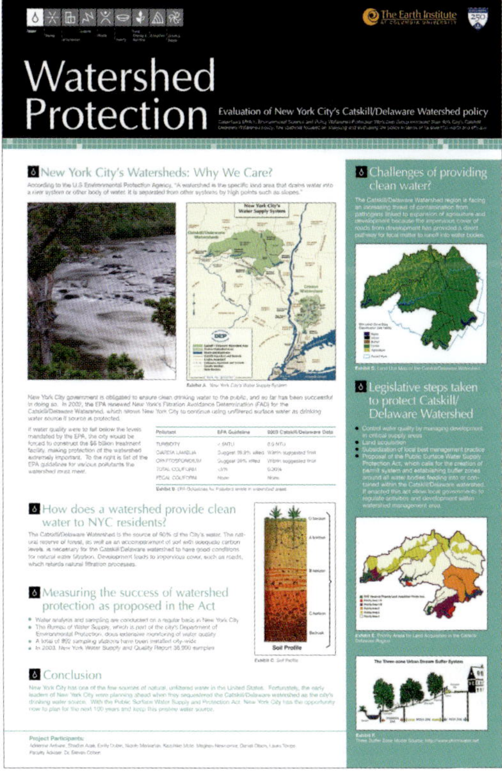

Auf seitlichen Leisten, ebenfalls mit Farbcode für jede Rubrik, werden Kategorien wie „Experimente und Forschung" hervorgehoben.

63. Geschickt aufteilen

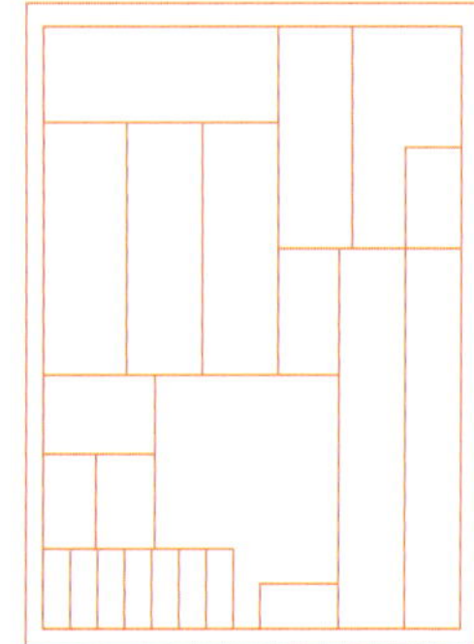

Was machen Sie mit zu vielen Geschichten, Bildern, Diagrammen, Cartoons, Statistiken, Schwafelei und Tiraden? Sie werfen alles in eine lockere Gesamtstruktur. Dann teilen Sie sie auf.

Seite gegenüber: Der Entwickler und Designexperte einer großformatigen Zeitung, die den Bürgern ein taktiles Gefühl des Lebens in einer übervölkerten Stadt vermitteln soll, sagt, dass die Zeitung kein Raster benutzt. Es ist aber klar, dass die Designer Raster verstehen und benutzen.

New York's Stories & Secrets & Supermarkets & Sex & Metadata & Terrorists & Herbalists & Scie... & Lovers

Civilization

A FUTURE HISTORY OF NEW YORK

PROTECT THE CHILDREN ● ISSUE 1 / VOL 1

Father, Mother, God, Bathe Me In Light

Darcie Wilder

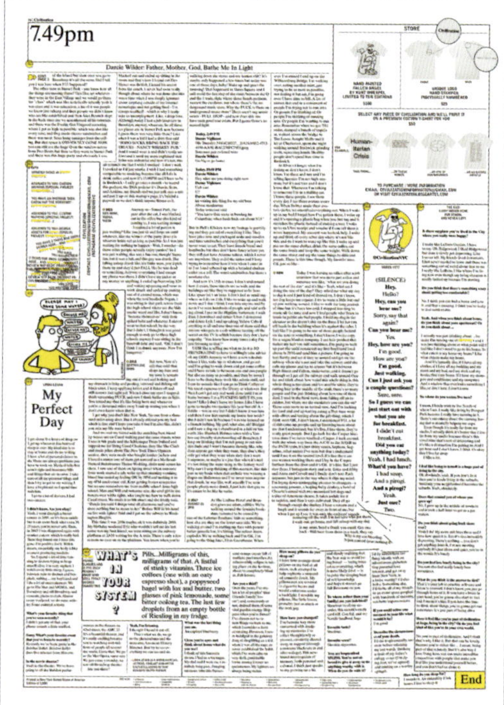

7.49pm

Darcie Wilder: Father, Mother, God, Bathe Me In Light

My Perfect Day

WHAT'S IN YOUR SYSTEM ?

End

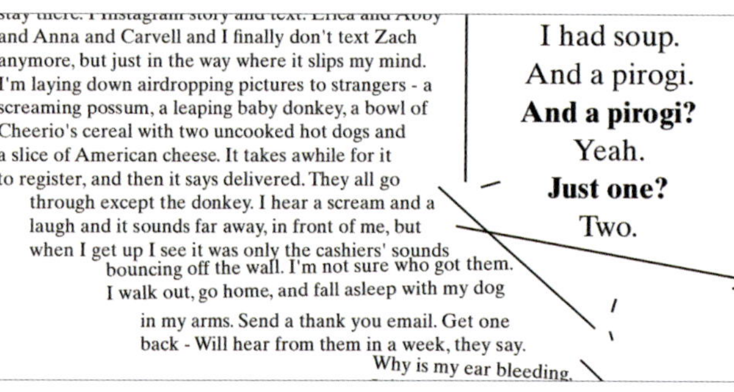

stay there. I Instagram story and text. Erica and Abby and Anna and Carvell and I finally don't text Zach anymore, but just in the way where it slips my mind. I'm laying down airdropping pictures to strangers - a screaming possum, a leaping baby donkey, a bowl of Cheerio's cereal with two uncooked hot dogs and a slice of American cheese. It takes awhile for it to register, and then it says delivered. They all go through except the donkey. I hear a scream and a laugh and it sounds far away, in front of me, but when I get up I see it was only the cashiers' sounds bouncing off the wall. I'm not sure who got them. I walk out, go home, and fall asleep with my dog in my arms. Send a thank you email. Get one back - Will hear from them in a week, they say. Why is my ear bleeding.

I had soup.
And a pirogi.
And a pirogi?
Yeah.
Just one?
Two.

living in the city.
Oh absolutely, ye then you're kinda Suddenly your ne suburbs. Yeah, ab

Does that remind grew up?
No. I grew up in t and it took a half station.

Do you think abo ever?
Yeah I do! My mo love how quiet it i

Schweizer System trifft konkrete Poesie, wie dieser Ausschnitt in Reproduktionsgröße zeigt.

PROJEKT
Civilization

KUNDE
Richard Turley

HERAUSGEBER/DESIGNER/TEXTER
Richard Turley, Lucas Mascatello, Mia Kerin, with other Beiträge,

Die im Eigenverlag erschienene New Yorker Zeitung ist voller Träumereien, Fakten und Halbwahrheiten wie die Stadt selbst. Beim Start im Jahr 2018 entschied sich ihr Entwickler bewusst für eine viktorianische Technologie, um das digital überfütterte Publikum zu erreichen.

3.1

Darcie Wilder

Cover story: We used to pass this park all the time on the bus, Dead Dog Park. There was a sign and everything. They were trying to up the number of Manhattan parks so they walked looking for squares of grass that could technically be parks. They ran out of names by the time they got to 168th and saw a dead dog and now it's Dead Dog Park, there's a sign and everything.

Father, Mother, God, Bathe Me In Light

○ 2018
HELL'S KITCHEN
1PM
15 NOTIFICATIONS

So I wake up in bed underneath a 25 pound weighted blanket but I just wanted the heaviest thing to put me out and the first thing I touch and look at is this blazing screen, and my eyes are still opening and they're making shapes adjusting to the light, I'm scrolling and looking for my dog and just scrolling and putting on the same podcasts as every morning before and trying to go back to sleep and failing. And then I'm using the new body wash that smells like pears and it's still so cold out, and I just found out my shower curtain was in the wrong place, years of making daily puddles on the floor, except who showers daily, more of a weekly puddle. I just got a new toothbrush that zaps me every thirty seconds, all I do it make it touch my teeth, moving around to a different tooth. My teeth were so soft they were stinging me when I brushed, a shooting pain up my gums. I'm trying to figure out what. I'm trying to. Every morning I can't fall back asleep but can't stay awake doing the things I want to do.

It's not that it's especially bad but that it just keeps never getting better, and I wake up around 1pm today so the kids at the high school next door are on their way home. Two recognizable character actors are talking on a bench, I'm throwing a tennis ball with a chihuahua mix and some puppy named Lemon comes up trying to steal it and does for a second. It's the kind of weather that could be fall but is March and taking it's time and even when it finally gets good enough to walk it'll be too hot, too much, too heavy, and I'll be wearing sleeves even when I shouldn't.

App goes off PUSH notification - assault 5 blocks away. Last night, suspicious package. Tonight a car'll jump the curb and a manhole a few miles away will blow off, like I used to imagine happening when I stood on them. The helicopter in the East River, I was a few blocks away. Repeated footage over and over, New York 1 in doctor's office, this is footage of people dying. This is the last moments of five peoples' lives.

Taking her off-leash even though she could die. But everyone, all the time. Aren't we all just trying to keep ourselves alive? It's crazy, they should teach kids about aneurysms earlier. But she's too scared of the curb, too scared of cars, too scared of the brushing whizzing noises to ever go near. Sometimes I worry she'll start chasing a rat, kill a rat with her bare paws and I'll have her, covered in blood, to clean up. My friend's dog killed a rat before they had time to stop her. Rats on 43rd run next to the parking lot bricks, behind the trees, burrowing deep. I used to sit on the wooden benches with Chiquita till my neighbor Kathy told me. Also the homeless sit there sometimes and you never know about bedbugs. My stomach itches. Peeling, flaking.

Why do croissants always stain the bag? So many greasy patches and they just soak through and I ate a croissant every day, I don't even like them. I guess I crave them when I go without but those bagels from the carts, you know they're either stale or not boiled, they're just straight up bread cut in a circle, they've got to be. Once I saw a delivery guy wipe out, spill all of them on Amsterdam Ave, scattered. Straight up on the pavement. Cart guy picked them up, dusted them maybe a little bit, sold them all day. Like when this girl in high school bought cart coffee, milk and sugar - I only did sugar then, and now I do nothing - and it started splitting out the seams of the cup. In her hand, driblets and then you know it was just all going to pour out. So then the guy's like, doing that hand motion like c'mere, come on, and he takes it from her and says like, hey you had gotten wet and he let them dry and tried to use them. Soggy cups.

A few weeks ago the G wasn't running and I was stranded at that PS 1 station. I ended up walking over that bridge to Greenpoint thinking I was going to be late because I was the only one that had to deal with it and I walked into the room and no one was there. It was one of those moments where the sun was setting, sky was some color maybe pink or even red, the water was making it colder and I was listening to that Jimmy Eat World song that sounds like a stomach ache when you're on the subway home with too much energy and not enough, and you're just turning everything, everything is about to bloom, it's a fade out that just fades into something else, I was the only one on the bridge taking a video and we were all going against the wind from the exhaust of the cars.

Coco comes when I call and I force the slimy tennis ball out her mouth. She's stripped it bald, tearing out neon green fuzz. I gnawed my teeth down to bits over the years and now they fit like puzzle pieces. This week I couldn't find her paperwork and had one of those spiraling meltdowns, started praying to anyone or no one or something or whatever, and then thought that I had to convince myself I'd never find it because you only get what you want when you don't want it anymore. Found it in a pile of loose papers in the closet.

○ 4PM
TIMES SQUARE

I walk on 42nd by my old office, the corporate office in a huge tower I used to walk there every morning during the Polar Vortex. When I was only wearing skirts and the same torn tights every day, and had to buy new tights and change before any of the bosses saw me. The wind would whip against my legs and my thighs got this stinging feeling, I wore Doc Marten dress shoes that gave me blisters and read the internet all day. Sometimes I had busy days, making coffee, ordering from Staples, time sheets. I remember phases by the lunches, brief stint doing halal but then the weeks I only craved guacamole, mushrooms, chicken strips. I would order Lenny's egg wraps and dunk them in a tiny tub of Tabasco hot sauce just to feel anything, just so bored and craving anything, and the salt would rush into my, my eyes watered, my tongue hung out of my mouth. I used to drink as much water as I could as fast as possible, Poland Spring after Poland Spring, out of boredom. Then I'd google "water poisoning."

She came from LA, surrendered 10/26/2016, up for adoption for nine months until I found her. The time I spent in LA is a blur, I can't remember if I worked in retail for 3 weeks, 6 weeks, or 3 months. They checked my bag going in and out so I left it in my car and had to pay $20 for a parking spot. Parked next to a matte'd out black sedan and dissociated after work circling the aisles of Target till I drove home.

They tore down the gas station on 44th and now it's a pit. Nearly empty, construction equipment moves around piles of dirt and drills in the middle of the day when I'm checking DMs. It says "Seen." But he hasn't written back and the real read receipts don't have periods but should, to imply that it's the end of the conversation. When did we stop using periods? I don't. The drill goes on and on so I got the blackout curtains and wake up in the single digit afternoon hours, after I wake up around 5 or 6 and look at my phone to avoid evil spirits. A ghost can't get you if you're looking at a screen, so I spent hours and hours staring at my phone in my bedroom, which used to be grandma's bedroom, where I cried and wept next to her bed when they said I should say goodbye to her for the last time. She laid there, and dead bodies look different in every stage, from dying to dead to been dead to on display.

So I light a lot of sage and whisper to myself and say the same non-denominational prayer my uncle told me when I was sleeping on the floor in LA, too scared to sleep in my own room alone, I closed my eyes and a thing, I don't know, something, gripped my throat and cut off my breath and I couldn't speak, couldn't breathe, couldn't, and so now I reflexively repeat "Father Mother God Bathe Me In Light" and ignored the God part for years, and forgot I was saying it for brief stretches of time, the times when I was doing molly on weeknights and coming down when the sun was coming up and sitting in a rocking chair waiting while they met the guy and got more, doing more and feeling dead or dying and scared of all of it suddenly. I forgot the words but they came back, and now when I close my eyes to say FUCK I accidentally find myself with them, repeating them over and over like a lotion that absorbed all it can and now it's just swimming on top of pores, it's just too much, like everything I want is always over the surface trying to get in but there's no room, there's no space, we're all just waiting for it to combine, to infuse, to germinate and it never does anything but slick the surface.

○ 2017
9AM
MARCH, APRIL, MAY, JUNE

It takes me awhile to get out of bed and I finally make coffee, Cafe Bustelo, from the grinds too small, too many of them, and the french press mesh, pushing down the plunger it always flies up, it splashes every day but today burns my fingers, grazes my hand and the counter is filled with puddles of grime and grinds that stay there for weeks, maybe just short of a month, because I feel guilty buying paper towels but it's more that I feel better avoiding tasks, evading errands.

I'm trying to come up with a way where I do what I need. I walk down the West Side Highway, I miss when it was April and May and June, the exhaust would mix with the humidity for something so unpleasant and rank but in that way where you destroy something, you keep on flossing too hard and keep touching a cut, you keep stinging yourself. Walk all the way down listening to headphones so much the soles of your foot hitting the pavement reverb up, bouncing up into your ears against the headphones. Taking selfies on West Street, whatever it's called. Walk until you reach the cobblestones. Up until a certain point and on weekends it's so crowded and I can't leave my house. In high school I rode my bike down Sixth Ave. on the fourth of July listening to the cracks of fireworks and avoiding it all, street so clear. Bike rides every month, Critical Mass, used to bike up the highway outnumbering the cars. I gave up when there were too many white bikes, ghost bikes, memorials of hits and dead people. I got hit once or twice.

Remember that guy I got in the cyber fight with? The twitter feud shit thing a few weeks ago? He was biking in Central Park and his handlebars shook, apparently, and he wiped out lost control and his face slid across the pavement.

There was this time right before I kept stealing stuff - but only from Whole Foods and Barnes and Noble - where, for a whole year, I wore these fake glasses. And I never talk about it but a whole year, and no one ever said anything when I stopped wearing them except Carla, but they were $8 glasses from St. Marks

I like the hot when the diner booths stick to the backs of your thighs and it hurts to move. It stings to stand up. Sometimes they don't even have AC, or maybe they all got AC now. Last time I was at Waverly Diner this guy flashed a blade, maybe just a scissor but still a metal edge, between his cotton glove and the glass window, maybe it's plastic? Flashed a blade while we ate omelets and we told the waitress and she said, "What do you want me to do about it?" He crossed the street, approaching the NYPD van, seemed like he was trying to get arrested. The van revved up, engine on, and pulled out into traffic, ignoring him. And I thought I knew rejection.

We used to come here in the mornings before we both got fired. I asked Dad to tell me about the scar again. Leaving Waverly Diner, where I cried in front of Steve that time in 2010. He tells me about Stuytown again, growing up on 15th and 1st, and climbing the metal fence in the 60s. A group of adults watched him throw himself over and catch his skin on the spike, sliced open, huge scar running down. He tells me about the crying girl in the amputation ward, who just lost her arm. I've never heard this part of the story but each time I find something else new out it's like, yeah.

○ 2014
MALT LIQUOR, COCAINE, RED WINE

I started to worry about myself when the grocery store employee, the guy stocking the milk, asked if I was OK. I thought I could just keep circling, no phone service just listening to music, and that no one would bother me but sometimes it's like, the city buffer gives out, it wears down like a rubber sneaker sole, and I end up touching the floor, the cement ground, the bottom, end up having a conversation with someone.

○ 2007
ALEX
QUEENS

There was that time ▇▇▇ ran out of the bookstore because I casually mentioned I made out with Alex - who I think moved to Portland and had a kid? And writes books on anarchist theory? He was living in Queens and no one else was really there, and his Greek landlady would give him liquor and he just stood in his doorway and maybe, looking back ten years, I should've left? Was that a social cue? It's like, weird with social cues because I have enough of my own brain telling me I'm a piece of shit and that I should just go, that like, I'm fighting that enough that I think normal things are social cues to leave and then I end up missing the real social cues because I'm talking myself down from the imagined stuff, you know?

○ 2014
IPHONE 5
RED WINE, WHISKEY, VODKA, WEED

There was that cat, on my way home when I was in Sunset Park, the year before my bag was locked up. When I was on my way home I saw a orange cat, infested and dirty, these huge black splotches of grime, and I kept petting it, talking to it, calling it a rat. My little orange rat. I thought it'd bite me eventually and I didn't want to deal with the paperwork so I kept walking. An hour and a half home on a yellow line train but no one calls them that but that's what they are. On the way home a guy told me, he was like, 40s maybe? That I should be a teacher. That I'm a "nice girl." It was the same subway station where I called that ambulance the last day of high school. She was nodding out on a bench, poisoned, and the cop asked her why her birth year was scratched out. Ended up going to Methodist and getting an IV and then I took a $60 cab ride home. The air is different at the top

CONT. PAGE 16

3.2

HERE:
11.09AM, CLOUDY
COULDN'T GET UP
DAY RATE
WELCOME TO ASANA

YOU HAVE 2 TASKS DUE: 2 INFOGRAPHIC... AND NEW LAYOUT * ... - ▇▇▇ CREATIVE

YOU HAVE 2 TASKS DUE: 2 INFOGRAPHIC... AND NEW LAYOUT * ... - ▇▇▇ CREATIVE

YOU HAVE AN OVERDUE TASK: 2 INFOGRAPHICS - NOT YET STARTED - ▇▇▇ CREATIVE

REMINDER: ▇▇▇ INVITED YOU TO JOIN ▇▇▇ CREATIVE

YOU HAVE AN OVERDUE TASK: 2 INFOGRAPHICS - NOT YET STARTED - ▇▇▇ CREATIVE

SPECIAL PROJECT: ▇▇▇ [▇▇▇ BRAND CUE]

YOU HAVE AN OVERDUE TASK: 2 INFOGRAPHICS - NOT YET STARTED - ▇▇▇ CREATIVE

DESIGN: ▇▇▇ [SPECIAL PROJECT: ▇▇▇]

ASSIGNED TO YOU: DESIGN: ▇▇▇ [SPECIAL PROJECT: ▇▇▇]

YOU HAVE AN OVERDUE TASK: DESIGN: ▇▇▇ - BUFFY CREATIVE

WELCOME TO ASANA

REVIEW DECK AND MAKE NOTES [#▇▇▇FACE]

ASSIGNED TO YOU: CHECK OUT THE ▇▇▇?

HOLD: ▇▇▇ CARD ON ▇▇▇ PAPER [▇▇▇]

YOU HAVE A TASK DUE TODAY: CHECK OUT THE ▇▇▇? - ▇▇▇ CREATIVE

SET LAUNCH DATE [#▇▇▇]

YOU HAVE AN OVERDUE TASK: CHECK OUT THE STICKERS? - ▇▇▇ CREATIVE

YOU HAVE AN OVERDUE TASK: CHECK OUT THE STICKERS? - ▇▇▇ CREATIVE

BRIEF SIDE PROJECT ON COMMERCIAL SCRIPT [SPECIAL PROJECT: ▇▇▇...]

UNBOXING UPDATE [▇▇▇ BRAND CUE]

OUTLINE DECK: PRODUCT FOCUS CONTENT [▇▇▇ BRAND CUE]

YOU HAVE AN OVERDUE TASK: CHECK OUT THE STICKERS? - ▇▇▇ CREATIVE

BRAINSTORM ALTERNATIVES TO THE ▇▇▇ [SPECIAL PROJECT ▇▇▇]

YOU HAVE AN OVERDUE TASK: CHECK OUT THE STICKERS? - ▇▇▇ CREATIVE

IDEAS FOR ▇▇▇ VIDEO? [SPECIAL PROJECT: ▇▇▇...]

ASSIGNED TO YOU: EDITOR INVITE IMAGE [SPECIAL PROJECT:

[CONT 16.2]

3.3

EMILY SEGAL

CLOSE YOUR EYES. THINK ABOUT HOME. DESCRIBE WHAT YOU SEE

"The living room of my mom and step dad's apartment on 91st Street. Oriental rug from my great grandpa, kitschy drippy orange and green glass vases from the 60s, a colossal plant that holds the spirit of my step dad's late mother Evelyn, heavy wrought iron coffee table my parents had made from a bank grate in Philadelphia when they were still in their 20s. Squeaking and squawking cars and sirens from Amsterdam Avenue, coming in from the street."

How many pillows do you sleep on?
Two.

Are you a thief?
Unfortunately I am no longer a thief, at least as far as physical objects are concerned. As a teen I was a consistent, if not terribly ambitious, shoplifter – bent on amassing the sum total of makeup available at a constellation of local CVS's and Duane Reades. I got caught taking a straightening iron (particularly unnecessary, because I have straight hair) one afternoon when I was fourteen.

I remember planning the heist during English class that afternoon. The manager told me she would call my mother, and I looked withering…"my mother is dead". Then, once they threatened to call my school (something I found humiliating beyond measure). I recanted and revealed my mother was still alive. Once she arrived to pick me up they told her I'd said she was dead, which she said she thought was pretty ingenious, given the circumstances. I was in huge trouble.

For years later I limited my shoplifting to the costume jewelry at Top Shop – it was kind of my stealing methadone. Eventually I gave that up, too.

To whom are you indebted?
Literally all of my friends (spiritually, not financially)

Changing the names to protect the guilty, what's the worst thing anyone's ever done to you?
Besides a couple gnarly heartbreaks and more-or-less typical divorced-family trauma – the worst things have been creative or collaborative betrayals…like when former collaborators had gone behind my back to talk trash or change an artwork, or bosses made announcements about me to large groups without talking to me first. I'm a Scorpio, obsessed with my work and horrified by loyalty transgressions!

Favorite soup?
Vichyssoise

Favorite scent?
Hinoki; Musty Basements

How do you deal with guilt?
I FaceTime my best friends Alexa and Rachel repeatedly until one of them tells me it's all right, or I write about it, or I psychically bury it!

If you could relive one moment in your life what would it be?
Toss up between a particular lecture and a particular fuck – both experiences of "flow" (blech)

Where do you sit on the many-worlds interpretation of quantum mechanics and the probability of infinite variants of every single moment in time?
Staunchly "pro"

Starsign?
Scorpio, Aries Rising, Leo Moon (<– Profoundly 'unchill' combination)

How long do you sleep for?
Eight hours. Sleep is the greatest determining factor in my sometimes disturbingly volatile mood landscape.

Tell me about the last time someone held your hand?
My long lost friend Elizabeth, dragging me through the streets of Hollywood to pluck an orange off the tree in someone's backyard!

Have you got your dues?
Not even close ○

— EMILY SEGAL IS A STRATEGIST & FOUNDER OF K-HOLE/NEMESIS

3.4

CONTAMINANTS DETECTED IN NEW YORK CITY DRINKING WATER

For the latest quarter assessed by the US Environmental Protection Agency (July 2017 - September 2017)

CONTAMINANTS DETECTED **ABOVE** HEALTH GUIDELINES:

- BROMODICHLOROMETHANE
  Cancer forming
- CHLOROFORM
  Cancer forming
- CHROMIUM (HEXAVALENT)
  Cancer forming
- DICHLOROACETIC ACID
  Cancer forming
- TOTAL TRIHALOMETHANES (TTHMS)
  Cancer forming
- TRICHLOROACETIC ACID
  Cancer forming
- OTHER DETECECTED CONTAMINANTS
  - Chlorate
  - Chromium (total)
  - Haloacetic acids (HAA5)
  - Monochloroacetic acid
  - Nitrate
  - Nitrate and nitrite
  - Strontium

From October 2014 to September 2017 this water utility was in **VIOLATION** of health-based drinking water standards

Source: ewg.org

3.5

SNEAKERS SEEN AT HOME DEPOT IN BED STUY.

NIKE	AIR JORDAN 1	RED/BLACK
ROCHE	RUN	RED
NEW	BALANCE	BLACK
NIKE	AIRMAX 180	WHITE
REDBOX	CLASSICS	WHITE
NEW BALANCE	(UNKNOWN)	GREY
ADIDAS	(UNKNOWN)	ORANGE
REEBOK'S	(UNKNOWN)	BLACK
NEW BALANCE	(UNKNOWN)	BLACK
CONVERSE	HI	ORANGE
NIKE	PEGASUS	GREY
SAUCONY	(UNKNOWN)	GREY
NIKE	AIR MAX LOW	BLACK
PUMA HI	(UNKNOWN)	BLUE
NIKE	HUARACHE	ORANGE
CHAMPION	(UNKNOWN)	BLACK

3.6

MY NEW YORK

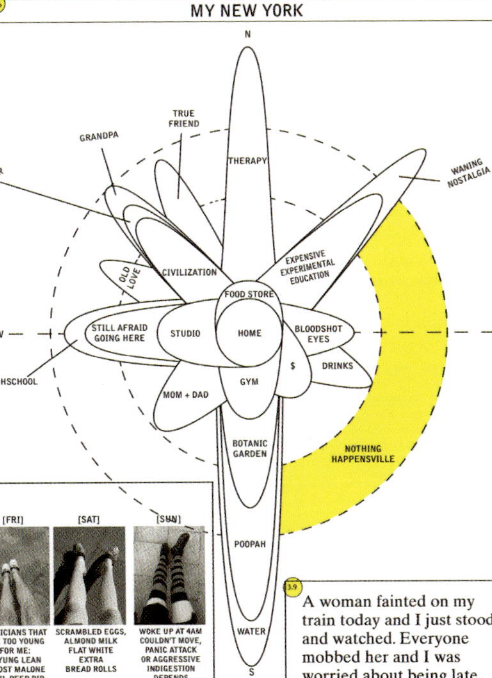

3.7

@CivilizationNYC

MOSTLY BELOW 59TH, USUALLY ABOVE DELANCEY

WEAR US → 16

BUSKER
WEST 4TH SUBWAY
SAT, MARCH 17, 2018
11.45PM

I'll tell y'all, first of all, you guys is looking all sad. You guys don't know what sad is. You don't know the half. I'm more than sad, **I'm dismayed**. Have a wonderful night y'all, this song is for us.

[inaudible 00:00:21]

This happened to me the other day, too, exactly what happened to me.

[inaudible 00:00:27] do you [inaudible 00:00:31]

[inaudible 00:00:32] especially with [inaudible 00:00:34]

(singing)

3.8

THE WEEK

[MON]
SLEPT IN A HOODIE LAST NIGHT. HAD A NIGHTMARE.

[TUES]
I HAVEN'T OPENED MY I-D INBOX IN 3 YEARS BECAUSE I'D RATHER NOT DEAL WITH THESPAM

[WED]
YOUR HAND CURLED INTO A FIST RESTING ON MY BACK AS YOU SLEEP

[THURS]
BRAN FLAKES, HALF A BOWL OF CAULIFLOWER SOUP, 2 CIGARETTES, VEGAN NOODLES, HALF AN EASTER EGG, ANOTHER NIGHTMARE

[FRI]
MUSICIANS THAT ARE TOO YOUNG FOR ME:
1. YUNG LEAN
2. POST MALONE
3. LIL PEEP RIP
4. LIL XAN

[SAT]
SCRAMBLED EGGS, ALMOND MILK FLAT WHITE EXTRA BREAD ROLLS

[SUN]
WOKE UP AT 4AM COULDN'T MOVE, PANIC ATTACK OR AGGRESSIVE INDIGESTION DEPENDS WHO U ASK

3.9

A woman fainted on my train today and I just stood and watched. Everyone mobbed her and I was worried about being late.

64. Ausgleich schaffen

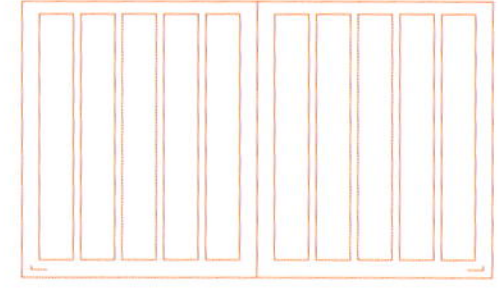

Manche Mitteilungen verlangen einen wahren Balanceakt. So ist in Newslettern, vor allem von gemeinnützigen Organisationen, häufig die Länge von größter Bedeutung. Um alles auf einer zuvor festgelegten Anzahl von Seiten (meist vier oder acht) unterzubringen, muss der Text verengt werden, was wiederum eine bestimmte Grobstruktur ergibt.

NEWSLETTER — The Cathedral Church of Saint John the Divine

Dean's Meditation

Mosaics

Die äußere Spalte dieser Seite enthält nützliche Informationen wie Quellenangaben, Serviceleistungen, Kontaktstellen und Adressen. Die übrigen Spalten sind von der äußeren durch eine senkrechte Linie getrennt und enthalten einen Essay. Der meditative Essay wird durch Abbildungen und ein Zitat ruhig unterbrochen.

The Cathedral Church of Saint John the Divine

Newsletter

Summer 2007 Volume 12 Number 37

ESTABLISHED in 1991, the Cathedral's Public Education & Visitor Services department welcomes all visitors to the Cathedral. From its Visitor Center in the West End of the Cathedral, the department provides hospitality by offering information on the Cathedral's history, liturgical services, performances, and events.

Cathedral Joins New York Transit Museum to Honor Architects Heins and LaFarge

Die Rasterstruktur wirkt auch auf der Rückseite. Alles lässt sich zu einem Mailing falten.

PROJEKT
Newsletter

KUNDE
Cathedral Church of St. John the Divine

DESIGN DIRECTION
Pentagram

DESIGN
Carapellucci Design

Ein Newsletter für eine gemeinnützige Organisation ist eine Hommage an die Vielseitigkeit des fünfspaltigen Rasters.

Winter 2007/2008

Der Veranstaltungskalender nutzt den Raster. Die Spalten für die Wochentage sind je nach Inhalt unterschiedlich breit. Trennlinien, dicke Linien als Behältnisse für die Schrift, Screens für Boxen und große Überschriften geben den Informationen Abwechslung und Textur.

Die Artikel und ihre Überschriften nehmen eine, zwei oder drei Spalten ein. Bilder füllen die Spaltenbreiten komplett aus, wobei die geordnete Struktur durch eine Vignette organisch aufgelockert wird.

65. Vollgepackt und doch dynamisch

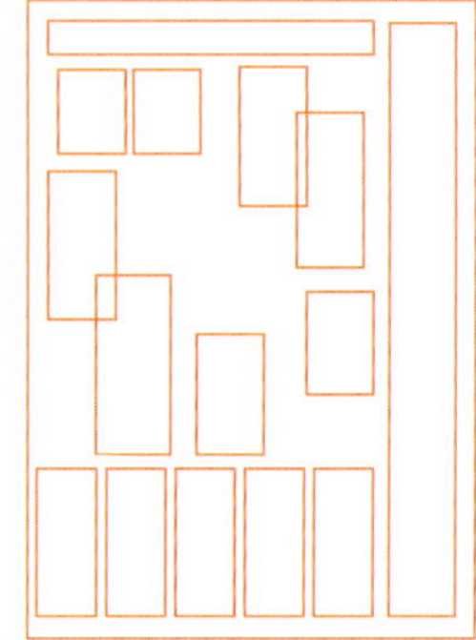

Oft besteht das Hauptziel darin, alles auf eine lesbare Art unterzubringen. Bei Verzeichnissen, Glossaren und Registern sollte man zunächst überlegen, wie alles hineinpasst.

"DO-IT-YOURSELF CULTURE MEANS TO LOOK AFTER YOURSELF AND THOSE ALL AROUND YOU... CREATE YOUR OWN FUTURE, WORK TO MANIFEST THE EQUALITY, LIBERTY AND KIND OF SOCIETY THAT YOU WOULD WISH TO BE A PART OF."

"IT'S A TIME FOR PEOPLE TO MAKE THEIR VOICES HEARD AND TO SAY NO."

GOOD TROUBLE

JUST SAY NON! P12

@-GOODTROUBLEMAG

INSIDE

LARA SCHNITGER / ADAM BROOMBERG P2
PETER KENNARD: LOUDER THAN BOMBS P3
EMEL: VOICE OF THE ARAB SPRING P4
DOSSIER: ART OF RESISTANCE P5
JARRETT GREGORY / JIMMY CAUTY P9
RAVE + RESIST: 90s UK YOUTH PROTEST P10
MALWARE: GRIZZLY STEPPE P11
CHINO AMOBI / TAUBA AUERBACH P12

"BEING A HUMAN IS NOT ABOUT VIOLENCE AND HATRED... ART IS HUMAN, A BEAUTIFUL THING THAT ONLY A HUMAN CAN DO."

P9

GOODTROUBLEMAG.COM

CUT-OUT-AND-KEEP! RUSSIAN CYBERWEAPON

GOODTROUBLEMAG@GMAIL.COM

1000

Issue Edición Édition Edizione

23

START AS YOU MEAN TO GO ON START AS YOU MEAN TO GO ON

PROTEST p2,3,8,9
ART p2,3,4,5,6,7
MUSIC p4,10,12
MORE ART p8,9,11,12
HACKING p11

CHANGE IS IN THE AIR

"WORK AS IF YOU LIVE IN THE EARLY DAYS OF A BETTER NATION"

NO ONE HAS THE RIGHT TO OBEY

"IT IS POSSIBLE FOR CULTURE TO BE AN INTEGRAL PART OF THE WORLDWIDE MOVEMENT FOR CHANGE."
(PETER KENNARD, P3)

PAY WHAT YOU WANT

WAKE UP FOR FREEDOM

"To resist is to exist in a way that directly opposes the powers that suppress your essence." – Matt Lambert (Photographer). Shot for *Good Trouble.*
"Being soft enough to fall with grace and to return standing each time, stronger and wiser." – MJ Harper (Dancer and model)

Diese Seite: Ein starker Zeitungskopf mit horizontaler Aufteilung verträgt auf der ersten Seite eines Protestmagazins einiges an visuellem Chaos.

Seite gegenüber: Die Spalten in diesem Magazin sind unregelmäßig angeordnet und bieten Fragen und Antworten in separaten, verstreut verteilten Kästen.

PROJEKT
Good Trouble

KUNDE
Ron Stanley

DESIGNER
Richard Turley

Die großformatige Protestzeitung bringt den wohl vorhandenen Plan durcheinander und schiebt ihn aus dem Raster.

NON WORLDWIDE is a group of African or diaspora musicians and artists. Their tagline is EXORCISE THE LANGUAGE OF DOMINATION. And they're here to shatter every illusion.

JUST SAY NON!

Even by the platform-agnostic standards of today, NON's activities are dizzying. They're a label, releasing stunning, radical music by musicians from Cape Town to Egypt to Virginia to Brixton. They're also magazine publishers who organise talks that run into all-night raves, who have opened a NON-branded range of travel merchandise in a duty-free store in downtown New York. Yet their ambitions go beyond these: NON is a borderless country open to all, a dissident political faction and a tight group of creative idealists…

Though the collective is sprawling and dozens of artists have released through their compilations and EPs, the core group is three DJ-producers-artists – the South African Angel-Ho, the Belgian-Congolese Nkisi, and Nigerian-American CHINO AMOBI. After a series of incredible mixes and mixtapes, Amobi has just released his debut album proper – the epic, collisionist double album Paradiso. It's an epic, complex, urgent, thrilling album, themed around an apocalyptic Edgar Allan Poe poem, and a radio station that flickers through moments of hellishness and total beauty. Kind of what life in America feels like right now.

"I like the chaos, throwing different variables in there, letting the chips fall where they may, shattering and breaking the canon in a way," Chino told us over Korean food in Berlin. "The depth and scale of the narrative is wider and deeper than one thinkpiece. The idea of NON is a constant rejection of definition. We're going to tell it ourselves."

RIGHT NOW

Your album is called Paradiso. Are you optimistic?
This time we're in has been growing. Trump is a benchmark, but a certain politicised feeling has been festering for time, with people like Black Lives Matter, the LGBTQ rights community, immigration, terrorism, home-grown terrorism and the way information is disseminated online,: it's all come to a head. it's like a boiling point.

Sometimes I think like the whole thing has to burn down in order for new life to be birthed. I say that optimistically. I'm not talking about masses dying – I don't want that – but destruction causes creation. It's always darkest before the dawn: in my life, the good things have happened directly after the bad.

I feel good about the future, about the youth, the spirituality in youth, the love. I think that the good will triumph. You can strengthen and pressure each other through productive measures. I'm all about shattering illusions, and the more you shatter, the better.

CHRIST

Your album is soaked in Christian allusion. Why is this so important today?
In times of strife which feel very dark, people go to faith to reconcile with what's going on, and [communicate] with something that's larger than themselves. Sometimes Christianity is represented like this fluffy thing, but the bible is super dark. It's gothic as hell. There's a mystery in those words. I'm just more malleable with data than some people are.

I identify as a follower of Christ, but I also identify as a queer body. That's often seen as a contradiction in the world, but the way I think about it, is it's about queering time and space. I really feel like The Bible did that.

Jesus entered time and space in a body that was queer, because only a queer body can transcend time and space, and change it in physical space. I believe that the body of Christ is here with us now and is changing who we are and our hearts. There's a sacred Blood that unites us together in that way, but that bond becomes more than just me, which connects me to other people, which is The Body. I know – it's a lot.

There's a certain magic element of faith that's important. A leap into that magic, I think, can change hearts. I put that into my music, and it's something that brings me closer to NON artists. Two become one. Transindividuation was something I was thinking about heavily on this album.

THE NATION

You've issued passports to concert attendees in the past. Is NON a nation?
Yes it is. It's a nation, it's a platform, it's an identification. We use the word NON because NON is everything and nothing. It's not limited to one thing. We can do anything. We can work with scientists, non-profit organisations, dancers, mathematicians, publications, designers. We can reflect our interests without things getting watered down. We have citizens all over the world, and I believe in multi-citizenship – so a NON citizen can be also be a citizen of the UK, or Nigeria, or the US, but NON-citizenship augments the citizenship of the location, where they are able to utilise that citizenship for creative intervention in their community, online, and the world.

I love that quote, "Work as if you are living in the early days of a better nation."
I love that too. NON is very nascent. I'm very concerned about giving ourselves space to develop. It's like a garden. Say there's water in the garden. If people hoard the water, the garden suffers. The power of the garden is its diversity. It's important to have that multiplicity of voices. The water is data. And the garden is the systems and infrastructures we work in.

AIRPORTS

You work a lot with air travel – you released an album called Airport Music For Black Folk, and opened a duty free-style shop of NON-branded travel accessories in New York. What's behind this?
I always come back to airports because of what they represent to me. It's a liminal space between cities and countries, and it's a trans space, where we literally are preparing to change our bodies, inside and out, by getting on a plane. It's a very democratic space, but there's so much class things. There's so many codes of society and ideology that's brought to surface this really transparent way. It's almost like a no space. It's like someone took white infinity and made a building out of it. I'm very drawn to that in a very tactile way.

My parents are from Nigeria, and oftentimes Nigeria is on the list of countries for Americans not to visit. So, sometimes I've been questioned and searched heavily, as have other NON artists. I've also had really good experience at airports. I love to people watch. There's multiple things going on: migrants, workers, amazing-looking dogs, the richest people in the country. There's a lot of spontaneity. But spontaneity in this formal way. When statements are isolated in a way, sometimes very mundane actions are way more powerful, because there's some much space around them. Airports are a very "NON" space.

It's between countries, but it's also the only place you can literally point at what a country is. It's a man with a gun saying "You can enter, and you can't." Everything else is scenery. You can really tell a lot from a country by its airport.

DIASPORA

For many of the global south, long-distance air journeys are an integral part of life – not a luxury, as in the global north. This may be an obvious point, but it blows my mind.
The diaspora has given people of the global south this fluidity. This, I think, changes how we create. The ability for our creativity to cross cultures, and also have enough being to assimilate to where we are. People of the diaspora learn to speak in many languages and touch on countries they're in and where they're from. It forces you to think in a way that's multi-levelled, very abstract, and highly conceptual. It's a trans idea.

You've said before that you make music to reject passivity. If you're a migrant, you took the most incredibly active step a person can. Take a lot of guts.
Heavy guts. And urgency. And you can see that urgency, in the work and the conversations. Like, they have so much life. Because you have to have that life – and light, because it can get super dark. And you have to do it together, because your take your family and culture and identity to survive. if you fail a little bit, you have at least that. There's this double consciousness.

MONEY

You're a corporation, rather than non-profit. You had the Duty Free shop, work with Red Bull, and set up Buy-Black Friday. How does money fit in?
I always go back to Robin Hood, man. Steal from the rich to give to the poor. Divide it as equally as we can. We believe in walking in the building and saying "We here. We don't believe in everything you believe, but: We. Are. Here. You need us, we don't need you." We're not playing around, we're smart, you know. Infiltrate and subvert culture in whatever ways we see fit.

It's more honest to operate in these spheres and to politick in them, than go back into the echo chamber and only be around voices that agree with me. Nah. We need a multiplicity of voices, and we deserve to be heard ●

CHARLIE ROBIN JONES

Photography by Johnny Utterback, Live photography by Brian Whar

"INSTRUCTIONS FOR NON-CITIZENS"

1 — Volunteer at an organization which benefits the quality of life of marginalized people.
2 — Feed your friends. Share your resources with one another.
3 — Spread the message of The Non State.

GANG LIFE

Q&A: RICHARD CABRAL

Born in the mid-80s into a family of East LA gang members, RICHARD CABRAL did his first time aged 13, going back to jail every year until he was 25. His longest stint, for attempted murder, was his last. On getting out, he left the gang he had grown up in. With the help of Christian organisation Homeboys Industries, he began mentoring those still caught up in gang life and prison, and embarked on a new career as an actor. He secured an Emmy nomination for his portrayal of Hector Tontz, a former gang member struggling to go straight, in the excellent ABC series American Crime, now in its third season.

"People see me how they see me, and that's all they see," Cabral's character says at one point. And Cabral's own story is one of identity and acceptance – of how the marks of a tough, violent past impact the present. But his story is also one of of how hard history can be held close, and how loyalty – to himself, as well as his fellow former gang members – can allow radical honesty to help others. "I witnessed guns, and violence, and everything people growing up there witness," he says, as we speak for an hour about prison reform, power and acting. "I finally came home at 25. And then it turned to what it is now."

GOOD TROUBLE: Tell us about life in LA.
Richard Cabral: I'm a second-generation Mexican-American, raised by my mom in East Los Angeles. I grew up in a metropolis of just Mexicans. The inner cities of Los Angeles have been riddled with guns and drugs since the beginning – it was poor, and law enforcement just didn't care. I was born in 1983, when the crack epidemic hit. So, I guess you could say I was a product of that energy, that time, and that sickness. Gangs, murder, mass incarceration.

LA historian Mike Davies said this explosion of gang violence from the 80s onwards is the result of deindustrialization. You have places where jobs were disappearing, so people were hanging around instead of working. And this coincides with the arrival of crack…
It was like these two forces that coincided at the same time. Boom. In the south side and in East LA, you have these cities which are all industrial. Right along the LA River, it's all factories and warehouses. So, you those kids with the mind to work, but all you have are drugs. The knowledge now is methamphetamine, and has been for the last 15 years. And while it's not as visible as the crack epidemic, it's taken its toll on the communities. The craziness of the stories, mothers killing babies and shit, all that has to do with drugs. The drugs really fucked things up.

One thing I heard about solving gang violence was that only warriors can end the war.
Yeah, that's a good one. For sure, for sure. To talk about the war, you have to know the war. To talk about death, you have to know death. There's a normality to it. It's the philosophy of a warrior, or a man in the army. It's not abnormal to know you might die, because there's a gang of other motherfuckers that might die with you. They all get it: we talk about death, and we talk about jail. The first time I went to jail as a kid, I looked around and thought 'Oh! There's hundreds of others like me.' I remember being young and seeing my uncle go to prison. My uncle has been a gang member since before I was born. You look outside and see gang members. You know the violence and craziness it carries, but you know they're not bad people… They're people.

What was the thing that turned it round?
The truth was I didn't want to spend my life in prison. I spent a year in jail. I had a whole year to think. And through my prayers or whatever, I got five years. But for that whole fucking year, you're thinking you might never come home.

What is the effect of all these years getting handed down by the state on the various communities affected?
You fuck up the community by having kids grow up without their fathers and mothers. You destroy the community. My best friend was 15 when he got life. Fifteen! California gives you life.

Why do you not hide from the past you had?
If I don't stand behind it, and say this is what made me, I cannot be inspiration. I cannot go into prisons and talk to people. Embracing it has been the most powerful thing.

Was getting the Emmy nomination for your acting a validation?
Yeah, but a validation I wasn't seeking. I'm happy now. I was in a cell eight years ago. Now I'm out and working and seeing my kids. But it was a surprise, because I just concentrate on the work, and this just meant people recognised the work.

What are your feelings about Trump?
Well, during Obama's reign, he deported more than any other president in history, so we've always been in the shit in a way! But when the threat becomes real evident, it makes people united. If I let someone piss me off, I've given them power. This. Too. Will. Pass. As a prison reformer, we're in a good place. In California, laws are getting passed, and we just need to push on ●

CHARLIE ROBIN JONES

Photography by James Mooney

'VIOLENCE= NO CHANGE'

CALLIGRAPHY

Over the last few months, artist TAUBA AUERBACH has written out the word 'Persevere' thousands and thousands of times.

A series of posters and public installations are now aiming to raise money and awareness for organizations including the Committee to Protect Journalists and GEMS (Girls Education and Mentoring Services).

"My favorite exercise in Daniel T Ames' Compendium of Practical and Ornamental Penmanship shows the word persevere written in lowercase script. Each letter is surrounded by a loop, similar to the a in the @ symbol. The loops are all the same but the letters are different, so the exercise teaches you to maintain a rhythm amidst otherwise varying circumstances."

"Calligraphy has become the activity during which I reflect on what's happening in the world, what's at stake, and what I'm willing to do about it. Maybe I've just needed something to do with my hands while I think. Until now, my politics have manifest mostly in quotidian, domestic choices like being vegan, composting and riding a bike. Feel free to roll your eyes. I support a few organizations. Big deal. I've always spoken my mind, but probably too politely. Besides, all of these choices are luxuries, and none of them registers as a sacrifice because they actually make my life more enjoyable. They are also, clearly, not sufficient."

"While doing calligraphy I've listened to a lot of speeches made by activists and philosophers. I've asked myself frequently if revolutionary change can take place without violence, and I've heard many sound arguments for why it cannot. Nonetheless, I remain certain that violence = no change, and that it is a doomed methodology for achieving it. In my view, violent means not only don't justify but also don't result in peaceful ends because the notion of an "end" is flawed. Now is the end.

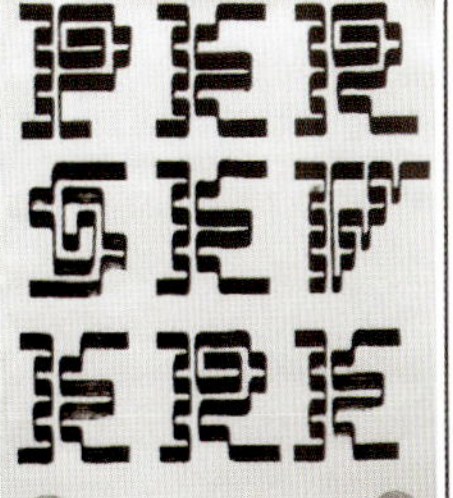

Every moment is the end. Civilization will always be in a state of becoming, so how we become what we want to be is what we are."

"Over the last few months, I've probably written the word persevere thousands times and in of hundreds ways. I've needed the time to think about what I can truly offer, about what a real contribution might be. I have some ideas, but I don't yet know if any of them are any good. In the meantime, I'm offering these drawings to support and thank some of the people I've held in my mind as I've written the word." ●

TAUBA AUERBACH

Persevere posters are available from diagonalpress.com for $25. 100% of profits benefit the Committee to Protect Journalists, GEMS (Girls Education and Mentoring Services), Chinese American Planning Council, and PLSE (Philadelphia Lawyers for Social Equity)

CLOSING SHOT

MARTIN SKAUEN

66. Den Leser an die Hand nehmen

Selbst das unwiderstehlichste Produkt profitiert von einem grafischen Design, das das Auge durch das Material führt.

Linien, hängende Initialen, fette Headlines und unterschiedliche (aber geregelte) Stärken und Farben können die Eintönigkeit vieler Seiten Fließtext durchbrechen und lassen den Leser interessante Themen leichter finden und alle übrigen überspringen. Durch Bilder, deren Größe und Platzierung wohlüberlegt ist, wird außerdem das Leseerlebnis erhöht.

NATIONAL

Tashnuba Hayder (who did not want to be photographed) lived on this street in Queens, N.Y.

A Girl in Exile

After the FBI pegged her as a potential suicide bomber, the 16-year-old daughter of Bangladeshi immigrants living in New York was forced to leave the United States

SUSPICIOUS CHAT ROOM

ALARM BELLS

By Nina Bernstein in Bangladesh

Agents later seized Tashnuba's diary, schoolwork, phone book—and the computer she had repeatedly tuned to Sheik Omar's sermons.

A KNOCK AT THE DOOR

HELD FOR QUESTIONING

UNLAWFUL PRESENCE

PROJEKT
Upfront

KUNDE
The New York Times und Scholastic

DESIGN DIRECTION
Judith Christ-Lafond

ART DIRECTION
Anna Tunick

Das Ziel dieser Zeitschrift, bei ihren jugendlichen Lesern Interesse für Nachrichten aus aller Welt zu erwecken und diese genauso „offen und ehrlich wie sich selbst" zu betrachten, wird durch das lebhafte Design unterstützt.

Große Initialen, fette Untertitel und Fußnoten geben den Seiten Farbe und Textur und machen sie interessant, eine Illustration aus weißen Linien über dem Foto schafft Textur und Tiefe. Die Seiten sind voll, wirken jedoch nicht überladen.

Linien mit negativ weißer Schrift bringen Elemente wie Taglines und Fußnoten zur Geltung. Eine fette Linie mit einer Bildunterschrift lenkt das Auge auf ein fesselndes Bild.

International

EMROZ KHAN IS HAVING A BAD DAY

Which is not unusual, and helps explain why Pakistan's youth are tinder for Islamic extremism.

By Peter Maass

Emroz Khan destroys for a living. He dismantles car engines, slicing them open with a sledgehammer, tearing out pistons, and throwing the metal entrails into a pile that will be sold for scrap. His hands and arms are stained a rich black, like fresh asphalt, and ribboned with scars.

He is 21 and has been doing this sort of work for 10 years, 12 hours a day, 6 days a week, earning $1.25 a day. Emroz rolls up his sleeve and puts my finger along a bulge on his forearm; it feels as hard as iron. It *is* iron, a stretch of pipe he drove into his body by mistake. He cannot afford to pay a doctor to take it out.

"We work like donkeys," Emroz says, a few paces from the tiny shop where he works in Peshawar, a city in northwest Pakistan. "That's what our life is like. It is the life of animals."

Javaid Khan watches with apprehension. Javaid, who is 17, began chopping up engines four months ago when he left school because he could no longer pay the fees. Javaid wishes he could be one of the clean-cut medical sales reps he sees at the nearby hospital. "I do not have the education," he acknowledges. "It makes me sad to think about it." →

AT WORK: Emroz Khan, center, chops engines into scrap for $1.25 a day.

CHRIS ANDERSON/AURORA

If you want to understand why young Muslim men line up to be suicide bombers, you would do well to stroll down Cinema Road, where Emroz and Javaid work. You would hear the chanting call to prayer, the shouts of peddlers selling bruised bananas, the groan of buses so overloaded that passengers ride on the roofs, and the cries of mutilated beggars pleading for a few cents. And all around, you would notice young men for whom life is abuse. The population of Peshawar (pronounced puh-SHAH-wuhr) reflects the population of Pakistan as a whole—63 percent are under the age of 25.

Most of these young men are not burning effigies of President George W. Bush or fighting Pakistani riot police. Their anger is only loosely expressed, often because they are struggling to survive and cannot afford the luxury of taking an afternoon off to join a demonstration.

They believe, or can be led to believe, that America is to blame for their misery. Many are adrift, cut off from their social foundations. Perhaps they moved to the city from dying villages, or were driven there by war or famine. There is no going back for them, yet in the city there is not much going forward; the movement tends to be downward. As they fall, they grab hold of whatever they can, and sometimes it is the violent ideas of religious extremists.

WORK OR PLAY: Children scoop up ash at a Peshawar brick kiln.

AN ANCIENT CITY PLAGUED BY WAR

Peshawar, once conquered by Alexander the Great and Genghis Khan, is one of the oldest cities in Asia. The city has long been the gateway to Afghanistan—a designation that became a curse 22 years ago when Afghanistan entered an era of warfare that has yet to end. Nearly half of Peshawar's 2 million inhabitants are Afghan refugees, most of them living in squalid camps. The local economy revolves around the smuggling of guns and ammunition, of VCRs and TVs, of heroin and hashish.

Aziz ul Rahman is a product of Peshawar. He is 18 and works in the mornings at a tire shop. In the afternoons, he studies the Koran at a madrassa, or religious school. The one he attends is of the extreme variety, as most are these days. I meet him at a protest organized by a pro-Taliban religious party.

"The American leaders are very cruel to Muslims, so that is why I am taking part in the demonstration today," he says politely. What he means is that America supports Israel, which is seen in the Muslim world as oppressing Palestinians, and supports certain Arab regimes, such as the one in Saudi Arabia, which are regarded as corrupt and oppressive.

In the background, a speaker is railing against Pakistan's military government, which supports the U.S. anti-terror campaign. "The generals are stupid," the speaker shouts. Then, like a rock star inviting crowd participation, he calls out, "Generals!" and the crowd roars back, "Stupid!" They are quick learners.

Aziz did not fall into religious extremism by choice; his preferred path, of becoming an engineer, was closed off by poverty. This is common in Pakistan. Poor families do their best to send a son to school, but in the end they cannot manage. The son will get a backbreaking job or maybe keep the donkey's life at bay by enrolling at a madrassa, most of which offer free tuition, room, and board. That's where they learn to think it's honorable to blow yourself up amid a crowd of non-Muslims and that the greatest glory in life is to die in a holy war.

1,000 BRICKS A DAY, SIX DAYS A WEEK

On the outskirts of Peshawar is Dabaray Ghara, an expanse of pits in which several thousand men, mostly Afghan refugees, make bricks. This labor, literally backbreaking, pays next to nothing and takes place outdoors, no matter how hot or cold.

Bakhtiar Khan began working in the pits when he was 10. He is now 25 or 26. He isn't sure, because nobody keeps close track. He works from 5 in the morning until 5 in the afternoon, making 1,000 bricks a day, six days a week, earning a few dollars a week. He is thin, wears no shirt or shoes, and he cannot believe a foreigner is asking about his life.

"Life is cruel," Bakhtiar says. "You can see for yourself. You wear nice clothes and are healthy. But look at us. We have no clothes to wear, and we are not healthy. Your question is amazing."

The youths at Dabaray Ghara are illiterate, and the world of politics is beyond their grasp. They can be led to rally behind any person or idea that promises to improve their lot. "I don't know about politics, but for our problems, I blame the world community," Bakhtiar says. "All humans should be equal, but we are not. . . . We arrived from Afghanistan 15 years ago. Since then I blame America, because it used to support us, but now it leaves us in a place like this. So if someone is fighting a jihad against America, I would support them. But if America is willing to help us, we support that, too."

PETER MAASS is the author of *Love Thy Neighbor: A Story of War*, his memoir of the conflict in Bosnia. Copyright 2001 Peter Maass.

CHILD LABOR: This 7-year-old works at a brick factory outside Peshawar. About 3.3 million Pakistanis under 14 work full-time.

The youths at Dabaray Ghara are illiterate, and the world of politics is beyond their grasp. They can be led to rally behind any person or idea that promises to improve their lot.

VIDEO GAMES & A FARAWAY FATHER

Ihsan u-Din is enrolled at a civil engineering college in Peshawar. Ihsan, 18, speaks good English, and he has the ultimate luxury in Pakistan—pocket money, which is why I ran into him at a video parlor. Compared with Emroz and the brick makers and most youths here, Ihsan has it good. But there's a catch. Pakistan is one of the poorest countries in the world. Even with a degree, it's very hard to get an engineering job. You need connections and money. Ihsan's family doesn't have enough of either.

"It is a game of money," he explains. "Even if you are a good engineer, you will not get a positive response when you apply, unless you pay. This has been the truth for 20 years."

The second catch is this: Ihsan's father is staying in the United Arab Emirates, where he works as a taxi driver earning infinitely more than he could in Pakistan. He sends money back to his family so that his children can eat well and go to school, but he doesn't earn enough to buy a plane ticket home.

"I have not seen my father for eight years," Ihsan says. "Is that right? He sends pictures and calls. But we don't want

20 The New York Times UPFRONT • upfrontmagazine.com

JANUARY 21, 2002 21

Ein geschickt ausgewähltes Foto und weiße Boxen, die das Bild durchbrechen, beleben die kraftvolle Struktur der Seite.

Farbe, Großbuchstaben, Linien und Boxen lenken das Auge des Lesers zum Textanfang. Gut zusammenwirkende typografische Elemente führen zu einem ergreifenden Foto.

VERSCHNAUFPAUSEN

67. Immer mit der Ruhe

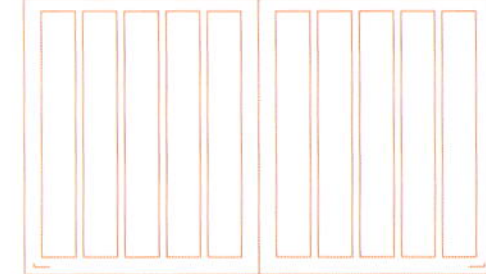

Ein Layout zu erstellen, das ist wie Geschichten erzählen, vor allem wenn es sich um ein Buch mit vielen großformatigen Illustrationen handelt. Für viele Projekte, speziell für Bücher oder Zeitschriftenartikel, muss das Layout für viele Seiten oder Screens gestaltet sein.

Große Doppelseiten bieten Gelegenheit für vollblutige Layouts. Wie der Titel eines Filmes macht diese Doppelseite gespannt auf das, was folgt.

Einzel- oder Doppelseiten mit unterschiedlichen Schriftgrößen, Formen, Spalten, Bildern und Farben lenken den Fluss der Geschichte und schaffen Spannung.

PROJEKT
Portrait of an Eden

KUNDE
Feierabend

DESIGN
Rebecca Rose

Ein Buch, das das Wachstum und die Geschichte einer bestimmten Gegend ausführlich behandelt, führt den Leser auf verschiedenartigen Doppelseiten durch die Zeitläufte.

Opposite:
Gertrude leaning against a coconut palm in Lummus Park wearing a playsuit, 1938. A hedge of *Malvaviscus arboreus*, a relative of the hibiscus, is in the background.

A Bermuda grass lawn was immediately planted with the hope that its aggressive root system would supply strong underground runners to hold the sandy soil in place. Coconut palms were planted as well, to provide inviting shade and a sense of site.

Left:
Barbara June Oka poses by the Shower of Gold *(Cassia fistula)*, late 1940s. Her right arm mimics the smooth barked limb. Joints of movement and growth, the elbow and node are parallel structures.

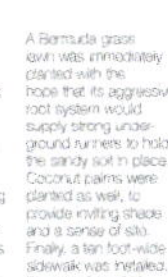

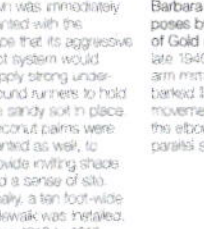

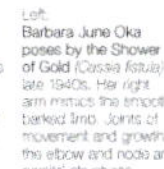

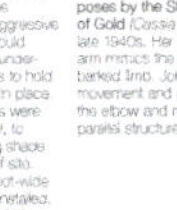

Healing Plant
The **Shower of Gold tree** was critical to early people dependent on the properties of nature to heal. *Cassia fistula* was valued by ancient Egyptians for its long cylindrical pods or fruits. Ripening from fresh green in hue to shiny black, these pods grow up to two feet in length.

Miami Beach of the Orient

Ink drawing by Gertrude Oka, c. 1960

68. Inseln schaffen

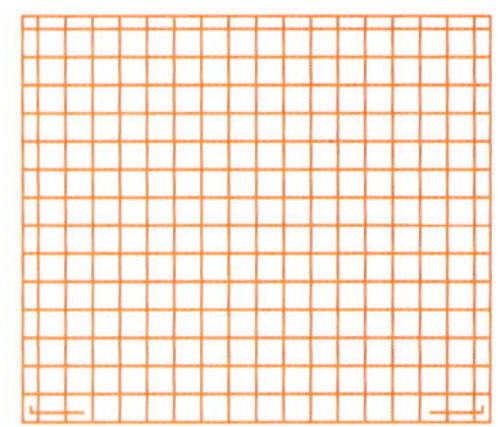

Um Glaubwürdigkeit zu vermitteln und das Interesse zu fokussieren, ist weniger wirklich mehr. Freie Fläche ermöglicht es dem Leser, sich zu konzentrieren.

Siehe auch Seite
138/139

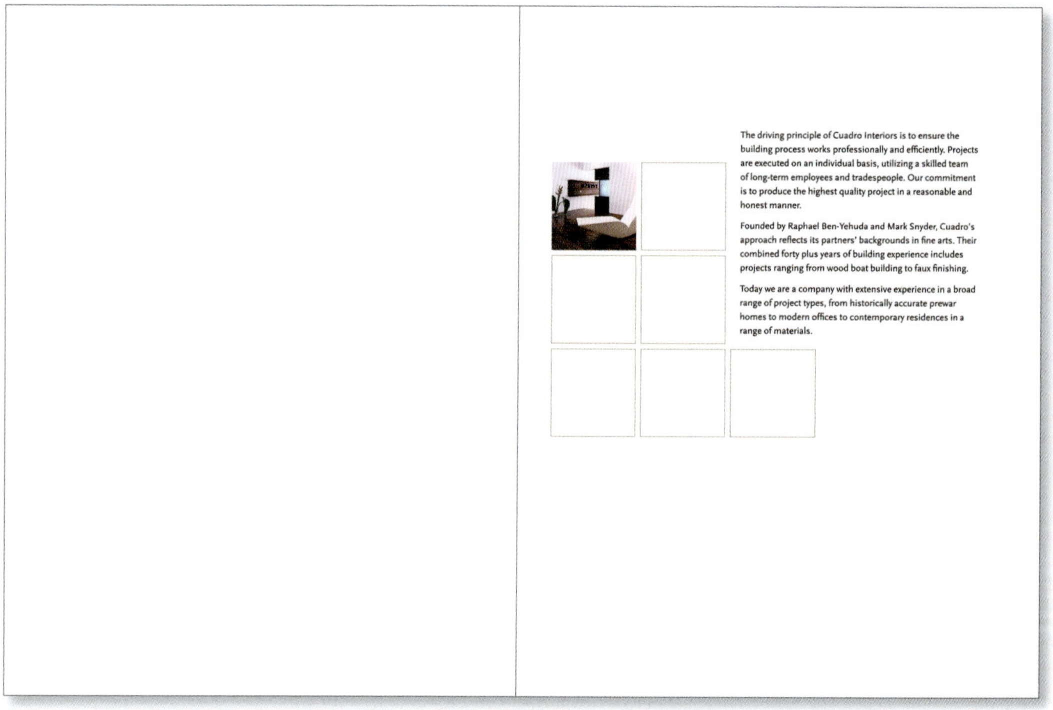

The driving principle of Cuadro Interiors is to ensure the building process works professionally and efficiently. Projects are executed on an individual basis, utilizing a skilled team of long-term employees and tradespeople. Our commitment is to produce the highest quality project in a reasonable and honest manner.

Founded by Raphael Ben-Yehuda and Mark Snyder, Cuadro's approach reflects its partners' backgrounds in fine arts. Their combined forty plus years of building experience includes projects ranging from wood boat building to faux finishing.

Today we are a company with extensive experience in a broad range of project types, from historically accurate prewar homes to modern offices to contemporary residences in a range of materials.

Ein Motiv in einem Modul führt in das Produkt ein.

PROJEKT
Cuadro Interiors
Produktbroschüre

KUNDE
Cuadro Interiors

DESIGN
Jacqueline Thaw Design

DESIGNER
Jacqueline Thaw

PRIMARY PHOTOGRAPHERS
Elizabeth Felicella,
Andrew Zuckerman

Die Broschüre mit den Projekten einer Firma für Interior Design ist so reduziert, dass sie die groß herausgestellten Wohn- und Büroräume in den Mittelpunkt rückt. Der Raster besteht aus Modulen.

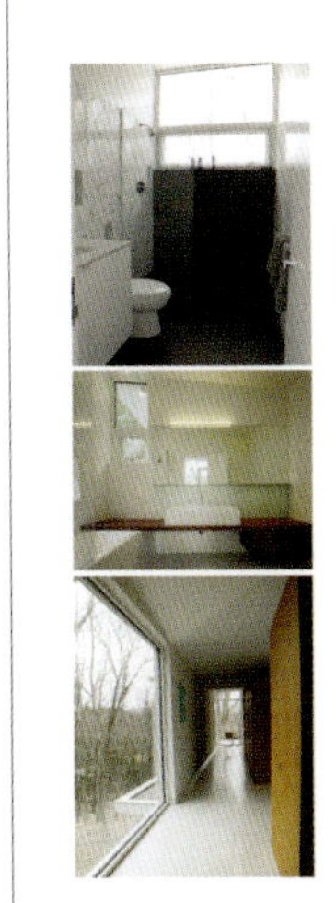

Ein Meer von Weiß gibt dem Leser Gelegenheit, die eingeklinkten Bilder und Informationen von vielen Standpunkten aus zu betrachten.

69. Bilder brauchen Luft

Eine sonst freie Seite rückt ein Foto oder eine Illustration darauf zwangsläufig in den Mittelpunkt. Der Betrachter kann die Hauptattraktion so ungestört auf sich wirken lassen.

FLÄCHE SCHAFFEN

Stets entscheidet der Inhalt eines Produkts darüber, wie der Designer die Fläche zwischen Text und Bildern verteilt. Bezieht sich der Text auf bestimmte Fotos, Kunstgegenstände oder grafische Darstellungen, ist es für den Leser am übersichtlichsten, wenn Bild und Text dazu möglichst nahe beieinanderstehen. Auf der Suche nach dem Text vor- oder zurückblättern zu müssen, ist kontraproduktiv.

Auch der Maßstab der Bilder ist wichtig. Ein vergrößerter Ausschnitt eines Kunstgegenstands verleiht einer Doppelseite Energie. Geht es darum, Interesse zu wecken, wird ein von viel Weißfläche umgebenes Bild den Betrachter eher anziehen als Bilder, die mit zahlreichen anderen Elementen gruppiert sind.

Siehe auch Seite
136/137

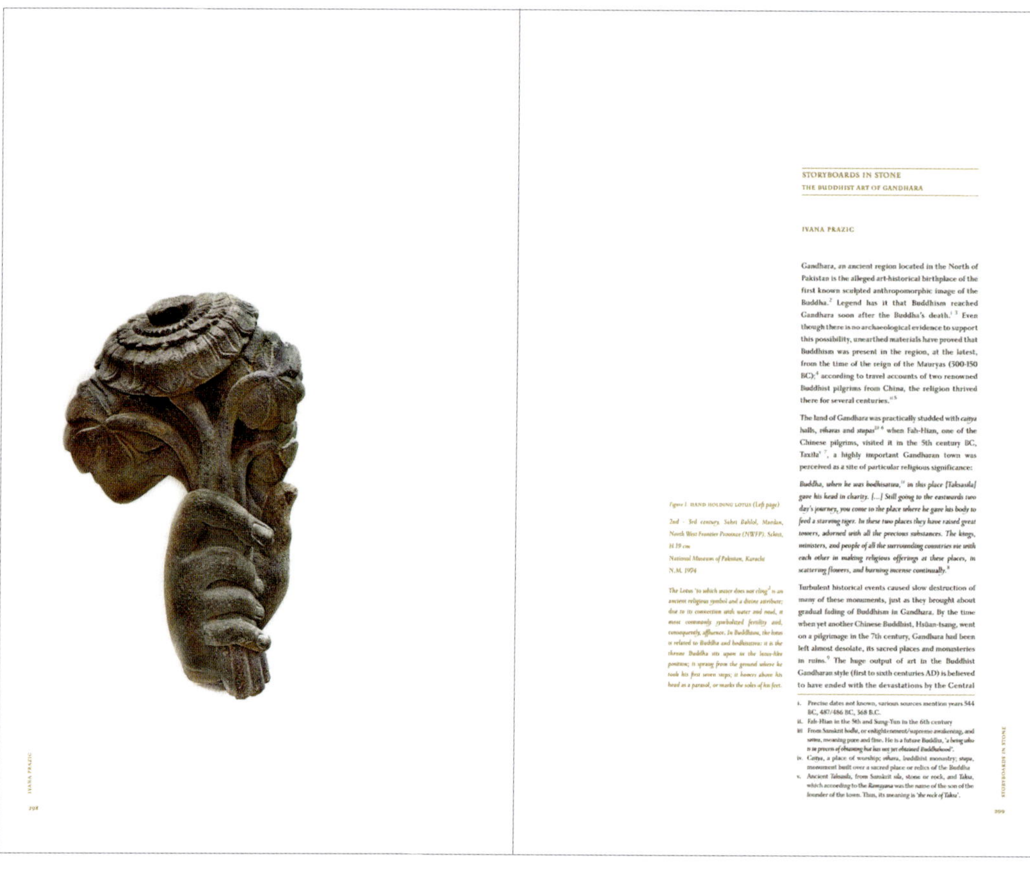

STORYBOARDS IN STONE
THE BUDDHIST ART OF GANDHARA

IVANA PRAZIC

Gandhara, an ancient region located in the North of Pakistan is the alleged art-historical birthplace of the first known sculpted anthropomorphic image of the Buddha.[2] Legend has it that Buddhism reached Gandhara soon after the Buddha's death.[i 3] Even though there is no archaeological evidence to support this possibility, unearthed materials have proved that Buddhism was present in the region, at the latest, from the time of the reign of the Mauryas (300-150 BC);[4] according to travel accounts of two renowned Buddhist pilgrims from China, the religion thrived there for several centuries.[ii 5]

The land of Gandhara was practically studded with *caitya* halls, *viharas* and *stupas*[iv 6] when Fah-Hian, one of the Chinese pilgrims, visited it in the 5th century BC, Taxila[v 7], a highly important Gandharan town was perceived as a site of particular religious significance:

Buddha, when he was bodhisatva,[iii] *in this place [Taksasila] gave his head in charity. [...] Still going to the eastwards two day's journey, you come to the place where he gave his body to feed a starving tiger. In these two places they have raised great towers, adorned with all the precious substances. The kings, ministers, and people of all the surrounding countries vie with each other in making religious offerings at these places, in scattering flowers, and burning incense continually.*[8]

Turbulent historical events caused slow destruction of many of these monuments, just as they brought about gradual fading of Buddhism in Gandhara. By the time when yet another Chinese Buddhist, Hsüan-tsang, went on a pilgrimage in the 7th century, Gandhara had been left almost desolate, its sacred places and monasteries in ruins.[9] The huge output of art in the Buddhist Gandharan style (first to sixth centuries AD) is believed to have ended with the devastations by the Central

Figure 1. HAND HOLDING LOTUS *(Left page)*
2nd - 3rd century, Sahri Bahlol, Mardan, North West Frontier Province (NWFP). Schist, H 19 cm
National Museum of Pakistan, Karachi
N.M. 1974

The Lotus 'to which water does not cling'[3] *is an ancient religious symbol and a divine attribute; due to its connection with water and mud, it most commonly symbolized fertility and, consequently, affluence. In Buddhism, the lotus is related to Buddha and bodhisattva: it is the throne Buddha sits upon in the lotus-like position; it sprang from the ground where he took his first seven steps; it hovers above his head as a parasol, or marks the soles of his feet.*

i. Precise dates not known, various sources mention years 544 BC, 487/486 BC, 368 B.C.
ii. Fah-Hian in the 5th and Sung-Yun in the 6th century
iii. From Sanskrit *bodhi*, or enlightenment/supreme awakening, and *sattva*, meaning pure and fine. He is a future Buddha, *'a being who is in process of obtaining but has not yet obtained Buddhahood'*.
iv. *Caitya*, a place of worship; *vihara*, buddhist monastry; *stupa*, monument built over a sacred place or relics of the Buddha
v. Ancient *Taksasila*, from Sanskrit *sila*, stone or rock, and *Taksa*, which according to the *Ramayana* was the name of the son of the founder of the town. Thus, its meaning is *'the rock of Taksa'*.

STORYBOARDS IN STONE

399

Ein Essay mit dem Titel „Storyboards in Stone“ zeigt eine Hand mit einer Lotusblume. Die Bildunterschriften, ein Essay und Fußnoten auf der Seite rechts bilden einen Ausgleich zur großen Fläche.

PROJEKT
Mazaar Bazaar: Design and Visual Culture in Pakistan

KUNDE
Oxford University Press, Karachi, with Prince Claus Funds Library, The Hague

DESIGN
Saima Zaidi

Auf einem strengen Raster wird die Geschichte des Grafikdesigns in Pakistan erzählt. Neben zahlreichen Schätzen des pakistanischen Grafikdesigns bleibt genügend Platz für Weißfläche übrig.

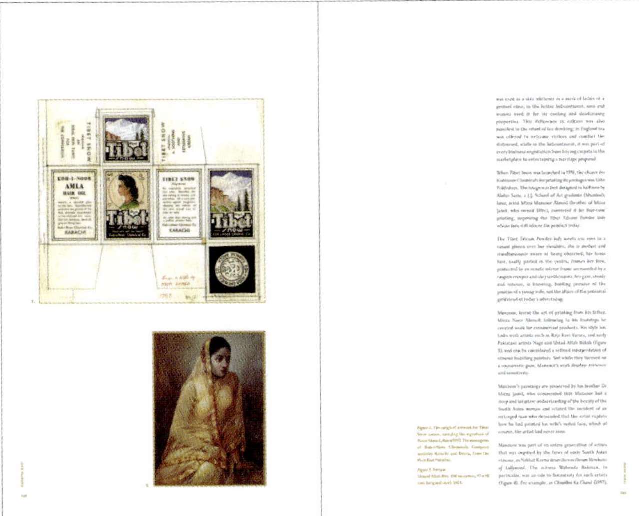

Die Kombination einer Verpackung für Haaröl und einem Porträt lässt viel Raum für Betrachtungen.

Gemälde und Ornamente, eines von der Rückseite eines Lastwagens, schaffen ein bunt texturiertes Layout.

Aina: PUBLICITY DESIGN AS A REFLECTION OF LOLLYWOOD FILM CONTENT

HASAN ZAIDI

Buxom belles in scarlet and gold, *gandasa*[i]-clutching heroes in beige and black, blood—or sweat-drenched villains in diabolical green and purple: nothing captures the essence of commercial film in the subcontinent better than its medium of publicity. Literally larger than life, cinema hall hoardings depict as much action and emotion on canvas as film-makers squeeze on to three hours of celluloid. As a result, they inevitably evoke the visual chaos we have come to associate with Lollywood,[ii] the Pakistani film industry.

Hoardings and posters are the first point of contact between the film and its target audience. The 'reel 'em in' splash of colours depicting dramatic scenes of sex and violence is calculated to have the masses lining up outside ticketing booths. Before television promotions and mass marketing made film publicity itself an art, these melodramatic hoardings were solely responsible for enticing audiences into darkened theatres screening black and white films. Even now, no cinema district would be worth its name if it did not offer an array of exaggerated, multihued interpretations of silver screen fantasy.

Much like film posters on the back of horse-drawn *tongas*[iii] and motorised rickshaws—which long preceded mobile advertising in the West—cinema hoardings came to be identified as a uniquely Subcontinental medium of film promotion, even though the films being promoted were often Hollywood musicals adapted to suit the indigenous taste for *nautanki*.[iv] Echoing the diversity of the region, the singular film poster—a staple of film publicity worldwide—was translated into two- or three- storey-high works of art for the street. As such, cinema hoardings or street art in the real sense of the phrase became popular. Inviting, arousing, brash, and disposable, the hoardings survived only as long as the films they represented screened

Figure 1. (Left page) Poster

FILM Inspector YEAR 1964 DESIGN Galib MEDIUM Lithography PRINTERS Nawa-i-Waqt, Lahore.

Aina meaning mirror, the title of an epic film from the 1970's.

i. A wooden baton

ii. The Pakistani cinema industry primarily based in Lahore is popularly termed Lollywood, referring to Bombay's Bollywood derived from the Hollywood Studios.

iii. Horse drawn carriage

iv. Folk theatre

Ein eindrucksvolles Bild als Auftakt zu einem Essay.

70. Von Hand entwerfen

In Skizzen nehmen Ideen Gestalt an und Skizzen erleichtern die Planung eines Seitenlayouts. Zwar gleichen erste Skizzen natürlicherweise oft eher Kritzeleien als durchdachten Vorschlägen, aber sie können einem Gesamtplan oder Konzept Form geben. Enthält ein größeres Konzept ein oder mehrere Bilder, empfiehlt es sich, Schablonen und Raster zu erstellen, um herauszufinden, wie die verschiedenen Elemente eines Kunstwerkes zueinanderpassen und zusammenwirken.

Eine Idee grob zu umreißen oder eine Schablone herzustellen kann viel Arbeit ersparen. Nur wenige Designer haben Zeit genug, um einzelne Schritte zu wiederholen. Ein Entwurf ist daher von entscheidender Bedeutung, ganz gleich, ob es sich um ein Layout mit Schrift, Bildern oder einer Kombination aus beidem handelt.

Dieser Entwurf zeigt Denk- und Planungsprozesse und eine mögliche Anordnung der vielen Bilder, die in diesem Kunstwerk enthalten sind.

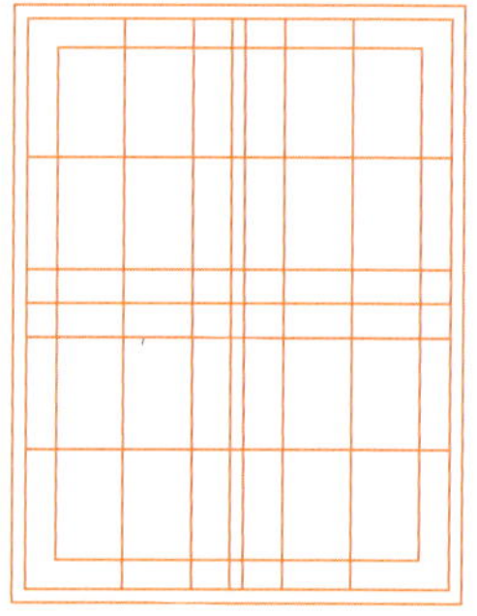

Oben und unten: Mit dem ausgearbeiteten großen Bild lassen sich alle Einzelteile gestalten.

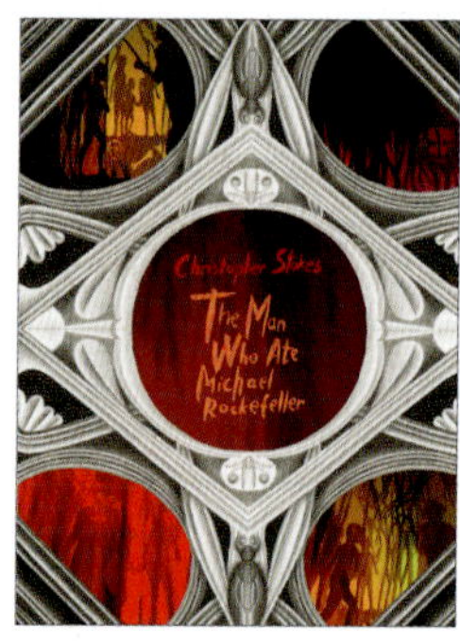

PROJEKT
McSweeney's 23

KUNDE
McSweeney's

DESIGN
Andrea Dezsö

MANAGING EDITOR
Eli Horowitz

Die handgezeichneten, sich spiegelnden und wiederholenden Muster von Andrea Dezsö auf diesem Umschlag für *McSweeney's 23* vereinen unterschiedlichste Kunstwerke: Bleistiftzeichnungen, Handstickereien, Fotos handgefertigter dreidimensionaler Schattenfiguren und Eitemperamalereien existieren innerhalb des Rahmenwerks nebeneinander. Dezsö hat den Computer nur zum Scannen und Zusammensetzen eingesetzt.

Bei dem Projekt geht es um Muster und Planung sowie um die künstlerische Gestaltung des Einbands für mehrere Bücher innerhalb eines großen gemeinsamen Schutzumschlags.

Rahmen innerhalb von Rahmen enthalten Illustrationen für zehn Vorder- und Rückseiten, eine für jede Geschichte in *McSweeney's 23*. Alle zehn Einbände finden sich darüber hinaus in einem Schutzumschlag wieder, der sich zu einem großformatigen Plakat entfalten lässt, das sich sogar zu Ausstellungszwecken eignet. Das handgezeichnete visuelle Rahmenwerk ist ein hervorragendes Mittel, um die einzelnen Kunstwerke sinnvoll und stimmig zu einem größeren Ganzen zusammenzufügen.

71. Anarchie oder Hierarchie?

Das durchgängige Band ermöglicht die Navigation durch die Informationen zum Kompostieren.

Zu viele Daten wirken kompliziert und unübersichtlich. Ein simples horizontales Band mit klaren Erläuterungen und Anweisungen bringt Übersicht in das Chaos.

PROJEKT
compost/r

KUNDE
Dopodomani

DESIGN
Suzanne Dell'Orto

ILLUSTRATOR
Nina Lawson

compost/r, eine App, die den Vorgang des Kompostierens widerspiegelt, verwandelt die gelöschten Daten in Gedichte, Muster, Klingeltöne und Musik.

Composting!

Composting!

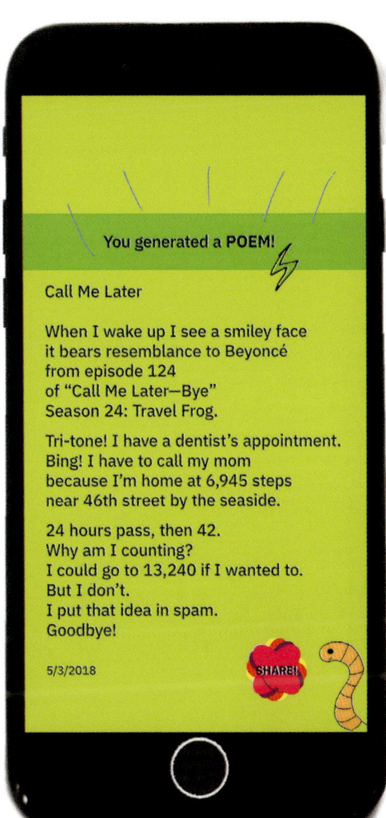
You generated a POEM!
Call Me Later
When I wake up I see a smiley face
it bears resemblance to Beyoncé
from episode 124
of "Call Me Later—Bye"
Season 24: Travel Frog.
Tri-tone! I have a dentist's appointment.
Bing! I have to call my mom
because I'm home at 6,945 steps
near 46th street by the seaside.
24 hours pass, then 42.
Why am I counting?
I could go to 13,240 if I wanted to.
But I don't.
I put that idea in spam.
Goodbye!
5/3/2018
SHARE!

72. Durch Strukturen Ordnung schaffen

Manchmal nutzen Designer das Prinzip des Rasters rein instinktiv. Einigen Designs sieht man an, dass sie auf Rastern basieren. Bei anderen ist es eher optisch. Oder man findet nur eine zarte Andeutung eines Rasters.

Siehe auch Seite
27

PROJEKTE
Some Fun, *I'm Special*, and *American Nerd*

KUNDE
Simon & Schuster, Inc.
Scribner, ein imprint von Simon & Schuster

Some Fun
ART DIRECTOR
John Fulbrook
DESIGNER
Jason Heuer

I'm Special
ART DIRECTOR
Jackie Seow
DESIGNER
Jason Heuer

American Nerd
ART DIRECTOR
John Fulbrook
DESIGNER
Jason Heuer
FOTO-ILLUSTRATIONEN
Shasti O'Leary Soundat

Diese drei Buchhüllen zeigen den mehr oder weniger strengen Einsatz von Rastern.

Some Fun hat einen strengen Raster, der nur für den Titel durchbrochen wird ..., das ist der Spaß daran!

I'm Special ist so speziell, dass auf den Raster zugunsten eines organischen Äußeren verzichtet wird. Das angedeutete Linienpapier betont die rasterfreien organischen Elemente.

American Nerd setzt auf ein optisches Raster – also kein mathematisches –, wie es auch in Galerien benutzt wird. Ein cleveres Konzept, das sich vom Rücken zum vorderen Umschlag fortsetzt.

OPTISCHE RASTER: EHER INSTINKTIV ALS MATHEMATISCH

Der Designer dieser Buchhüllen meint, dass Künstler das Raster und den Goldenen Schnitt instinktiv und manchmal auch bewusst anwenden. Wie viele Designer setzt Jason Heuer bei der Arbeit auf einen optischen (rein instinktiven) Raster und wendet zum Schluss einen mathematischen Raster an, um die Designelemente aufzuräumen und anzuordnen.

73. Den Fluss bewahren

Ein gut strukturiertes Design hat einen soliden Unterbau, auch wenn dieser Rahmen nicht unmittelbar zu erkennen ist.

Charles Coiner

advertising art director

May/June **1963**

What of the next 25 years [in advertising]? Will the picture continue to tell more and more of the story—or will we revert to more emphasis on copy?

Will the average American accept longer copy once again, as his leisure time increases? I wish I knew the answers.

But I can predict that the reader's taste will continue to move upward, placing even more challenges before the advertising designer. The art director's position and responsibility in the advertising agency will increase greatly. And this will mean that he will have to be a well-rounded advertising man....

I believe the greatest challenge in the next 25 years will be in upgrading the great mass of advertising. Instead of a handful of great art directors, as we have today, we will have hundreds of art directors equal to the tasks and opportunities. And they will not be located in the larger cities New York, Philadelphia, and Chicago alone.

PROJEKT
Illustration in einer Zeitschrift

KUNDE
Print magazine

DESIGN
Marian Bantjes

In dieser Zeitschrift über Grafikdesign liegt der Zusammenhang zwischen den Händen und der Kunst der Typografie buchstäblich auf der Hand.

MARIAN BANTJES ÜBER GRAFIKDESIGN

„Ich arbeite mit visuellen Fluchtlinien und lege mit Fanatismus eine Struktur an. Ich ordne Dinge in einer Linie mit Teilen der Bilder oder kräftigen vertikalen Linien in Überschriften an und stelle viel an, damit das funktioniert. Ich bin auch fanatisch, wenn es um die logische Struktur, den Aufbau der Informationen und die Konsistenz geht. Grafikdesign und Typografie sind wie ein maßgeschneiderter Anzug: Wer sich nicht auskennt damit, achtet vielleicht nicht auf die von Hand angenähten Knöpfe (Unterschneidung), die maßgefertigten Abnäher (perfekte geradlinige Anordnung) oder den feinen Stoff (perfekte Schriftgröße) ..., erkennt aber instinktiv dessen Hochwertigkeit.

LINKE SEITE UND RECHTS: Mit einer gleichmäßigen Spationierung von Buchstaben und Wörtern sowie Schriftarten aus einer bestimmten Epoche schenkt Marian Bantjes typografischen Details, etwa logischen Absätzen, Beachtung. Was die Seite zum Klingen bringt, ist Bantjes' illustrativer, kalligrafischer Esprit.

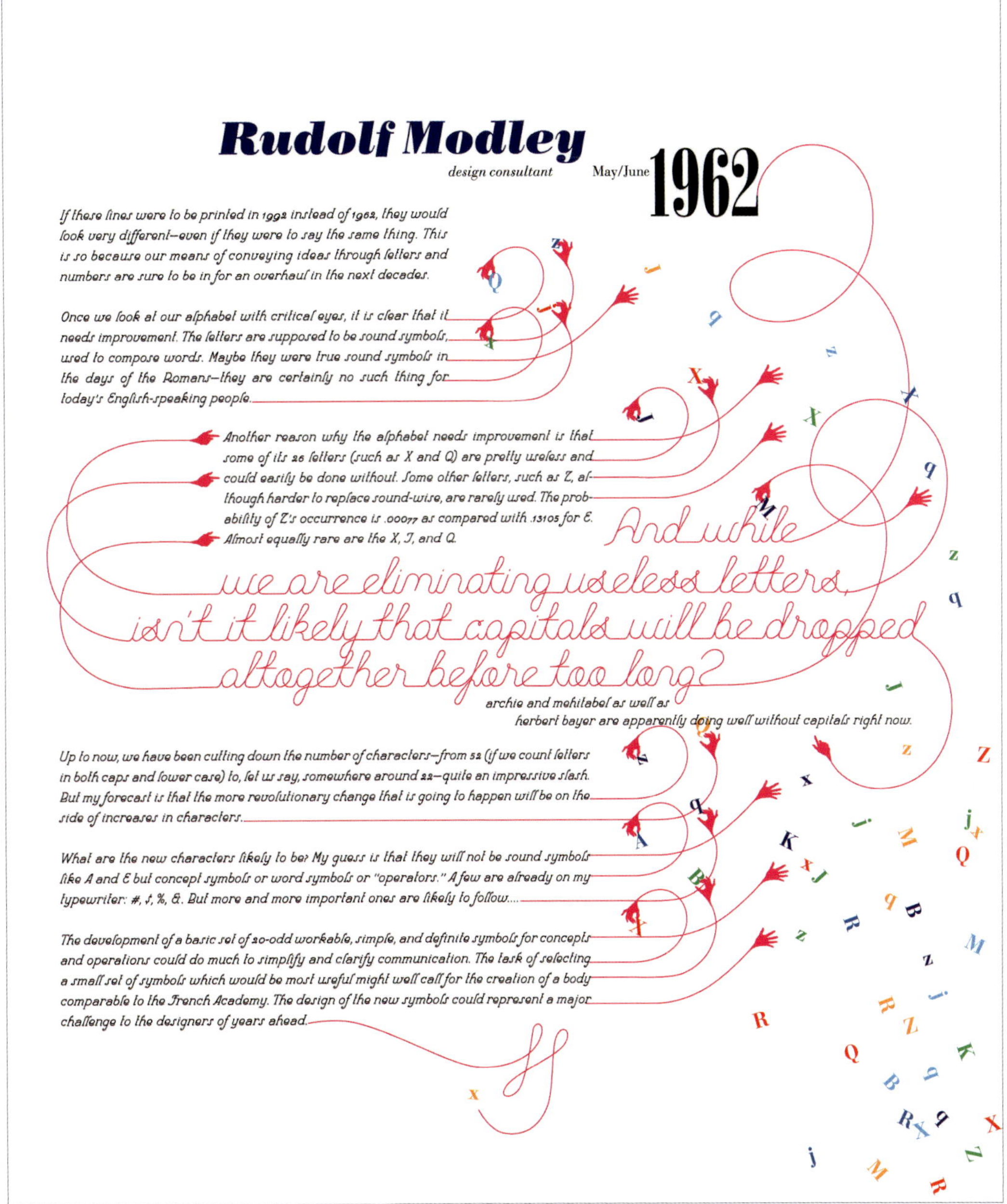

Rudolf Modley

design consultant

May/June **1962**

If these lines were to be printed in 1992 instead of 1962, they would look very different—even if they were to say the same thing. This is so because our means of conveying ideas through letters and numbers are sure to be in for an overhaul in the next decades.

Once we look at our alphabet with critical eyes, it is clear that it needs improvement. The letters are supposed to be sound symbols, used to compose words. Maybe they were true sound symbols in the days of the Romans—they are certainly no such thing for today's English-speaking people.

Another reason why the alphabet needs improvement is that some of its 26 letters (such as X and Q) are pretty useless and could easily be done without. Some other letters, such as Z, although harder to replace sound-wise, are rarely used. The probability of Z's occurrence is .00077 as compared with .13105 for E. Almost equally rare are the X, J, and Q.

And while we are eliminating useless letters, isn't it likely that capitals will be dropped altogether before too long?

archie and mehitabel as well as herbert bayer are apparently doing well without capitals right now.

Up to now, we have been cutting down the number of characters—from 52 (if we count letters in both caps and lower case) to, let us say, somewhere around 22—quite an impressive slash. But my forecast is that the more revolutionary change that is going to happen will be on the side of increases in characters.

What are the new characters likely to be? My guess is that they will not be sound symbols like A and E but concept symbols or word symbols or "operators." A few are already on my typewriter: #, ¢, %, &. But more and more important ones are likely to follow....

The development of a basic set of 20-odd workable, simple, and definite symbols for concepts and operations could do much to simplify and clarify communication. The task of selecting a small set of symbols which would be most useful might well call for the creation of a body comparable to the French Academy. The design of the new symbols could represent a major challenge to the designers of years ahead.

74. Geplante Störungen

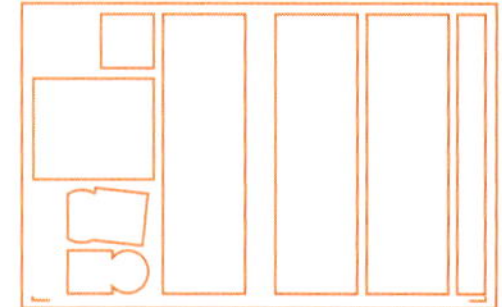

Planung ist eine der wichtigsten Prinzipien im Grafikdesign. Formate sind Pläne. Raster sind Pläne. Störungen können ein wichtiger Teil des Planes sein und die Typografie kann Teil eines sehr klaren Planes für Störungen sein. Wenn ein Designer festlegt, welcher Name oder welches Detail eine größere und fettere Schrift verdient, welches Element farbig gestaltet werden muss und ob eine Initiale empfehlenswert oder notwendig ist, dann trifft er Entscheidungen über das, was als typografische Störung betrachtet werden kann.

Auch Bilder von unterschiedlicher Größe können eine Seite oder Doppelseite gezielt stören und ihr genau damit Kraft und Spannung verleihen.

sylvia Tournerie revisite les formes des **avant-gardes.**

Douze ans de création graphique française : la production de Sylvia Tournerie est un échantillon symptomatique d'une génération, tout en étant singulière. En 2008, la graphiste peut s'enorgueillir d'une livraison clairsemée mais régulière. Douze ans d'indépendance, et cette liberté s'affirme aujourd'hui avec une même fragilité davantage décomplexée. Son dernier travail est un "plaisir" de simplicité, elle a composé un univers cohérent, tout en variation, pour le groupe Prototypes. Pour leur dernier album, *Synthétique*, Sylvia Tournerie manipule un répertoire formel basique, un triangle, deux couleurs, le noir et blanc. Peu d'éléments, mais l'infini s'esquisse. Son plaisir d'orchestrer une communication avec un minimalisme rythmé ravive ses envies sur ce métier, et confirme ses audaces typographiques.

retour sur un parcours

À sa sortie de l'école Penninghen, ses commandes s'enregistrent toutes sur le répertoire de la musique électronique. Non pas que le culturel ou l'espace public ne l'attirent pas – ils furent son sujet de diplôme[1] –, mais la musique lui offre l'occasion d'amorcer ses premières manipulations. Le groupe Bosco lui demande de créer son image et lui assure ainsi une prodigieuse liberté. Suivront d'autres invitations de labels indépendants qui lui ouvriront un espace d'expérimentations à même de construire son répertoire, de s'affranchir d'une certaine pudeur, puisqu'elle se rend compte que *tout est possible*. Les pochettes se suivent et donnent le jour à une palette variée, allant d'un registre construit et typographique (Bosco, *Novo Screen*) à des collages provocants et plus trash (Bosco, *New Pax*). Milieu des

42 | 3.2008

Pour l'art et la musique : **constructions et collages** par vanina Pinter ■ ■ ■

années 1990, la musique électronique est un territoire expérimental et propice aux compositions graphiques. À l'image des samples, les mariages cocasses de sons, le tempo électronique trouve une résonance en connivence avec les motifs géométriques, répétitifs et anonymes, et/ou les collages dissonants.

Transferts de savoir-faire

En 2004, la prise en charge de la revue *02*[1] marque un tournant dans le parcours de Sylvia Tournerie. Sa mise en forme de la revue gratuite et pointue d'art contemporain la fait remarquer par d'autres commanditaires, une brèche s'ouvre dans le monde de l'art contemporain. Cette brèche rappelle la difficulté pour les graphistes à aborder d'autres genres s'ils n'ont pas été invités et repérés. Pour le deuxième numéro, elle décide de travailler avec son ancien camarade d'école Gilles Poplin, avec lequel elle avait déjà créé des clips. *02* est ainsi à l'origine d'une fructueuse collaboration, un travail en duo, une joute en complicité, donnant naissance à une dizaine d'alphabets. En effet, pour chaque numéro, ils dessinent une typo, cette dernière fournira une allure et une unité à la revue, qui, pour le reste, affiche une maquette sobre, une intégration classique des images. *Nos typos de titrage sont un outil pour construire un mot et pour que visuellement, elles aient la force d'une forme.* Leur travail du signe est leur réponse pour ne pas ajouter du visuel à la création artistique contemporaine. Réponse qui, pour des catalogues d'expositions, lui fera adopter des postures de retrait : la graphiste manie les lettres et cherche la lisibilité des images des artistes.

1. disque promotionnel. 2. verso. 3. intérieur de la pochette cd. 5. stickers pour synthétique. sylvia compose autour d'un triangle des motifs aux évocations "péruviennes ou constructivistes". elle dessine le "typotype", sur la variation de sa gamme. 4. sylvia habille une série de vêtements, qu'elle souhaite voir prendre son indépendance et réfléchit à les autoproduire. ces motifs prendraient ainsi leurs aises dans l'espace quotidien.

3.2008 | 43

Große farbige Bilder stehen einem klaren Raster gegenüber.

PROJEKT
étapes: magazine

KUNDE
Pyramyd/*étapes:* magazine

DESIGN
Anna Tunick

Doppelseiten aus der französischen Zeitschrift *étapes* zeigen, wie mit einem großen Bild, einem Freisteller oder viel Weißfläche vermieden werden kann, dass eine Doppelseite oder Geschichte nur mechanisch wirkt.

6. Pochette du maxi-vinyle "Novo Screen" pour le groupe Bosco, 2002.
7. Pochette CD pour Panti Will, album "H.E.L.L.", 2005.
8. Pochette de "Fastback" maxi-vinyle pour Sodex, (pour le label client 2000st), utilisation d'une typo originale, la Copland.
9. Pochette CD pour Experience, album "Hémisphère gauche", 2004.

ses "gimmicks"

À l'incontournable – et douloureuse – question sur l'autodéfinition de son style, Sylvia Tournerie évoque deux éléments signifiants. L'école s'étant équipée d'ordinateurs à la fin de ses études et le recours à la photocopieuse étant également plus facile, cela a entraîné un style repérable, économique, un jeu de découpes. *Mon travail est marqué par des grosses masses noires avec des couleurs primaires.* Difficile de ne pas faire allusion à l'empreinte de Cieslewicz. Sylvia Tournerie a étudié à l'ESAG-Penninghen au temps où Roman Cieslewicz y enseignait[3]. Il fut son maître de thèse. De lui, elle se souvient d'un rire qu'il eut, durant un stage, alors qu'il manipulait des formes et concevait un hors-série pour *Le Monde*. Cette excitation, cette légèreté, qui ne s'essouffle pas malgré les années, cette ouverture d'esprit face aux étudiants, n'excluant pas la sévérité, sont les "outils" qu'il lui légua. L'attitude de Cieslewicz, entre détachement et jouissance personnelle d'une affirmation, semble être une aspiration, comme un moteur pour la graphiste. Son style se forgea aussi en raison des contraintes financières qu'elle subit. Les labels n'ayant pas de budgets pour une production photo, jugeant que ses propres photos ne peuvent se suffire à elles-mêmes, elle transforme celles qu'elle reçoit ou qu'elle prend en paysages. Ainsi, ses photos sont-elles plus à l'aise avec l'esprit décalé provoqué par les collages. Dans ces conditions naît la mémorable et si furtive identité de *Point éphémère*, où elle transforme en une toile de Jouy, *les acteurs de la musique*.

émergence

Sylvia Tournerie ne compose que sur ordinateur, et parle de la légèreté de l'outil, puisque, au propre comme au figuré, les données ne pèsent rien. Sur son Mac, un dossier vrac regroupe ses premières sessions de travail peu organisées, *une étape de vidage, suite à ma rencontre avec le commanditaire*. Dans un état presque hypnotique, où *l'important est de se laisser aller*, elle façonne une matière formelle abstraite. Elle la pétrit jusqu'au moment où se manifeste la première émotion, cette émotion, qu'elle peut perdre en cours de route, mais qu'elle *n'a de cesse de faire vivre, de conserver jusqu'au bout du projet*. Tout est dans le doigté et dans ces ressentis impalpables. Sylvia Tournerie parle avec sensibilité, avec intelligence de cette étape de travail, capitale, qui l'interroge douloureusement aussi. Elle évoque son incapacité à décrypter ses convictions. Cette étape est de l'ordre de l'émotion. *J'ai rarement une idée avant de faire les choses*. Ainsi, l'objet graphique émerge-t-il de son façonnage. *Je justifie les formes une fois qu'elles sont là*. Pendant longtemps, il lui fut difficile d'assumer cette prétendue gratuité, aujourd'hui, Sylvia Tournerie se dit plus sereine face à sa façon de composer[4]. Ses formes ne sont pas le fruit du

44 : 3.2008

Poster pour la styliste Andrea Crew (avec la participation de Leslie David, 2006). Au recto, les mannequins présentaient la collection de la saison et des motifs géométriques auréolaient chaque modèle et accentuaient leurs postures irrévérencieuses. Au recto, le processus de travail d'Andrea Crews se révèle dans un vaste désordre recollé : la styliste élabore ses pièces uniques à partir d'habits récupérés et recyclés.

hasard, avec l'expérience, toutes relèvent d'un choix. Sylvia Tournerie agit dans la traduction – le graphisme avec ses composants parle à l'âme directement de la même manière que la musique parle avec ses notes et ses gammes –, elle n'est pas sur le territoire des intentions. Ses identités visuelles ne sont pas des chartes, mais des pulsations, des vibrations, concentrées ou fragmentées.

Peu d'affiches, pas de théâtre, ni d'identité institutionnelle (excepté sa participation avec Gilles Poplin à l'identité du CNAP *é : 126*), pas de gros chantiers, ni de régularité (cette situation qu'on retrouve chez d'autres de ses contemporains devrait inciter les commanditaires à défier ces graphistes sur ces terrains balisés). Pourtant, les gammes de Tournerie marquent leur empreinte dans le

3.2008 : 45

Freisteller und eine geschickte Auswahl von Gegenständen machen eine geordnete Doppelseite zu einem kraftvollen Kunstwerk.

75. Für Drama sorgen

Beschneiden bringt Drama in die Bilder. Das originale Bild kann die Geschichte erzählen, aber der Beschnitt kann Punkte herausheben, Standpunkte vermitteln und Furcht oder Aufregung erzeugen. Er kann auch die Aussage des Fotos verändern, das Auge auf einen bestimmten Aspekt lenken und überflüssige Informationen entfernen.

URHEBERRECHTSBESCHRÄNKUNGEN BEACHTEN

Manchmal ist ein Beschnitt aufgrund des Urheberrechts nicht möglich. Viele Fotografen, Bildagenturen und Museen haben strenge Vorschriften über die Reproduktion von Kunstwerken. Einige Bilder, besonders von berühmten Gemälden oder Skulpturen, dürfen nicht angetastet werden.

Seite gegenüber: Auf diesem Buchumschlag sind die Textboxen kreisförmig (wie Spotlights). Collagen können die Aussage verwässern, aber der Beschnitt der Bilder hat hier einen atemberaubenden Effekt.

Diese Seite rechts: Die runden Textboxen bringen Bewegung in die grobkörnigen Bilder, die im Raster angeordnet sind.

PROJEKT
Broadway: From Rent to Revolution

KUNDE
Drew Hodges, Autor;
Rizzoli, Verleger

CREATIVE DIRECTOR
Drew Hodges

DESIGN
Naomi Mizusaki

Energie, Formen, Beschnitt und eine tolle Typografie bringen Drama in ein Design, das auf einem klaren Raster basiert, aber nichts auslässt.

MAX VADUKUL
PHOTOGRAPHER

AT THE TIME, I WAS DOING lots of black-and-white work for French *Vogue*, working with models like Helena Christensen and Linda Evangelista—women who are full of personality and charisma. To do black-and-white photography was almost taboo at the time. Everything had to be in color. So when the request came in to do *Chicago*, I remember the layouts were actually swipes of my editorial work, narrow slits of photos like they ended up being in the ads. And I thought, "Wow, this design looks so good." It was fantastic, and I had to do it.

I originally wanted to style it to be very edgy like my editorial work, but we realized the photographs had to have staying power like the show. So William [Ivey Long, the show's costume designer] came in with a series of looks. And I shot the girls in relation to the theater—sitting on the stage, all over the place. I'd call out to people to get a look—and with very simple light. It was the biggest blast ever. The enthusiasm on the set was incredible.

CHICAGO WAS MY FIRST lesson in embracing a possible negative and turning it into a positive from the very beginning. I now know if you don't correct a perception right from the start, you will never be able to do so.

We began as if the show had never existed before on Broadway, and we banished the word "revival." But someone wrote a column asking how would audiences pay top dollar ($75 at the time) for a glorified concert. I decided the real perception challenge here was to make it clear from the get-go that with all the budget in the world, our bandstand and bentwood chairs were all the spectacle we needed. But how?

ALL THAT GRIT
AROUND THE THEATER AND IN THE STREETS—IT WAS JUST THERE. AND WE'D GO STRAIGHT TO PRINT. NO RETOUCHING.
MAX VADUKUL

above
Ute Lemper giving you legs for days in the aisle, and above right, the shoot goes out of control.

below
A calendar from *Chicago* featuring SpotCo friends and family, and die-cut bullet holes on the cover.

21

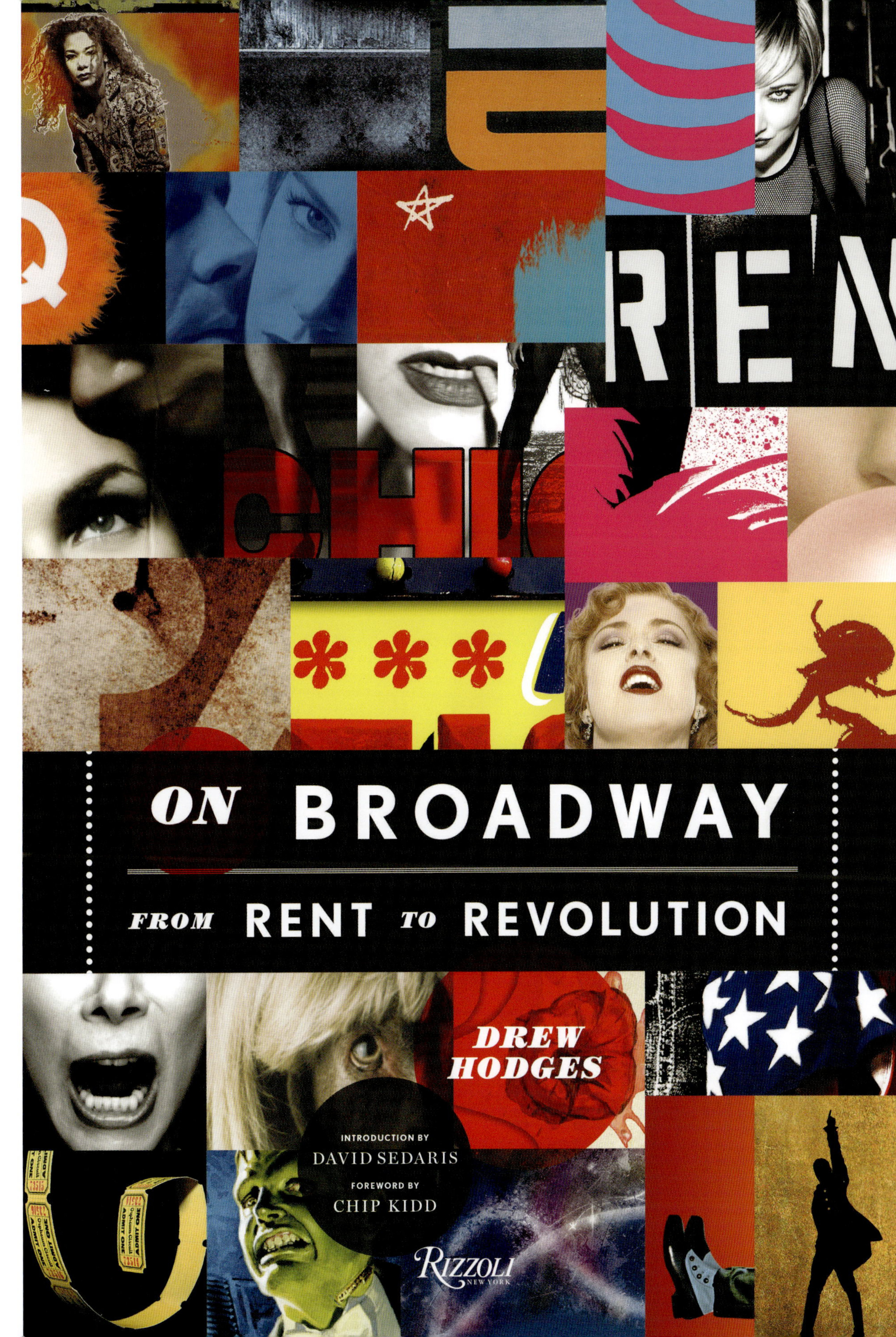
ON BROADWAY
FROM RENT TO REVOLUTION
DREW HODGES
INTRODUCTION BY
DAVID SEDARIS
FOREWORD BY
CHIP KIDD
RIZZOLI
NEW YORK

76. Silhouetten

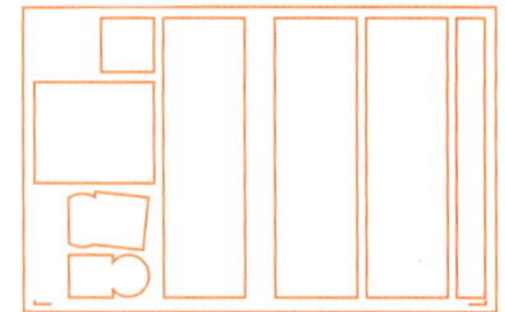

Freisteller sind ein wunderbares Werkzeug, um zu verhindern, dass eine Doppelseite zu systematisch oder wie ein einziger Block erscheint.

Im Zusammenhang mit einem Layout ist ein Freisteller ein Bild, dessen Hintergrund entfernt wurde und nur noch selbst in seinen Umrissen zu sehen ist. Das kann eine organische Form wie ein Blatt sein oder eine gleichmäßige wie ein Kreis. Freisteller mit einer weichen, fließenden Form geben einer Doppelseite mehr Bewegung.

旬の、相性のいいもの同士が醸しだす味わいの深いことといったら。

新じゃがの花まぶし

ふきの炊いたんと鯛の子

若ごぼうの炊いたん

糸こんちりめん

わけぎとイカの辛子酢味噌和え

若竹汁

Vertikale und horizontale Linien begrenzen die Flächen für Überschriften, Einleitungen und Informationen. Der nüchterne Lehrcharakter dieser Seiten wird von den organischen Formen der Freisteller belebt.

PROJEKT
Croissant magazine

ART DIRECTOR
Seiko Baba

DESIGNER
Yuko Takanashi

Diese Doppelseite aus einer japanischen Fachzeitschrift macht deutlich, wie eine geordnet und strukturiert dargestellte Geschichte von Freistellern profitiert. Das Magazin heißt MOOK, eine Spezialausgabe der *Croissant*-Herausgeber. Der Titel *Mukashi nagara no kurashi no chie* bedeutet so viel wie „Altehrwürdige Lebensweisheiten".

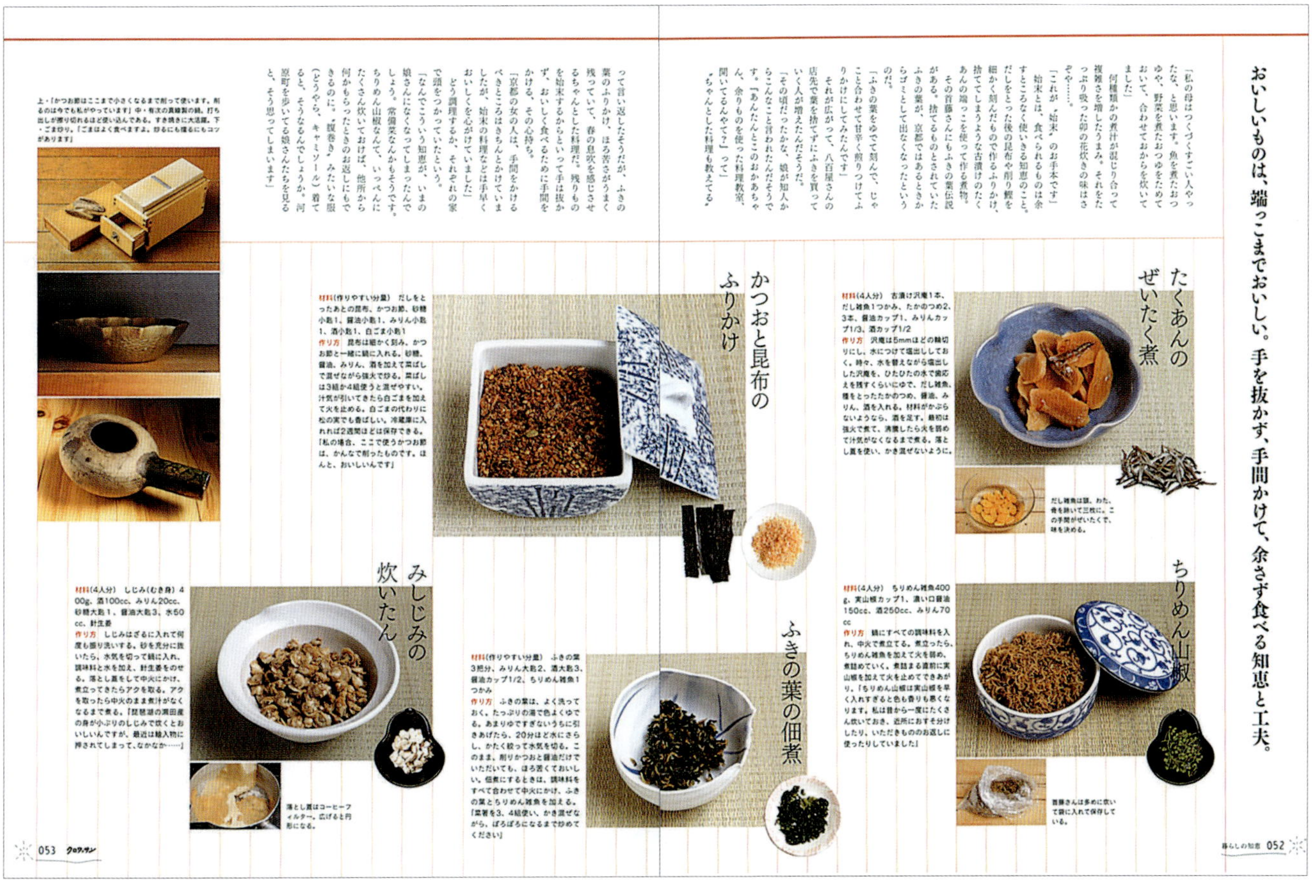

おいしいものは、端っこまでおいしい。手を抜かず、手間かけて、余さず食べる知恵と工夫。

たくあんのぜいたく煮

ちりめん山椒

かつおと昆布のふりかけ

ふきの葉の佃煮

しじみの炊いたん

Linien erzeugen einen zusätzlichen Raster innerhalb des Zeitschriftenrasters. Die geradlinige Anordnung ist klar und sauber. Unterschiedliche Formen verleihen dieser geordneten und hierarchisch strukturierten Doppelseite Dynamik.

77. Ein Platz für die Kultur

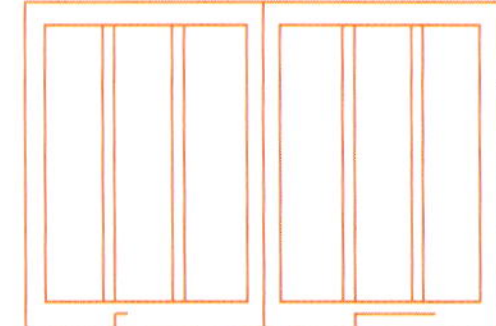

Man kann einen starken, deutlichen Rahmen mit zahlreichen Blickfängen schaffen und doch Aspekte beibehalten, die das Projekt einzigartig machen und den Betrachter oder Leser vielleicht sogar weiterbilden.

Durch Kulturgeschichliches, Mythologien oder Symbole kann der Designer sein Werk viel eindrucksvoller gestalten. Gleichzeitig wird die Welt mit dem Blick auf andere Kulturen kleiner und dennoch reicher.

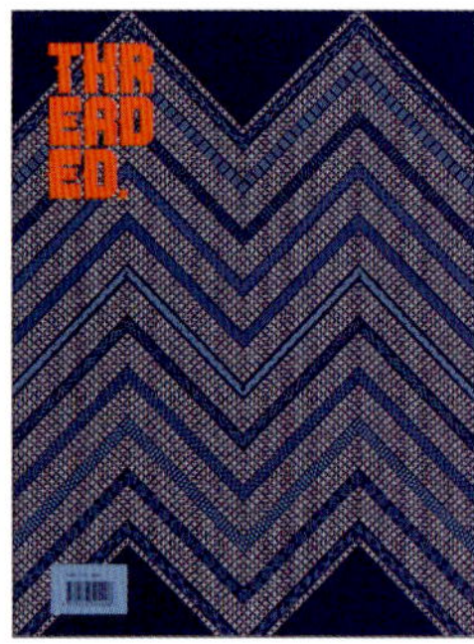

PROJEKT
Threaded, Zeitschrift

KUNDE
Threaded Studio, Verleger

DESIGN
Threaded Studio

DESIGNER
Kyra Clarke,
Fiona Grieve,
Reghan Anderson,
Phil Kelly,
Desna Whaanga-Schollum,
Karyn Gibbons,
Te Raa Nehua

BILDNACHWEIS
Threaded Media Limited, 2016/17

New Beginnings, Ausgabe 20 eines internationalen Magazins, das von einer Designagentur aus Auckland, Neuseeland, gestaltet wurde, befasst sich ausschließlich mit dem *Kaupapa* (Thema) der Kunst der Maori und ihrer Designer.

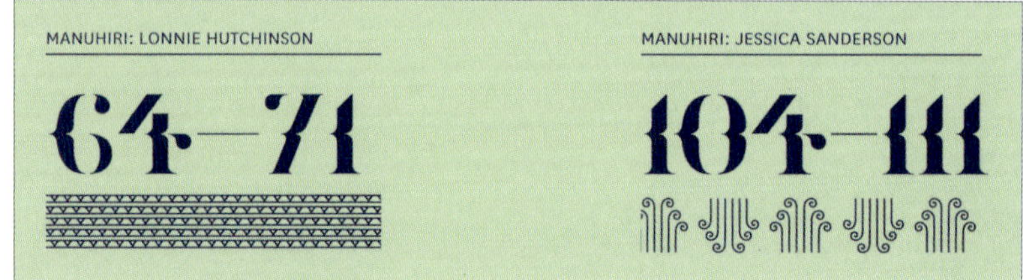

Die serifenlose Schrift und die fantasievollen Zahlen arbeiten mit Symbolen aus der Maori-Kultur. Jedes Muster ist individuell entwickelt, um den *Korero* (Vortrag) jedes *Manuhiri* (Gast) des Magazins widerzuspiegeln. So wird die Grafik zu einem einzigartigen kulturellen Boten jedes einheimischen Designers.

SEITE GEGENÜBER: Ein einfacher Raster und viel Raum für die atemberaubende Kunst der Maori. Die Ornamente feiern die Kultur und meiden alles Banale oder Stereotype. Zwischentitel und Text in schlichter serifenloser Schrift bilden einen Kontrast zu den spektakulären Bildern.

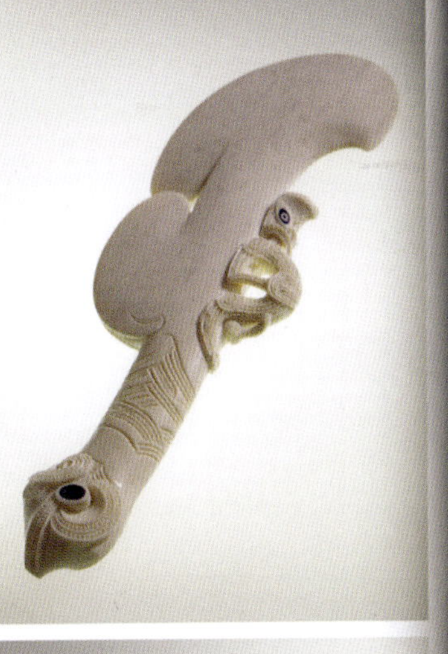

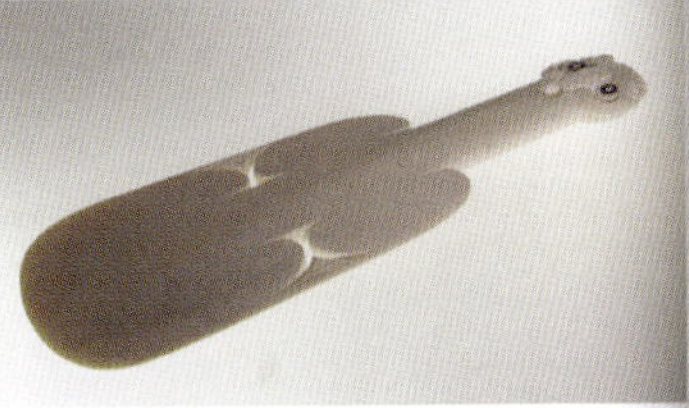

ON LINE:

I'm acutely aware of how much the power of a line can influence how people read visual material. For me, when I'm working in a sculptural sense I'm analyzing everything by the mana of those lines; if they're transported into low relief or 3D we're talking about edges. Edges are everything. You create a deep or powerful sense of space, direction and form with something that's relatively shallow. The edge of the line enables you to use light to give the impression of depth. I'm aware of it and I just try to exploit it I suppose. There's a beauty in line that's difficult to explain, but I get seduced by the ability to reduce down to linear forms and play with it, there's so much you can do it's really just up to your imagination and over a period of time you get to a point where you can master it and then people, they follow it, they get it. They're not necessarily able to interpret it or explain it but they get it.

The energy contained in the line is no different than the principle of physics. It's the same as how you use your arm to develop a centrifugal force to throw something. The line can do that as well. You can use that energy to influence and to give the impression (of that force) in the same way whether in 2D and 3D.

TOP
—
Wahaika, whalebone
•
BOTTOM
—
Kotiate, whalebone

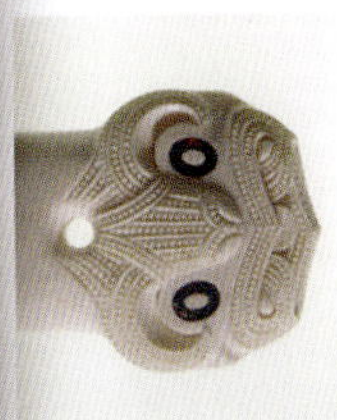

REGARDING SCALE:

I work in so many different genres and scales. There are fundamental principles to design but there are so many different ways to exploit them, they have weaknesses and strengths. And they can change as soon as you change scale or your relationship – for example physical proximity or moving from 2D to 3D. There are so many ways in which you can change the nature of the game and it forces you to engage differently with those elements or principles. As I get older in my craft and sense of self (if you're fortunate enough to be given the freedom in your practice to pursue your own sense of design truth) you find that as you journey along eventually as you refine based upon your own sensibilities, as you refine your craft you find the sweet spots. You find what works and what doesn't. What works for me in moko, using line at that proximity – you're working within one foot of your hand guiding the gun as its laying ink in the skin – works differently when you're standing away doing a large mural or a 6 metre bronze sculpture. So you can't use and engage the same principles in the same way, you're forced to renegotiate your own axis or your own sense of gravity to your work.

TOP LEFT
—
Heru, whalebone
•
TOP RIGHT
—
Hei tiki, whalebone
•
BOTTOM LEFT
—
Kotiate, (detail), whalebone
•
BOTTOM RIGHT
—
Madonna and Child, whaletooth

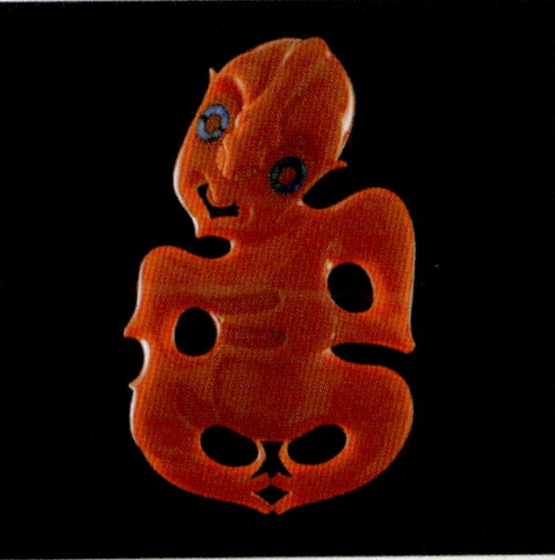

CONNECTIONS AND PATHWAYS:

My artistic practice is strongly influenced by my political belief that we need to be relevant and that we have a role to bridge the past and the future. I'm lucky enough to grow up in my tribal area. I've had strong cultural connections to my community so I have a sense of allegiance to my culture and community that manifests itself in the work. But I can't change the past. I'm trying to visualize a pathway in the future and trying to use my art as a tool to help lay down some of that pathway. I'm trained as a social scientist as well so if you connect that to my cultural background it's part of my imperative to drive my art to always be forward focused. I believe we have a role to visualize the future and make it happen. If you look over my practice over the past 20-30 years, moko was like that, taonga puoro – Māori musical instruments – was like that, my role in waka in Taranaki was like that. So by kicking those things off and by continuing to push them – the same with taonga whakarakai (adornment arts) – restoring those art forms but restoring them in a way that continues to have relevance not only for now and being present in your work but also into the future. I'm trying to push my work so far ahead that it actually looks futuristic, that people can go 'wow – I really like that'. They can see the footprint of our old world in it but it's also out there tugging on them so the materiality, the aesthetics and the cultural imperative that's subtly locked in there pulls at them.

NEW MATERIALS:

The interesting thing was about seeing how the Māori community would respond to media that wasn't seen as valuable, that didn't have a valuable (cultural) attribution, so creating beautiful stuff out of it, it enables itself to be relevant. The beautiful thing about Corian is you can get a great sense of colour – and if you're really good at finishing the work – it's as seductive as what whale tooth, whale bone or pounamu can be. That was the beauty of that exercise, seeing that material being adopted as a taonga. These approaches to materials are interesting journeys, they're not necessarily answers to questions but they're part of the journey.

> “… you're working within one foot of your hand guiding the gun as its laying ink in the skin… ”

TOP
—
Moko peha
•
BOTTOM
—
Poroto, Corian

OPPOSITE PAGE

TOP
—
Hei tiki, Corian
•
BOTTOM
—
Mourei, whaletooth

78. Ein vielseitiges System

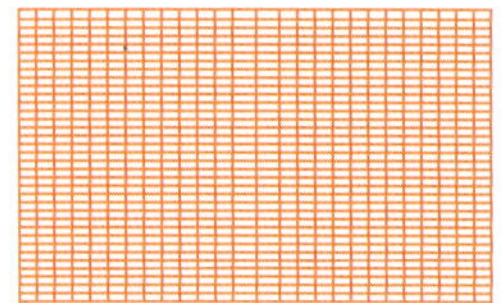

In einem vielseitigen System können verschiedene Größen, Formen und Informationen in zahlreichen Konfigurationen zusammenwirken.

PIONIERE

Ellen Lupton sagt, dass die Pioniere des Schweizer Grafikstils, Josef Müller-Brockmann und Karl Gerstner, ein Design-„Programm" als Satz von Linien zur Konstruktion einer Auswahl an visuellen Lösungen definierten. Lupton fasst das so zusammen: „Die Schweizer Designer nutzten die Grenzen einer sich wiederholenden Struktur, um Abweichung und Überraschung zu erzeugen. Ein System lässt innerhalb eines Projekts sowohl dichte als auch weiträumige Seiten zu."

josef müller-brockmann, certitudes visuelles

Les affiches de concert

Durch dieses Rastersystem lässt sich die Seite in Hälften, Drittel und Viertel aufbrechen. Sie kann auch horizontal unterteilt werden.

Der starke Raster reguliert die Bildgrößen und unterstützt Abweichungen.

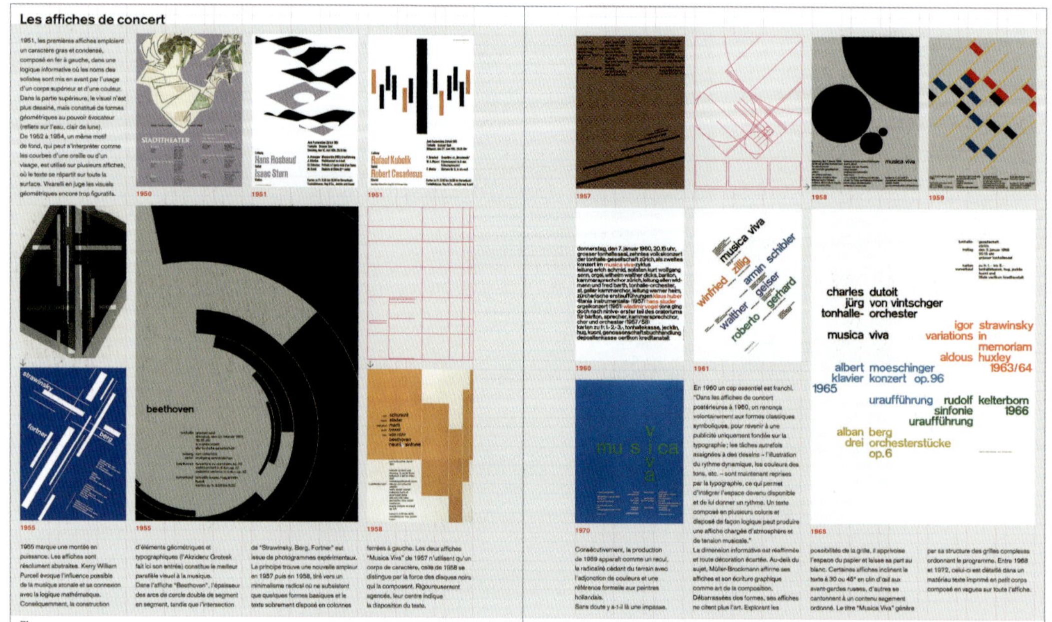

Les affiches de concert

PROJEKT
étapes: magazine

KUNDE
Pyramyd/*étapes:* magazine

DESIGN
Anna Tunick

Dieser Zeitschriftenartikel über die Arbeit des berühmten Schweizer Designers Josef Müller-Brockmann wendet flexible Raster an.

Strenge Raster schließen Spannung nicht aus. Fesselnde Bilder und eine rhythmische Anordnung schaffen Variation und Überraschung.

Auf dieser Doppelseite wird deutlich, dass auf dem Raster sowohl für eine Box als auch für eine Seite mit einer großen Weißfläche Platz ist.

« En une fraction de seconde, l'affiche doit agir sur la pensée des passants, les contraignant à recevoir le message, à se laisser fasciner avant que la raison n'intervienne et ne réagisse. Somme toute, une agression discrète, mais soigneusement préparée. »

rationalité, objectivité et efficacité

Philosophie de la grille et du design

L'usage de la grille comme système d'organisation est l'expression d'une certaine attitude en ce sens qu'il démontre que le graphiste conçoit son travail dans des termes constructifs et orientés vers l'avenir.

C'est là l'expression d'une éthique professionnelle, le travail du designer doit avoir l'évidente, objective et esthétique qualité du raisonnement mathématique.

Son travail doit être la contribution à la culture générale dont il constitue lui-même une partie.

Le design constructiviste qui est capable d'analyse et de reproduction peut influencer et rehausser le goût d'une société et la façon dont elle conçoit les formes et les couleurs.

Un design qui est objectif, engagé pour le bien-être collectif, bien composé et raffiné constitue la base d'un comportement démocratique. Un design constructif signifie la conversion des lois du design en solutions pratiques. Un travail accompli de façon systématique, en accord avec de stricts principes formels, permet ces exigences de droiture d'intelligibilité et l'intégration de tous les facteurs eux aussi vitaux pour la vie sociopolitique. Travailler avec un système de grille implique la soumission à des lois valides universellement.

L'usage du système de grille implique
- la volonté de systématiser, de clarifier ;
- la volonté de pénétrer à l'essentiel, de concentrer ;
- la volonté de cultiver l'objectivité au lieu de la subjectivité ;
- la volonté de rationaliser les modes de production créatifs et techniques ;
- la volonté d'intégrer des éléments de couleur, de forme et de matière ;
- la volonté d'accomplir la domination de l'architecture sur l'espace et la surface ;
- la volonté d'adopter une attitude positive et visionnaire ;
- la reconnaissance de l'importance de l'éducation et les effets du travail conçu dans un esprit constructif et créatif.

Tout travail de création visuelle est une manifestation de la personnalité du designer. Il est marqué de son savoir, de son habileté et de sa mentalité. – *Josef Müller-Brockmann*

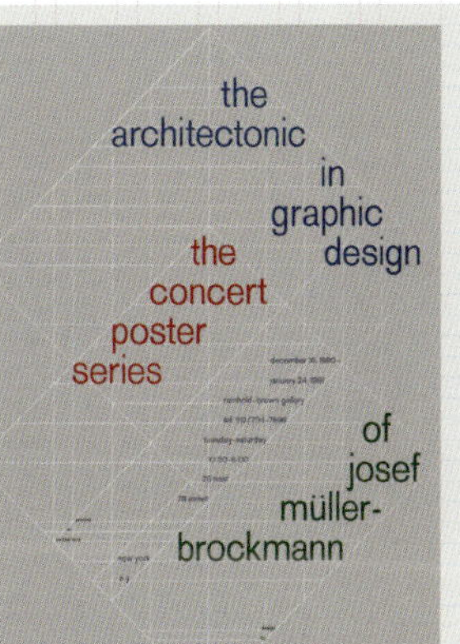

Unis, visite le Mexique et prend des contacts à New York, où il songeait à s'établir, devant la difficulté pour la Suisse à reconnaître et à laisser s'épanouir ses talents, du fait de son *esprit de villageois et de paysans*. Il retourne finalement à Zurich, où il prend la suite de son professeur à l'école des arts et métiers, Ernst Keller, et met en place la revue qu'il songeait à monter depuis 1955: *Une publication pour un graphisme rationnel et constructif pour contrer les excès d'une publicité irrationnelle et pseudo-artistique que je voyais autour de moi.* Animée et éditée avec Richard Paul Lohse, Carlo Vivarelli et Hans Neuberg, la revue *Neue Grafik ("Graphisme actuel")*, éditée en allemand, anglais et français approximatif, comptera dix-huit numéros publiés jusqu'en 1965. D'abord approchées, des personnalités comme Armin Hoffman ou Emil Ruder sont écartées, leurs productions étant jugées trop diversifiées par le quarteron de puristes. Une idéologie formelle et fonctionnelle se met en place. Les trois mots-clefs en sont rationalité, objectivité et efficacité: *J'en suis venu à apprécier l'Akzidenz Grotesk davantage que ses successeurs Helvetica et Univers. Il est plus expressif et ses bases formelles sont plus universelles. La fin du "e", par exemple, est une diagonale qui produit des angles droits. Dans le cas de l'Helvetica et de l'Univers, les terminaisons sont droites, produisant des angles aigus ou obtus, des angles subjectifs.* Après la Seconde Guerre mondiale et le désordre nazi, le graphisme espère un retour à l'harmonie et ambitionne un rôle constructeur. La subjectivité du dessin est écartée au profit de l'objectivité de la photo et de la construction. Les règles de la nouvelle typographie constituent avec le fer à gauche une dynamique vers le progrès technique et social: *La symétrie et l'axe central sont ce qui caractérise l'architecture fasciste. Le modernisme et la démocratie rejettent l'axe.* Le savoir-faire du designer se précise et quitte la théorie pour passer à l'épreuve du réel au service des entreprises: *Un design constructif signifie la conversion des lois du design en solutions pratiques.* C'est dans ce sens que s'oriente son premier livre *Problèmes d'un artiste graphique*, dont la publication en 1961 correspond à son départ de l'école des arts et métiers de Zurich, où il n'est pas parvenu à installer son enseignement. Dix ans plus tard, il publie une *Histoire de la communication visuelle* et (avec sa seconde épouse) une *Histoire de l'affiche*, qu'il organise de nouveau avec l'affiche constructiviste en ligne de mire et l'efficacité en lieu et place de l'expressivité: *En une fraction de seconde, l'affiche doit agir sur la pensée des passants, les contraignant à recevoir le message, à se laisser fasciner avant que la raison n'intervienne et ne réagisse. Somme toute, une agression discrète mais soigneusement préparée.* Quatre ans plus tôt, Müller-Brockmann a fondé avec trois associés l'agence Müller-Brockmann & Co, qui intègre la publicité dans son activité régulière, aux côtés de l'identité visuelle, la signalétique et la communication culturelle. Au terme de dix années supplémentaires, en 1981, il publie son ouvrage de référence: *Raster systeme für die visuelle Gestaltung.* Ses expérimentations dans les affiches du Tonhalle ainsi que son récent travail pour les chemins de fer suisses lui ont permis de forger une théorie mais aussi une éthique de la grille. Derrière son apparence de manuel technique, l'ouvrage est un manifeste. Le livre est introduit par un texte sur la *philosophie de la grille et du design (voir encadré)* qui conclut par un renvoi à l'individualité du créateur: *Tout travail de création visuelle est une manifestation de la personnalité du designer. Il est marqué de son savoir, de son habileté et de sa mentalité.* Las, les progrès qu'il contient et propose ne seront pas perçus comme les choix déterminés d'un graphiste ou comme des règles parfois compassées proposées à la profession, mais plus souvent

programme d'identité, de signalétique et d'informations visuelles des chemins de fer suisses (SBB). Assorti de recommandations typographiques (un Helvetica modifié), le gabarit permet de garantir l'uniformité du système dans le temps et d'en tirer bénéfice sur une multiplicité de supports. Projet réalisé par Müller-Brockmann & Co et Peter Spalinger, primé en 1993 par le Swiss Design Prize.

79. Stark und groß

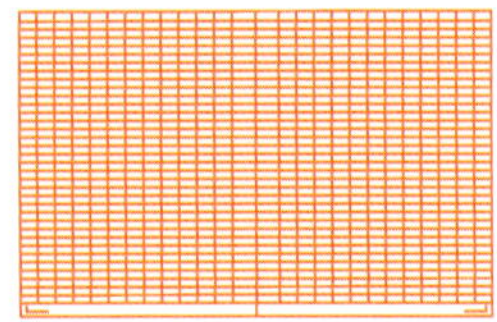

Ein Raster, dessen Grundlage der Schweizer Grafikstil bildet, kann aus viel Text ein Lesevergnügen machen. Dieser Raster übermittelt visuell Informationen, die klar und deutlich lesbar sind. Mehrspaltige Raster können reichlich Informationen sowie Bilder und farbige Boxen für Teilinformationen enthalten. Auch Abweichungen sind möglich: Was ausgelassen wird, bringt das zur Geltung, was hineingepackt wird.

7 GREAT SERIES. 7 GREAT EXPERIENCES!

1
LCJO SERIES
Lincoln Center Jazz Orchestra with Wynton Marsalis
4 Concerts
Rose Theater, 8pm

COLTRANE
SEPTEMBER 14, 15 & 16, 2006
Blue tranes run deeper. Ecstatic and somber, secular and sacred, John Coltrane's musical sermons transform Rose Theater into a place of healing and celebration with orchestrations of his small group masterpieces "My Favorite Things," "Giant Steps," "Naima," and more. Join us as the **LCJO** with **Wynton Marsalis** marks the 80th year since the birth of one of

2
JJ SERIES
Jazz Jam
4 Concerts
Rose Theater, 8pm

WYNTON AND THE HOT FIVES
SEPTEMBER 28, 29 & 30, 2006
Hearts beat faster. It's that moment of pure joy when a single, powerful voice rises up from sweet polyphony. Louis Armstrong's Hot Five masterpieces—"West End Blues," "Cornet Chop Suey," and others—quicken the pulse with irresistibly modern sounds. **Wynton Marsalis, Victor Goines, Don Vappie, Wycliffe Gordon**, and others re-imagine the recordings that defined jazz, and then bring that pure joy to the debut of equally timeless new music inspired by the original.

RED HOT HOLIDAY STOMP
DECEMBER 14, 15 & 16, 2006
Tradition gets fresher. When Santa and the Mrs. get to dancin' the "New Orleans Bump," you know you're walking in a *Wynton Wonderland*—a place where joyous music meets comic storytelling. **Wynton Marsalis, Herlin Riley, Dan Nimmer, Wycliffe Gordon, Don Vappie**, and others rattle the rafters with holiday classics swung with Crescent City style. *Bells, baby. Bells.*

THE LEGENDS OF BLUE NOTE
APRIL 26, 27 & 28, 2007
Bop gets harder. The music is some of the best ever made—Lee Morgan's *Cornbread*, Horace Silver's *Song for My Father*, Herbie Hancock's *Maiden Voyage*—all wrapped up in album cover art as bold and legendary as the music inside. The **LCJO** with **Wynton Marsalis** debuts exciting and long-overdue big band arrangements of the best of Blue Note, complete with trademark cracklin' trumpets, insistent drums, and all manner of blues.

IN THIS HOUSE, ON THIS MORNING
MAY 24, 25 & 26, 2007
Tambourines testify. It's that sweet embrace of life—sometimes celebratory, sometimes solemn—rising from so many houses on so many Sundays. We mark the 15th anniversary of Wynton's first in-house commission, a sacred convergence of gospel and jazz that

3
MM SERIES
Music of the Masters
4 Concerts
Rose Theater, 8pm

FUSION REVOLUTION: JOE ZAWINUL
OCTOBER 27 & 28, 2006
Grooves ask for mercy, mercy, mercy. Schooled in the subtleties of swing by Dinah Washington, keyboardist **Joe Zawinul** brought the fundamentals of funk to Cannonball Adderley, the essentials of the electric to Miles Davis, and carried soul jazz into the electric age with his band Weather Report. Now the **Zawinul Syndicate** takes us on a hybrid adventure of sophisticated harmonies, world music rhythms, and deeply funky grooves. *Mercy.*

BEBOP LIVES!
JANUARY 26 & 27, 2007
Feet tangle and neurons dance. Fakers recoil, goatees sprout, and virtuosos take up their horns. Charlie Parker and Dizzy Gillespie set the bebop revolution in motion, their twisting, syncopated lines igniting the rhythms of jazz. Latter day fakers beware as the legendary **James Moody** and **Charles McPherson**, the alto sax voice of Charlie Parker in Clint Eastwood's *Bird*, raise battle axes and *swing*.

CECIL TAYLOR & JOHN ZORN
MARCH 9 & 10, 2007
Souls get freer. Embark on a sonic voyage as the peerless **Cecil Taylor** navigates us through dense forests of sound—percussive and poetic. He is, as Nat Hentoff proclaimed, "a genuine creator." The voyage banks toward the avant-garde as **John Zorn's Masada** with **Dave Douglas** explores sacred and secular Jewish music and the "anguish and ecstacy of klezmer." Musical wanderlust *will* be satisfied.

THE MANY MOODS OF MILES DAVIS
MAY 11 (Kisor/blanchard) &
MAY 12 (Payton/Miller), 2007
Change gets urgent. "I have to change," Miles said, "It's like a curse." And so his trumpet voice—tender, yet with that *edge*—was bound up in five major movements in jazz. The LCJO's **Ryan Kisor** opens with bebop and the birth of the cool. GRAMMY®-winner **Terence Blanchard** interprets hard bop and

Ausschnitt (oben) und rechte Seite: Die wechselnden Stärken, Zeilenabstände, Beschriftungen, Rubriken und Taglines sind geregelt, der Aufbau ist klar und deutlich. Farbmodule weisen auf die sieben Serien hin. Die Typografie innerhalb eines jeden Farbmoduls ist klar und ausgeglichen, die Schriftgrößen und -stärken kennzeichnen die Informationen. Die Farbmodule erscheinen als gelungene sekundäre Layouts innerhalb des Gesamtlayouts der Broschüre. Innerhalb der Module dienen gut gewählte Schriftarten als Minibanner.

PROJEKT
Abo-Broschüre

KUNDE
Jazz at Lincoln Center

DESIGN
Bobby C. Martin Jr.

Erfolgreicher Wettstreit zwischen Typografie und reichhaltigem Programmangebot.

JAZZ AT LINCOLN CENTER'S 06-07 SEASON

From Satchmo's first exuberant solo shouts to Coltrane's transcendent ascent, we celebrate the emotional sweep of the music we love by tracing the course of its major innovations. Expression unfolds in a parade of joyous New Orleans syncopators, buoyant big band swingers, seriously fun beboppers, cool cats romantic and lyrical, blues-mongering hard boppers, and free and fusion adventurers. From all the bird flights, milestones, and shapes of jazz that came, year three in the House of Swing is a journey as varied as the human song itself, and the perfect season to find your jazz voice.

4 A-LJO SERIES

Afro-Latin Jazz Orchestra with Arturo O' Farrill
3 Concerts
Rose Theater, 8pm

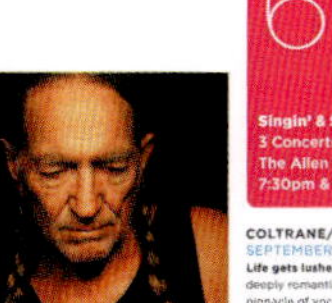

BEBO VALDES
OCTOBER 13 & 14, 2006

Mambo migrates. Bebo Valdes is a true legend. The *Los Angeles Times* calls him "elegant and restrained, a product of the golden era of Cuban music, when delicate *danzones* and tasty cha-cha-chas reigned supreme." When Bebo first lit up the Tropicana with his irresistible mambo beat, a musical blaze spread from Havana to the dance halls of New York City. Today, after over 40 years of exile in the cooler climes of Sweden, Bebo still brings the heat of his native island like no one else. Joined by **Arturo O'Farrill** and the **Afro-Latin Jazz Orchestra**, the 87-years-young Afro-Cuban pioneer serves up *el ritmo caliente* with the United States premiere of his GRAMMY®-winning *Solo Cubana*. *¡Viva Bebo!*

5 SM SERIES

Singers Over Manhattan
4 Concerts
The Allen Room
7:30pm & 9:30pm

STEPHANIE JORDAN & THE WESS ANDERSON QUARTET
OCTOBER 20 & 21, 2006

Standards get fresher. Every so often a new voice stands up and proclaims itself, but few do so with such supreme depth and understated soul. Emerging from the New Orleans jazz family Jordan, **Stephanie Jordan** was last year's *Higher Ground Concert's* "real discovery" (*JazzTimes*). She's joined in sweet counterpoint by **Wess "Warmdaddy" Anderson**, whose buoyant saxophone voice continues to proclaim itself with unmatched joy and warmth.

WILLIE NELSON SINGS THE BLUES
JANUARY 12 & 13, 2007

Blues get democratic. He's got the right to sing the blues. As folk legend **Willie Nelson** told the great B.B. King and the late, great Ray Charles, "Gentlemen, I think I'm the only one here who actually had to pick cotton." Country outlaw Willie Nelson and Crescent City son **Wynton Marsalis** come together on two stellar evenings in The Allen Room to demonstrate soulfully why the blues should be our national anthem.

6 SS SERIES

Singin' & Swingin'
3 Concerts
The Allen Room
7:30pm & 9:30pm

COLTRANE/HARTMAN
SEPTEMBER 15 & 16, 2006

Life gets lusher. Between the grooves of a deeply romantic recording that stands at the pinnacle of vocal and instrumental harmony, John Coltrane and Johnny Hartman were relaxed and electric, muscular and gentle. Tenor saxophonist **Todd Williams** with his "commanding tone" (*The New York Times*) and vocalist **Kevin Mahogany**, who makes "you want to just sit down and listen" (*USA Today*), recast the sophisticated balladry of a recording that flows like an "April breeze on the wings of spring..." All against a backdrop of Coltrane's most beloved instrumental ballads.

PAQUITO D'RIVERA
NOVEMBER 17 & 18, 2006

Streams converge. The jazz-meets-classical clarinet tradition of Benny Goodman is carried into a new century by the seven-time GRAMMY®-winner and 2005 National Medal of Arts recipient **Paquito D'Rivera**. "A formidable musician... with lovely, clear low registers." (*The New York Times*) Paquito and company explore the dynamic Third Stream convergence of classical and jazz with interpretations of the music of Ravel, Bartók, and Stravinsky.

THE BIRTH OF COOL
MARCH 30 & 31, 2007

Whispers shout louder. Cool. Lester Young invented the word as his saxophone angled toward the heavens—*original cool*. Count Basie set Kansas City ablaze without breaking a sweat—*slow hand cool*. Billie Holiday relaxed each syllable and clichés wilted—*tragic cool*. And Miles Davis caressed melody like a plate with a blowtorch—*blue flame cool*. Pianist **Bill Charlap**, who "approaches a song the way a lover approaches his beloved" (*TIME* magazine), leads an ensemble in a celebration of the classics of cool, and as the last note drops, we'll *all* be cool.

7 GREAT SERIES. 7 GREAT EXPERIENCES!

1 LCJO SERIES

Lincoln Center Jazz Orchestra with Wynton Marsalis
4 Concerts
Rose Theater, 8pm

COLTRANE
SEPTEMBER 14, 15 & 16, 2006

Blue tranes run deeper. Ecstatic and somber, secular and sacred, John Coltrane's musical sermons transform Rose Theater into a place of healing and celebration with orchestrations of his small group masterpieces "My Favorite Things," "Giant Steps," "Naima," and more. Join us as the **LCJO** with **Wynton Marsalis** marks the 80th year since the birth of one of the most admired, influential, and adventurous artists in the history of jazz.

GERSHWIN
NOVEMBER 16, 17 & 18, 2006

Rhapsodies get bluer. "Composers have been walking around jazz like a cat around a plate of soup," said legendary conductor Walter Damrosch, "waiting for it to cool so that they could enjoy it without burning their tongues." George Gershwin, however, had chutzpah. The **LCJO** with **Wynton Marsalis**, special guest **Marcus Roberts**, and the **American Composers Orchestra** perform Gershwin's groundbreaking *Rhapsody in Blue*, and then underscore vocal performances of Nelson Riddle's finest arrangements from the Gershwin songbook. The evening closes as composer **Derek Bermel** brings his innovative vision to an exciting new work.

JAZZ AND ART
FEBRUARY 22, 23 & 24, 2007

Sound bleeds color. The lipstick reds of a bebop fanfare and midnight blues of a low-down dirge syncopate our musical canvas. Inspired by the **Museum of Modern Art's** collection, the **LCJO** with **Wynton Marsalis** performs the music that moved Mondrian, Bearden, Pollock, and others to saturate their art with the rhythmic energy of jazz. Then in a debut work, *Down Beat* magazine 2005 "Rising Star" **Ted Nash** interprets in song the masterworks of Picasso, Chagall, and other twentieth century masters.

THE SONGS WE LOVE
MARCH 29, 30 & 31, 2007

Perfection endures. They are arranged to perfection—"April in Paris" arranged by Wild Bill Davis, "Summertime" by Gil Evans, and many more—and they're our life soundtracks, elevating the everyday, making the mundane magical. The **LCJO** with **Wynton Marsalis** plays some of the greatest arrangements of our favorite songs—swinging and supple, sophisticated and spirited—and remind us all over again how great music becomes legendary.

2 JJ SERIES

Jazz Jam
4 Concerts
Rose Theater, 8pm

WYNTON AND THE HOT FIVES
SEPTEMBER 28, 29 & 30, 2006

Hearts beat faster. It's that moment of pure joy when a single, powerful voice rises up from sweet polyphony. Louis Armstrong's Hot Five masterpieces—"West End Blues," "Cornet Chop Suey," and others—quicken the pulse with irresistibly modern sounds. **Wynton Marsalis, Victor Goines, Don Vappie, Wycliffe Gordon**, and others re-imagine the recordings that defined jazz, and then bring that pure joy to the debut of equally timeless new music inspired by the original.

RED HOT HOLIDAY STOMP
DECEMBER 14, 15 & 16, 2006

Tradition gets fresher. When Santa and the Mrs. get to dancin' the "New Orleans Bump," you know you're walking in a *Wynton Wonderland*—a place where joyous music meets comic storytelling. **Wynton Marsalis, Herlin Riley, Dan Nimmer, Wycliffe Gordon, Don Vappie**, and others rattle the rafters with holiday classics swung with Crescent City style. *Bells, baby. Bells.*

THE LEGENDS OF BLUE NOTE
APRIL 26, 27 & 28, 2007

Bop gets harder. The music is some of the best ever made—Lee Morgan's *Cornbread*, Horace Silver's *Song for My Father*, Herbie Hancock's *Maiden Voyage*—all wrapped up in album cover art as bold and legendary as the music inside. The **LCJO** with **Wynton Marsalis** debuts exciting and long-overdue big band arrangements of the best of Blue Note, complete with trademark crackin' trumpets, insistent drums, and all manner of blues.

IN THIS HOUSE, ON THIS MORNING
MAY 24, 25 & 26, 2007

Tambourines testify. It's that sweet embrace of life—sometimes celebratory, sometimes solemn—rising from so many houses on so many Sundays. We mark the 15th anniversary of Wynton's first in-house commission, a sacred convergence of gospel and jazz that gave rise to a new sound, at once modern and familiar. When **Marsalis, Riley, Gordon, "Warmdaddy,"** and **Williams** testify, the House of Swing will shake down to its drum skin floors.

Jazz at Lincoln Center proudly acknowledges:

Lead New York Sponsor

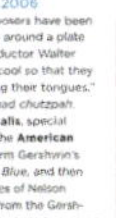

Cadillac · Bank of America · jazz · Bloomberg · Brooks Brothers · The Coca-Cola Company · Time Warner · Satellite Radio

3 MM SERIES

Music of the Masters
4 Concerts
Rose Theater, 8pm

FUSION REVOLUTION: JOE ZAWINUL
OCTOBER 27 & 28, 2006

Grooves ask for mercy, mercy, mercy. Schooled in the subtleties of swing by Dinah Washington, keyboardist **Joe Zawinul** brought the fundamentals of funk to Cannonball Adderley, the essentials of the electric to Miles Davis, and carried soul jazz into the electric age with his band Weather Report. Now the **Zawinul Syndicate** takes us on a hybrid adventure of sophisticated harmonies, world music rhythms, and deeply funky grooves. *Mercy.*

BEBOP LIVES!
JANUARY 26 & 27, 2007

Feet tangle and neurons dance. Fakers recoil, goatees sprout, and virtuosos take up their horns. Charlie Parker and Dizzy Gillespie set the bebop revolution in motion, their twisting, syncopated lines igniting the rhythms of jazz. Latter day fakers beware as the legendary **James Moody** and **Charles McPherson**, the alto sax voice of Charlie Parker in Clint Eastwood's *Bird*, raise battle axes and *swing*.

CECIL TAYLOR & JOHN ZORN
MARCH 9 & 10, 2007

Souls get freer. Embark on a sonic voyage as the peerless **Cecil Taylor** navigates us through dense forests of sound—percussive and poetic. He is, as Nat Hentoff proclaimed, "a genuine creator." The voyage banks toward the avant-garde as **John Zorn's Masada** with **Dave Douglas** explores sacred and secular Jewish music and the "anguish and ecstacy of klezmer." Musical wanderlust *will* be satisfied.

THE MANY MOODS OF MILES DAVIS
MAY 11 (Kisor/Blanchard) &
MAY 12 (Payton/Miller), 2007

Change gets urgent. "I have to change," Miles said, "It's like a curse." And so his trumpet voice—tender, yet with that edge—was bound up in five major movements in jazz. The LCJO's **Ryan Kisor** opens with bebop and the birth of the cool. GRAMMY®-winner **Terence Blanchard** interprets hard bop and the modal *Kind of Blue*. New Orleans adventurer **Nicholas Payton** conjures the great 60s quintet. And Miles vet **Marcus Miller** electrifies with fusion.

ROSE THEATER

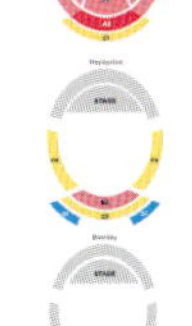

THE ALLEN ROOM

CUBANA BE CUBANA BOP
JANUARY 12 & 13, 2007

Cultures collide and rhythm explodes. "Dizzy no peaky pani," said conguero Chano Pozo. "I no peaky engly, but boff peak African." When they wed *Afro-Cuban* grooves to the frenetic language of bebop, their music—cubop—burst on the scene with the heat of an inferno. The **Afro-Latin Jazz Orchestra** with **Arturo O'Farrill** revisits Machito and Dizzy Gillespie masterpieces—"Panga," "Manteca," "Tin Tin Deo," "A Night in Tunisia"—along with new works that demonstrate that when jazz gets tropical, everybody moves.

TODO TANGO
APRIL 13 & 14, 2007

Swing gets sultry. Dancers step closer when jazz treads on South American shores. Tango crusader, composer, bassist, and arranger **Pablo Aslan** joins the **Afro-Latin Jazz Orchestra** with **Arturo O'Farrill** as they explore the dynamic intersection where jazz and tango converge. Join us as we celebrate the living tradition of tango with performances including the legendary music of Astor Piazzola, along with new works that pay tribute to the rich musical culture of Argentina.

DIANNE REEVES
APRIL 20 & 21, 2007

Satin shimmers divinely. At once shimmering and sultry, four-time GRAMMY® winner—most recently for the soundtrack to the Oscar®-nominated *Good Night, and Good Luck*—**Dianne Reeves** is in a class by herself—powerful when soft, intimate at a fever pitch, and agile at any tempo. With her trio, classics are rendered with timeless grace and style.

DARIN ATWATER GOSPEL
MAY 25 & 26, 2007

Spirits run deeper. The rhythms of the sanctified church will shatter The Allen Room windows. Masterful young pianist **Darin Atwater**, composer-in-residence of the Baltimore Symphony Orchestra and artistic director of the Soulful Symphony is joined by **Kim Burrell**, a talent "graced with the thunder of a gospel shouter and the sophistication of a classy jazz chanteuse" (*Billboard*). With guest singers—full-throated and gossamer—they raise voices in a divine confluence of jazz, classical, and gospel.

7 J4YP SERIES

Jazz for Young People®
3 Concerts
Rose Theater, 12pm & 2pm

What an amazing show!...every time I go to a Young People's concert they get better and better. The lessons were great, the pacing was perfect and the singing along was such a good way to get the kids involved...beautiful!!
—Maria Allerhouse, Vocal & General Music Hutchison Colonial Elementary Schools

WHAT IS AN ARRANGER?
DECEMBER 2, 2006

Brass and reeds reconcile. How do 15 strong-willed musicians come together in perfect harmony? How does the standard become fresh again? Thank the arranger, the unsung hero who musically choreographs every show, bringing order and imagination to the bandstand. The **LCJO** with **Wynton Marsalis** explore the techniques arrangers use to help everyone get along and sound good, good, good.

WHAT IS LATIN JAZZ?
MARCH 3, 2007

Rhythm becomes everything. What happens when you put a little Latin in your jazz? As Latin jazz pioneer Mario Bauza explained, "You walk with rhythm, you talk with rhythm, you eat your food with rhythm." **Arturo O'Farrill** and the **Afro-Latin Jazz Orchestra** show us how Dizzy, Machito, Bauzá, and others infected the lines of jazz with Latin grooves and started a rhythm epidemic. Watch out! Your knees will become bongos, your kid sister will turn into a timbale!

HOW DO WE CREATE JAZZ MOODS?
MAY 19, 2007

Moods mingle. Jazz is more than a mood indigo. In the beat of a heart it can take you from April in Paris to autumn in New York, from a stormy Monday to the sunny side of the street. Host **Wynton Marsalis** and friends show us how jazz's changing tones, colors, rhythms, and tempos take us from moanin' to groovin' high!

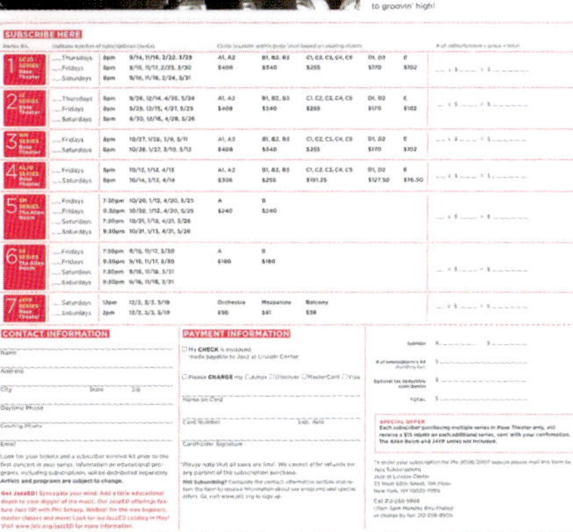

SUBSCRIBE HERE

CONTACT INFORMATION

PAYMENT INFORMATION

80. Verwendung von Helvetica

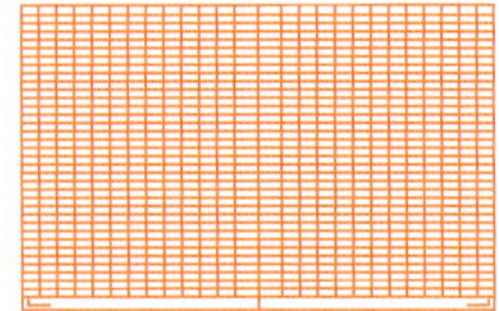

Im Jahr 2007 wurde die klassische, serifenlose Helvetica 50 Jahre alt und seitdem ist sie endgültig ein Star unter den Schriften. Warum wird die Helvetica aber immer gleich mit dem Schweizer Raster in Verbindung gebracht? Abgesehen von der Bezeichnung, die sich von *Helvetica*, dem lateinischen Namen für die Schweiz ableitet, besteht ein enger Zusammenhang zwischen den funktionalen Linien der Schrift, die ursprünglich Neue Haas Grotesk genannt wurde, und dem regelmäßigen Raster, der die Kunst der Moderne in den 1950er Jahren kennzeichnete.

Selbst eine feine, anmutige Helvetica kann ruhig und exquisit aussehen.

Verschiedene Verwendungen der Helvetica

KUNDE
- Designcards.nu by Veenman Drukkers
- Kunstvlaai/Katja van Stiphout

PHOTO
Beth Tondreau

Helvetica kann in verschiedenen Stärken und Größen benutzt werden. Medium und Fett signalisieren oft einen nüchternen Ansatz zur täglichen Information. Feinere Striche vermitteln Einfachheit, Luxus und eine Zen-artige Ruhe.

Mit ihrer Lesbarkeit ist Helvetica für die Typografie so wichtig wie Luft und Wasser.

81. Linienstärken

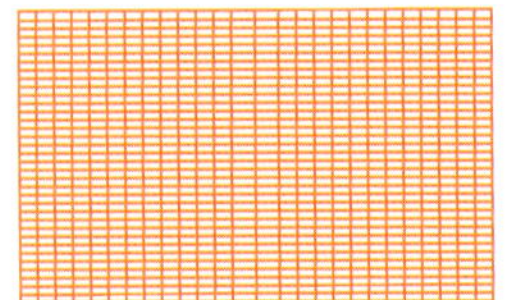

Linien sind vielseitig. Sie dienen u.a. als

- Navigationsleiste
- Behältnis für Überschriften
- Grundlinie für Bilder
- Trennmittel
- Leitlinie

Home	Recent	News	Clients	Awards	Contact

United States
National Parks Service
Washington, DC
Publications Program
1977

In order to achieve better identification and financial savings through standardization of every aspect of the publications program, we designed a modular system that determined everything from the paper size to graphics to cartography and illustration.

Corporate Identity 1 of 3 ▶

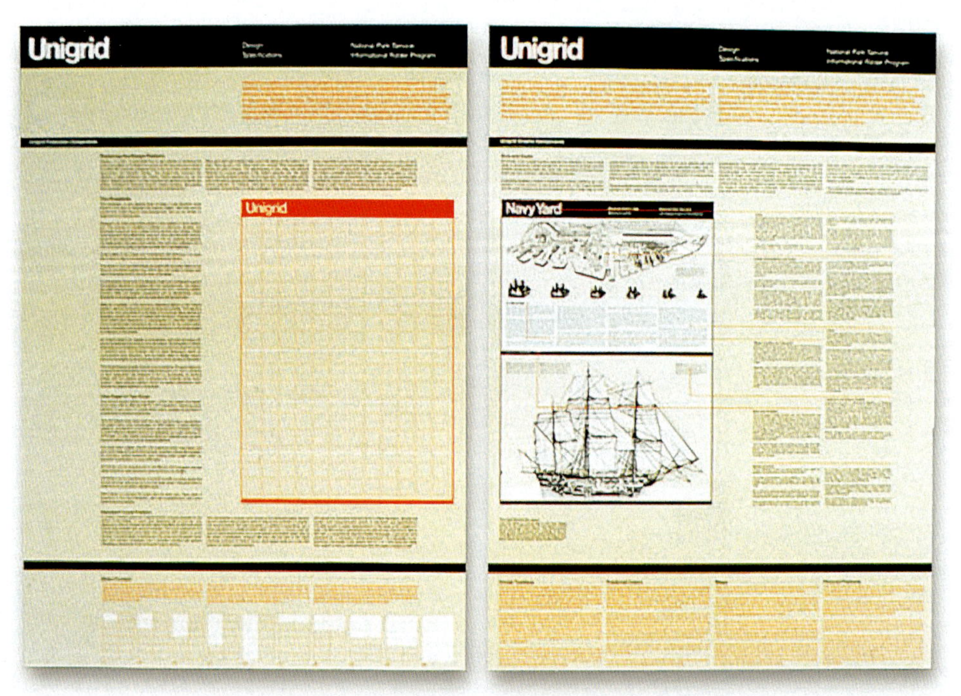

PROJEKT
www.vignelli.com

KUNDE
Vignelli Associates

DESIGN
Dani Piderman

DESIGN DIRECTOR
Massimo Vignelli

Der verstorbene Massimo Vignelli, ein Meister der Raster und Linien, zeigt sein Unigrid-System auf seiner Webseite. Diese Doppelseite in der aktuellen Ausgabe von *Layout Basics* ist eine Hommage an Vignelli und seine Mitstreiter.

Linke Seite oben: Die stets gleichmäßigen, klar strukturierten Arbeiten von Vignelli Associates lassen sich auch ins Internet übertragen.

Linien unterschiedlicher Stärke trennen und umschließen Informationen.

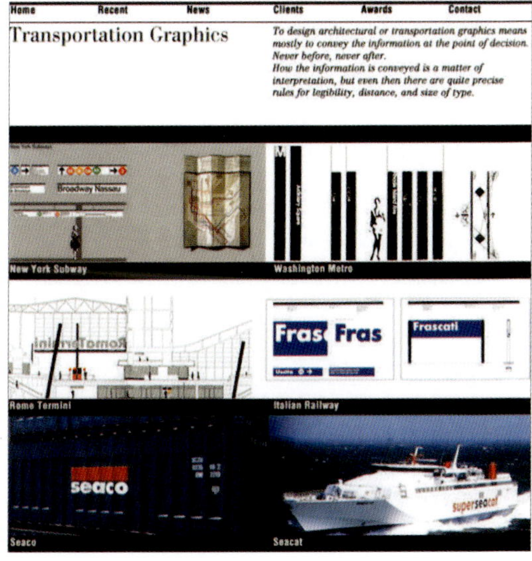

Linke Seite unten: Überschriften in der Franklin Gothic Bold stehen Elementen in Bodoni und Bodoni Italic gegenüber und werden von diesen ergänzt. Das Schweizer Design erhält dadurch eine italienische Note.

82. Vertikale und horizontale Ausrichtung

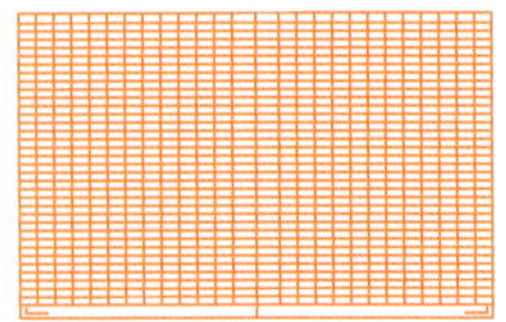

Die Unterteilung einer Seite in klar erkennbare Bereiche lässt Schreibwaren, Formulare und selbst Quittungen schön und zweckmäßig zugleich erscheinen. Horizontale und vertikale Raster existieren ohne Behinderung gut nebeneinander und Informationseinheiten werden zwar eventuell anders angeordnet als erwartet, enthalten aber alle notwendigen Elemente.

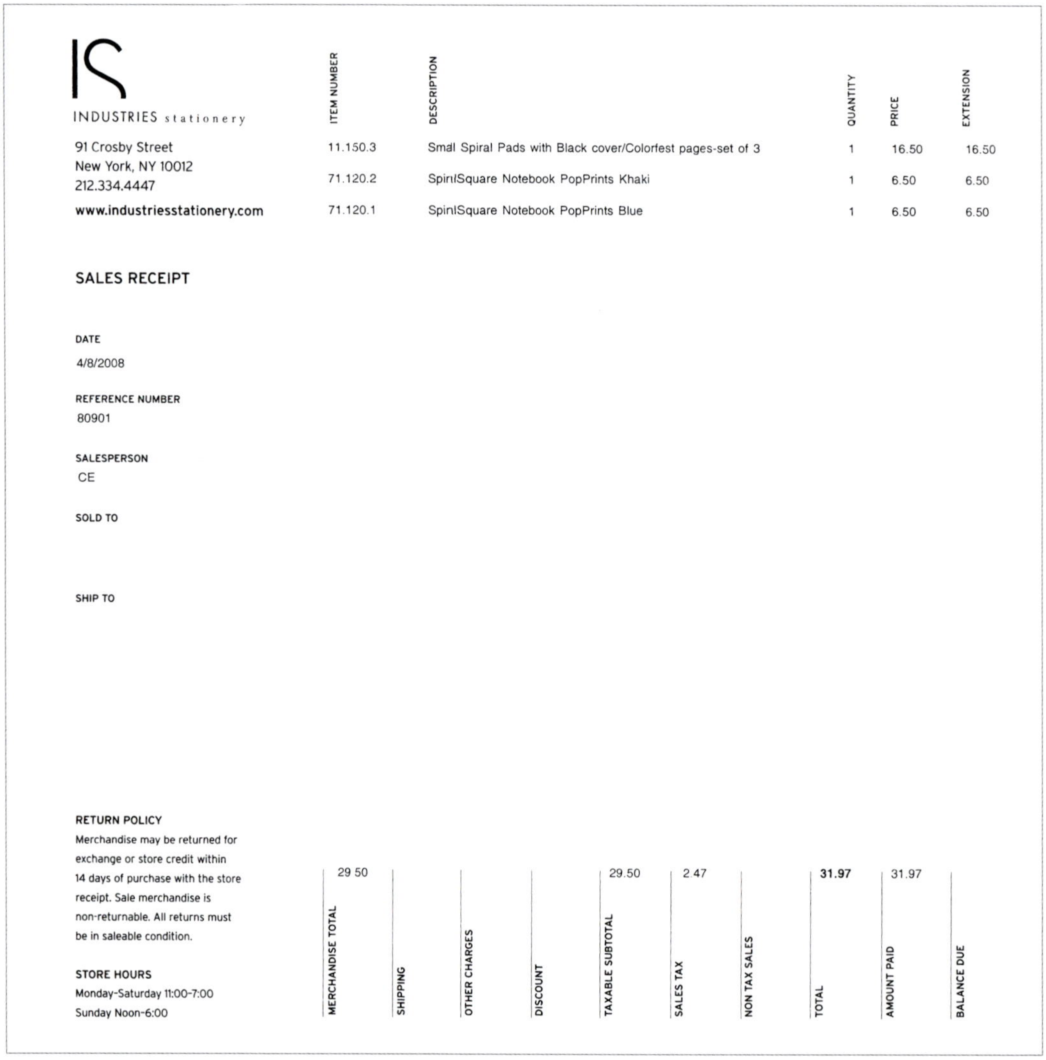

IS
INDUSTRIES stationery

91 Crosby Street
New York, NY 10012
212.334.4447
www.industriesstationery.com

ITEM NUMBER	DESCRIPTION	QUANTITY	PRICE	EXTENSION
11.150.3	Small Spiral Pads with Black cover/Colorfest pages-set of 3	1	16.50	16.50
71.120.2	SpinlSquare Notebook PopPrints Khaki	1	6.50	6.50
71.120.1	SpinlSquare Notebook PopPrints Blue	1	6.50	6.50

SALES RECEIPT

DATE
4/8/2008

REFERENCE NUMBER
80901

SALESPERSON
CE

SOLD TO

SHIP TO

RETURN POLICY
Merchandise may be returned for exchange or store credit within 14 days of purchase with the store receipt. Sale merchandise is non-returnable. All returns must be in saleable condition.

STORE HOURS
Monday-Saturday 11:00-7:00
Sunday Noon-6:00

MERCHANDISE TOTAL	SHIPPING	OTHER CHARGES	DISCOUNT	TAXABLE SUBTOTAL	SALES TAX	NON TAX SALES	TOTAL	AMOUNT PAID	BALANCE DUE
29 50				29.50	2 47		**31.97**	31.97	

PROJEKT
Quittung einer Schreibwarenhandlung

KUNDE
INDUSTRIES Stationery

DESIGN
Drew Souza

Das Design dieser Quittung schenkt Herbert Bayers Methode Beachtung, eine Seite als Fläche zu behandeln, die unterteilt werden muss.

LINKE SEITE UND UNTEN: Durch horizontale und vertikale Ausrichtung werden das Formular und die Quittung zu einem klar unterteilten Behältnis für zahlreiche Daten. Ohne die Verkaufsdaten ist die Quittung eine abstrakte Komposition. Mit den geschäftsbezogenen Daten ist sie ein funktionales System.

IS

INDUSTRIES stationery

91 Crosby Street
New York, NY 10012
212.334.4447

www.industriesstationery.com

SALES RECEIPT

DATE

REFERENCE NUMBER

SALESPERSON

SOLD TO

SHIP TO

ITEM NUMBER | DESCRIPTION | QUANTITY | PRICE | EXTENSION

SALES DRAFT

DATE

REFERENCE NUMBER

SALESPERSON

SOLD TO

DISCOUNT

MERCHANDISE TOTAL

SHIPPING

OTHER CHARGES

TAXABLE SUBTOTAL

SALES TAX

NON TAX SALES

TOTAL

AMOUNT PAID

BALANCE DUE

PAID BY

PAID BY

RETURN POLICY

Merchandise may be returned for exchange or store credit within 14 days of purchase with the store receipt. Sale merchandise is non-returnable. All returns must be in saleable condition.

STORE HOURS

Monday-Saturday 11:00-7:00

Sunday Noon-6:00

MERCHANDISE TOTAL | SHIPPING | OTHER CHARGES | DISCOUNT | TAXABLE SUBTOTAL | SALES TAX | NON TAX SALES | TOTAL | AMOUNT PAID | BALANCE DUE

83. Überraschung!

Die strenge Untermauerung des Magazins durch einen Raster sorgt für einen durchgehenden Plan, aber zu viel Struktur bringt schnell unangenehme Langeweile. Ein unerwarteter Einschub, die Unterbrechung des Rasters oder der Einsatz freier Räume sorgen für Aufmerksamkeit.

PROJEKT
No Man's Land

KUNDE
The Wing

DESIGN
Pentagram

CREATIVE DIRECTION
Emily Oberman

PARTNER
Emily Oberman

SENIOR DESIGNER
Christina Hogan

DESIGNER
Elizabeth Goodspeed

DESIGNER
Joey Petrillo

PROJECT MANAGER
Anna Meixler

Die Designer entwickelten einen starken Grundraster, den sie aber bewusst durch Poster, Sticker und typografische Bildspiele durchbrechen.

No Man's Land

Twelve miles from downtown and five miles from the sea, I-10 arrives to cross the I-405 at the most beautiful freeway interchange in L.A.—the Santa Monica-San Diego interchange. On a satellite map, these intertwined roads look something like a game of cat's cradle between fingers. The scale, from head-on, is vast and monumental, as striking as the Gateway Arch in repose. In a more pastoral setting, it might have been called land art. In a city defined by its lack of a center, it seems to both enable and pay homage to the sprawl.

72

Feature

Three stacked ramps spring forth from small hills, scattering cars in all four directions. This is the junction between normal L.A. and the parts of L.A. that we always see in movies—the beaches and hills and the very big houses. No doubt my fondness for its ramps has something to do with how they often lead to fun.

This past spring, driving on the 10, preparing to change to the 405, I noticed a light green highway sign with the junction's formal municipal name: the Marilyn Jorgenson Reece Memorial Interchange. Sitting in traffic, I typed the honorific into Google, hoping to learn what a woman must do in order to win a junction for herself. The search results were few—an obituary, some blog posts—and all seemed to draw from the same pool of facts: the first licensed female engineer in California, designed the Santa Monica-San Diego interchange, died at 77 in 2004. Reece was a readymade Woman of the Month, a frequent subject of those bland but nice blurbs that often say more about corporate obligation than the actual women they seek to reward.

Her image results were far more compelling, or at least lent some inadvertent texture to her life. Reece, it seems, was a snappy dresser—a fan of polka-dot blouses and smocks before the era of easy office separates. In one photo of the interchange under construction, she wears a linen skirt suit with a Peter Pan collar. Cranes and pylons rise from the dirt, casting her ballet flats as comically flimsy. It's hard to overstate the strangeness of the photo. The scene looks cribbed from a B-movie—a woman swept up from the pew of her church and set down unharmed at the helm of a job site. Archives are full of such photos of men, surveying big holes in their hard hats and suits. Reece looks more focused than photo-op ready. A gentle breeze ruffles her jacket but scarcely moves her hair-sprayed hair.

Back at my house, I tried to learn more. As the daughter of a working mother and the Spice Girls, I still feel a surge of neoliberal pep at the thought of a female-designed highway junction (or, if you prefer, a *she*-way int-*her*-change). Reece, like most private citizens of her time, barely makes a blip in the searchable

"The worst-designed freeways make you wrestle for this thrill; the best, like Reece's, induce it without effort."

news, despite the relative high profile of her work. The day-to-day details of her life and career are less readily available than the same information for any given ex-boyfriend's new girlfriend. I tracked down a colleague from her time at CalTrans who never responded to my interview request. I tweeted at the former Secretary of State who sponsored the light green freeway sign. Deep in the text of a press release, I learned of a $1,000 prize in her name, awarded to a Pomona middle school student for her "Empirical Analysis of Wooden Structures."

It's nice how women cross paths across time but not especially interesting. (I suppose this was the inadvertent message of *The Hours*.) Getting more serious, I filed a public records request. Dredging up some nouns that might pertain to engineering, I wrote: "Any correspondence, designs, photos, or proposals would be useful." The CalTrans librarian emailed me back, asking me to clarify the nature of my interest. The records form acknowledged litigation research but didn't have a box for vague curiosity relating to the life of a former employee. The librarian said she'd see what she could do, but she sounded overworked and confused by my request. I didn't expect to hear from her again.

73

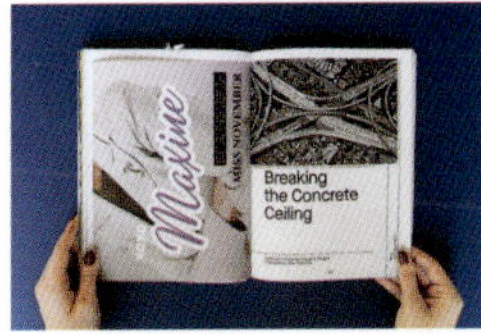

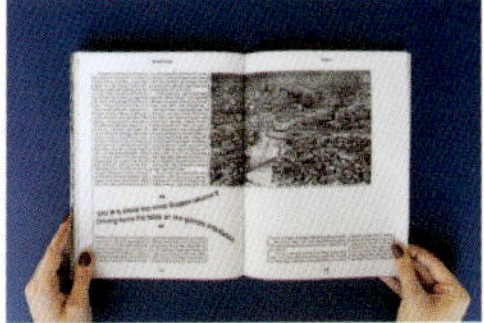

Um auf dem immer stärkeren digitalen Markt um Leser/Nutzer/Betrachter zu buhlen, setzen die Publikationen auf immer neue Taktiken. Der Aufbau einer Community über ein Printmagazin – hier eine Community von Frauen mit einer gemeinsamen Philosophie – ist im digitalen Zeitalter riskant. Da helfen taktile Belohnungen wie Sticker oder clevere Faltblätter. Dieses Poster zeigt eine etablierte Politikerin und spielt mit dem Klischee des Ausfaltblatts, indem es aus der „Miss" eine starke Frau macht.

Seite gegenüber und oben:
Der Bericht mit dem Titel „Die erste weibliche Überführungsentwicklerin Kaliforniens schuf Poesie aus Beton" spielt ein bisschen mit der Poesie, indem es aus dem Raster fällt und die Form der Überführung in der Typografie aufnimmt. (Ganz nebenbei: Dieser Artikel ist faszinierend und ein echter Spaß!)

84. Größen variieren

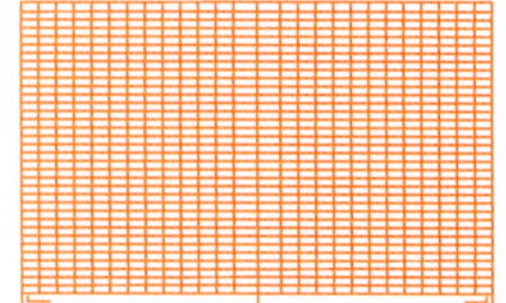

Sobald der allgemeine Raster festliegt, hat der Grafikdesigner Gelegenheit, mit Maßstab, Fläche, Größe und Typografie zu spielen. Ob Bild- und Schriftgrößen dynamisch oder einförmig gestaltet werden, hängt davon ab, welche Absicht der Text verfolgt, wie wichtig er ist und wie groß die Fläche ist, die für das Material benötigt wird.

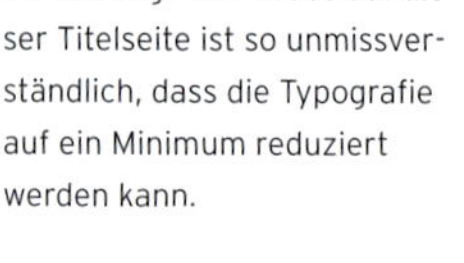

Die Aussage des Bildes auf dieser Titelseite ist so unmissverständlich, dass die Typografie auf ein Minimum reduziert werden kann.

PROJEKT
What Is Green?

KUNDE
Design within Reach

DESIGN
Design within Reach Design

CREATIVE DIRECTOR
Jennifer Morla

ART DIRECTOR
Michael Sainato

DESIGNER
Jennifer Morla, Tim Yuan

TEXTER
Gwendolyn Horton

„Grün sein" und Nachhaltigkeit sind heiße (global erwärmte) Themen, mit denen sich viele Unternehmen befassen, darunter auch die Agentur für Grafikdesign DWR, die seit Jahren ökologisch bewusst auftritt. Auf den ersten dreizehn Seiten dieses Projekts entsteht ein Gespür für eine Geschichte, in der es in einer Vielfalt von Layouts um ein einziges Problem geht.

Auf der ersten Seite wird mithilfe der Typografie eine – übermäßig lange – Erklärung abgegeben, die die gesamte Rasterfläche ausfüllt.

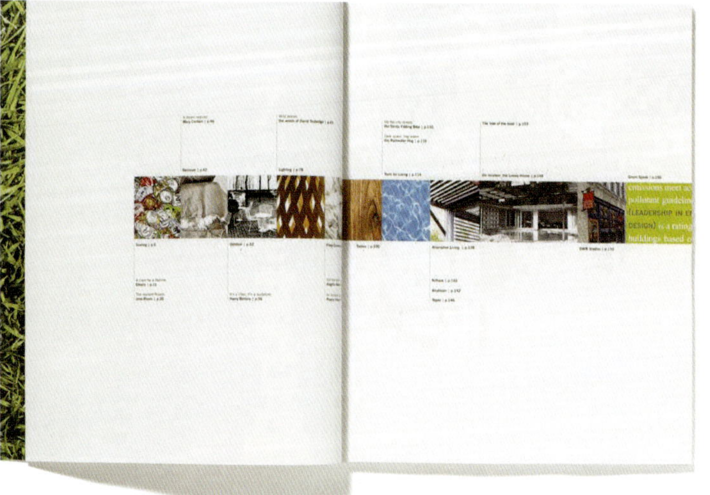

Mit einer drastischen Veränderung der Größenordnung und einem horizontalen Aufbau erscheint dieses Inhaltsverzeichnis angenehm fließend. Leitpunkte – etwa Linien – lenken das Auge auf den Inhalt. Kleine Abbildungen sind ein kurzer optischer Hinweis auf die Themen.

The hand-brushing department at Emeco, U.S.A.

At Emeco, all aluminum waste is recycled, even the aluminum dust that's filtered out of the air.

The upside of up-cycling aluminum: chairs for a lifetime or two.

When Emeco started making its aluminum chairs in 1944, you can be darn sure there wasn't a marketing brief that said, "Make it attractive to the eco-conscious community." Emeco had other things on its mind, namely how to make a chair withstand a torpedo blast. The irony is that Emeco chairs have become an outstanding example of what's commonly referred to as "green." To create the 1006 Navy Chair (1944), Emeco invented a 77-step process to satisfy the military's need for lightweight, corrosion-resistant chairs for destroyers and submarines. In the process, the company invented a method to make aluminum three times stronger than steel, and a chair so durable that it has an estimated lifespan of 150 years. Legend has it that Wilton Dinges, who founded Emeco in 1944, actually tossed a 1006 Navy Chair out the window of a six-story building. The people on the sidewalk below were a bit surprised, but the chair was fine, with the exception of a few scratches. Today, everything Emeco makes is still manufactured by hand using the same 77-step patented process. Emeco chairs and tables all begin with 80% recycled aluminum, which requires only 5% of the energy needed to produce virgin aluminum, and they're all made in Pennsylvania, U.S.A. Emeco's all-aluminum chairs and stools are built to last, and generations from now, when your great-great-grandchildren finally manage to wear out a chair that's tested to withstand 1,700 pounds of weight (big kids), the aluminum can be 100% recycled and made into something else. In recent years, Emeco has partnered with Philippe Starck, Norman Foster and others to create classic designs for a new century, and these collections are made in the same facility, using the same processes and by the same people who make everything else at Emeco. Perhaps Philippe Starck said it best when he explained that "working with Emeco has allowed me to use a recycled material and transform it into something that never needs to be discarded – a tireless and unbreakable chair to enjoy for a lifetime. It is a chair you never own, you just use it for a while until it is the next person's turn." On the next page you'll find Emeco chairs and stools, all of which contribute to LEED™ credit #4.2 Recycled Content (and credit #5.1 if shipped within 500 miles of Hanover, Pennsylvania). For the entire **Emeco Collection**, visit dwr.com.

DESIGN WITHIN REACH: APRIL 2008 | 11

Diese Layouts zeigen die Veränderungen in den Textgrößen. Das sehr breite Textmaß auf einer Doppelseite ist für den Satz im Allgemeinen nicht vorteilhaft. In diesem Fall jedoch werden die gewöhnlichen Regeln des Grafikdesigns von Stil und Mitteilung übertrumpft. Möchte man etwas über die Stühle aus recyceltem Aluminium lesen, kann man das, denn die Beschreibung der Stühle ist sehr prägnant.

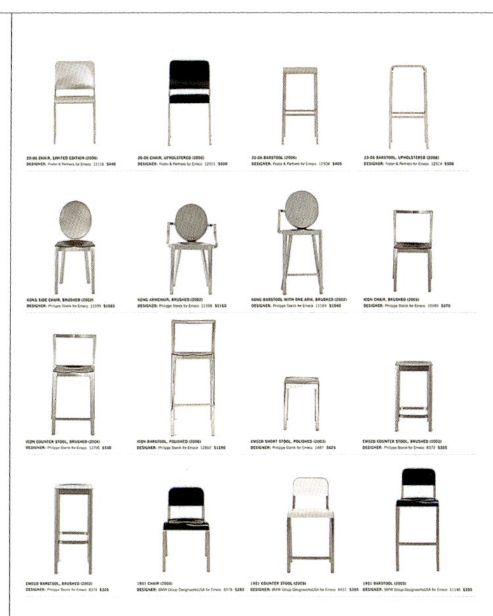

85. Was kann man weglassen?

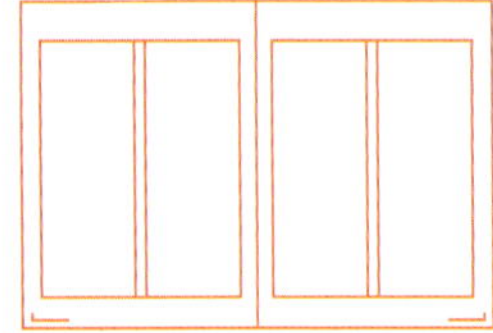

Ein tolles Foto sollte man nicht ruinieren! Manchmal ist es am besten, ein Foto so groß wie möglich zu machen, es nur sehr wenig oder überhaupt nicht zu beschneiden und über dem Bild weder Text noch Grafiken zu platzieren. Kurz: Übertragen Sie es auf Ihr Raster, aber lassen Sie es ansonsten in Ruhe.

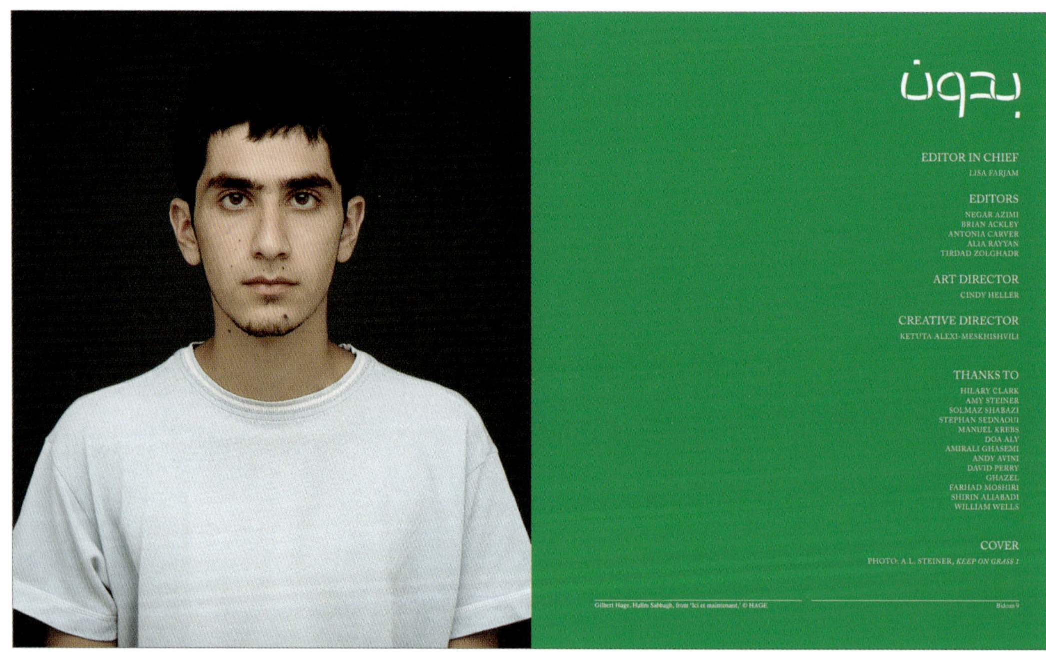

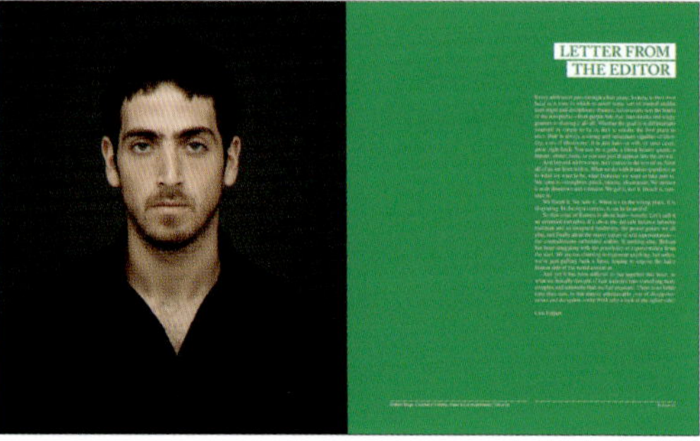

HIER UND RECHTE SEITE: An diesen Fotos muss nichts geändert werden. Auch ohne grafische Mittel sprechen sie Bände.

PROJEKT
Magazine

KUNDE
Bidoun

CREATIVE DIRECTOR
Ketuta-Alexi Meskhishvili

DESIGNER
Cindy Heller

FOTOS
Gilbert Hage (Porträts) und
Celia Peterson (Arbeiter)

82 Tourism

Cautious Radicals

Art and the invisible majority

By Antonia Carver

At the 2005 Sharjah Biennial, artist Peter Stoffel attempted to get himself banned. Taking inspiration from the notices placed by employers in local newspapers, featuring the names, nationalities, passport numbers and mug shots of ex-employees, Stoffel requested that the biennial's organizing body fire him and announce his occupational demise in the same way. Other potential employers—presumably those organizing another biennial in the UAE—would be hiring him "at their own risk and responsibility." At the same time, the biennial would write Stoffel a recommendation letter "acknowledging his reliable services as an artist," which would be freely available to visitors to the biennial.

The artist's conceit turned out to be more potent than the proposed work itself. In keeping with the generally taboo nature of discussion surrounding the rights of the Gulf's underclass of foreign maids and laborers, the biennial organizers declined to go along with Stoffel's ruse. During the exhibition, he showed two panels of text—one a narrative explaining his concept and the outcome, the other a page from a local newspaper with advertisements placed by "sponsors" of Sri Lankans and Pakistanis who had "absconded from duty" and were therefore now outside the employer's responsibility.

For Gulf-based biennial visitors, Stoffel's project was audacious in its attempt to query the region's strict racial and financial hierarchy of workers' rights. (Since the biennial, new legislation has begun to address both the rights of the employee in the transferral of sponsorship and the prerogative of sponsors to impose the customary six-month ban—from the country, and/or from working for a competitor company—on some employees.)

As he describes it, Stoffel attempted to establish a connection between the smallest minority in the UAE, that of the immigrant artist, and the largest, the immigrant laborer. (About two-thirds of the UAE's work force comes from abroad, and about a quarter of all expats work as unskilled laborers for construction companies.) Stoffel concluded that the "two parallel lines of the biennial artist and the Pakistani worker never cross, and that is the paradox of the paradox: that even at an imaginary point, within an artwork, it's impossible to establish a connection."

Despite being the largest segment within the UAE population, the foreign working class remains by and large a faceless majority, known only to the wealthy minority through increasingly ballsy local media stories. Every week, the usually self-censoring UAE newspapers detail gory tales of trafficking, suicide, and rape; of false promises made by dubious foreign employment agencies and mounting debts; of dehydration while working in extreme summertime heat and humidity; of industrial accidents and loan sharks; of depressed, desolate labor camps. The Indian Embassy's official list of its functions includes such grisly tasks as "processing applications received for providing free air tickets by Air India/Indian Airlines for transportation of dead bodies of destitute/stranded/absconded Indian nationals."

In many ways, the situation faced by the Gulf's legions of indentured laborers is mirrored worldwide, from Chinese cocklepickers in the UK to Mexican meatpackers in US abattoirs. But the particular state of affairs in Dubai, with its rapid growth and surface profligacy, takes a microscope to what's vaguely termed globalization.

Tourism 83

Photos of laborers in Dubai by Celia Peterson, 2005, courtesy of Celia Peterson and arabianEye

84 Tourism

Tourism 85

86. Information in der Box

Eine Box mit einer Nebengeschichte, die daneben im Haupttext ausführlicher behandelt wird, ist ein allgemein gebräuchliches Mittel, um Informationen hervorzuheben. Diese Informationen können sich auf den Haupttext beziehen, müssen dies aber nicht. Boxen lassen sich auch in den Raster integrieren. Im Gegensatz zu Unterbrechungen liefern sie zusätzliche Informationen.

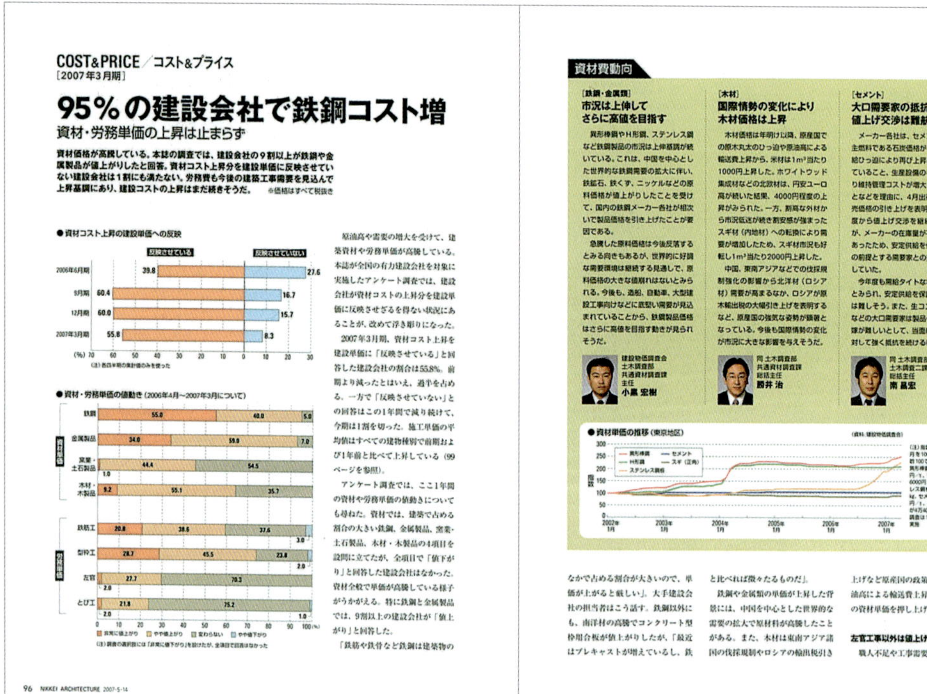

COST&PRICE／コスト&プライス
［2007年3月期］

95％の建設会社で鉄鋼コスト増

資材・労務単価の上昇は止まらず

資材価格が高騰している。本誌の調査では、建設会社の9割以上が鉄鋼や金属製品が値上がりしたと回答。資材コスト上昇分を建設単価に反映させていない建設会社は1割にも満たない。労務費も今後の建築工事需要を見込んで上昇基調にあり、建設コストの上昇はまだ続きそうだ。　※価格はすべて税抜き

●資材コスト上昇の建設単価への反映

資材費動向

［鉄鋼・金属類］
市況は上伸して
さらに高値を目指す

［木材］
国際情勢の変化により
木材価格は上昇

［セメント］
大口需要家の抵抗で
値上げ交渉は難航

Die Boxen in diesem klar strukturierten Raster können sämtliche Spalten oder auch nur zwei oder gar nur eine Spalte einnehmen.

PROJEKT
Nikkei Architecture

KUNDE
Nikkei Architecture, Magazin

DESIGN
ar

In dieser Fachzeitschrift für Architektur enthalten die Boxen und Tabellen technische Informationen.

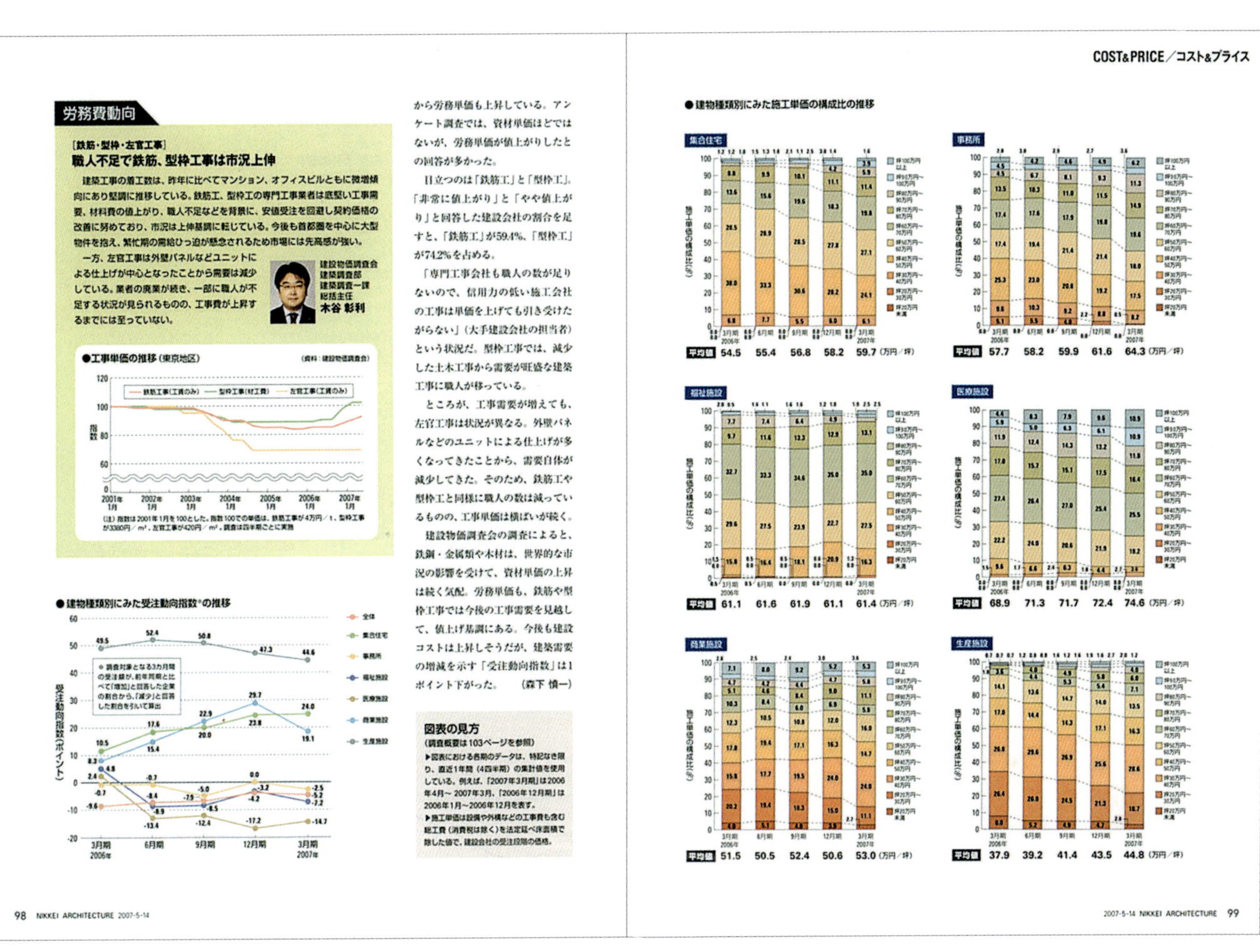

Boxen erscheinen oft wie getrennte Strukturen, aber durch den Einsatz gemeinsamer Farben, Schriftarten oder Linien sind sie grafisch immer mit der Hauptgeschichte verbunden.

87. Auf alte Meister hören

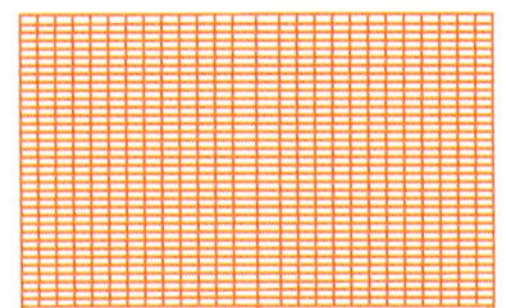

Eine genauere Betrachtung von Arbeiten der Pioniere des Grafikdesigns kann zu Layouts führen, die den Werken der alten Meister zwar gleichen, aber dennoch neue Interpretationen des Rasters darstellen.

Layouts, die als Hommage gestaltet sind und in denen die Schweizer Meister widerhallen, sind dank der tiefen und grundlegenden Kenntnisse der allgemeinen Regeln nicht zwangsläufig eine sklavische Kopie von Elementen, sondern können sehr erfrischend sein.

chronologie

1914 Naissance de Josef Müller le 9 mai à Rapperswil (Suisse).

1930 Apprentissage à Zurich.

1932–1935 Suit les cours de Ernst Keller et Alfred Willimann à l'école des arts et métiers de Zurich.

1934 Débute sa carrière comme illustrateur, scénographe d'exposition et décorateur de théâtre.

1939-1945 Lieutenant dans l'armée suisse.

1943 Épouse Verena Brockmann (violoniste) et adopte le nom de Josef Müller-Brockmann. Courante en suisse, cette pratique le distingue des autres Müller.

1944 Naissance de leur fils Andreas.

1950 Premières commandes du Zürich Tonhalle.

1951 Membre de l'AGI.

1952-1958 Affiches de sécurité pour l'Automobile-Club de Suisse.

1956 Voyages en Amérique : conférences à Aspen et New York, voyage au Mexique avec Armin Hoffman. Songe à s'installer aux États-Unis.

1957 Nommé professeur à l'école des arts et métiers de Zurich.

1958

1960 Quitte l'école de Zurich. Conférence à Tokyo.

1961 Second voyage au Japon.

1962 Contrats de design et de conseil en RFA. Enseigne à Ulm.

1964 Mort de Verena dans un accident de voiture.

1965 Fonde avec Eugen et Kurt Federer la Galerie 58 (puis Galerie Seestrasse) dans la maison familiale. Le lieu deviendra une référence de l'art concret. Conférences et expositions de Johannes Itten, Max Bill, Karl Gerstner… Fermeture en 1990.

1967 Fondation de Müller-Brockmann & Co avec Max Baltis, Ruedi Rüegg et Peter Andermatt et une quinzaine d'employés.

1971 Mariage avec Shizuko Yoshikawa.

1976 Brouille avec Baltis et Rüegg, qui quittent l'agence. Müller-Brockmann dirige seul l'agence jusqu'à sa fin en 1984.

1975–1980 Système de signalétique pour les chemins de fer suisses.

1981

1989

1994

1996 Meurt le 30 août.

Neue Grafik
New Graphic Design
Graphisme actuel

Revue *Neue*
(Graphisme a
avec Carlos V
Richard Paul
et Hans Neub

1958

J. Müller-Brockmann
Gestaltungsprobleme des Grafikers
The Graphic Artist and his Design Problems
Les problèmes d'un artiste graphique

1961

Les Problème
artiste graphi

Plakat
Affiche
Poster

Histoire de la
communicati
visuelle.

Histoire de l'a
cosigné avec
Yoshikawa.

1971

Grid systems
Raster systeme

Grid System
Graphic Des

1981

1989 *Fotoplakate – Von den Anfängen bis zur Gegenwart.*

1994 *Mein Leben: Spielerischer Ernst und ernsthaftes Spiel*, autobiogra

82:

PROJEKT
étapes: magazine

KUNDE
Pyramyd/*étapes:* magazine

DESIGN
Anna Tunick

Diese Doppelseite aus einem Zeitschriftenbeitrag über den Grafikdesigner Josef Müller-Brockmann ist eine wahre Fundgrube für Raster – von der Chronologie seines Lebens bis zu Bucheinbänden und originellen Bildern.

Die genaue Beobachtung der Arbeiten von Müller-Brockmann führt zu einem großartigen grafischen Design, das zugleich eine Hommage an den alten Meister als auch eine unabhängige Arbeit ist.

« Plus la composition des éléments visuels est stricte et rigoureuse, sur la surface dont on dispose, plus l'idée du thème peut se manifester avec efficacité. Plus les éléments visuels sont anonymes et objectifs, mieux ils affirment leur authenticité et ont dès lors pour fonction de servir uniquement la réalisation graphique. Cette tendance est conforme à la méthode géométrique. Texte, photo, désignation des objets, sigles, emblèmes et couleurs en sont les instruments accessoires qui se subordonnent d'eux-mêmes au système des éléments, remplissent, dans la surface, elle-même créatrice d'espace, d'image et d'efficacité, leur mission informative. On entend souvent dire, mais c'est là une opinion erronée, que cette méthode empêche l'individualité et la personnalité du créateur de s'exprimer. »

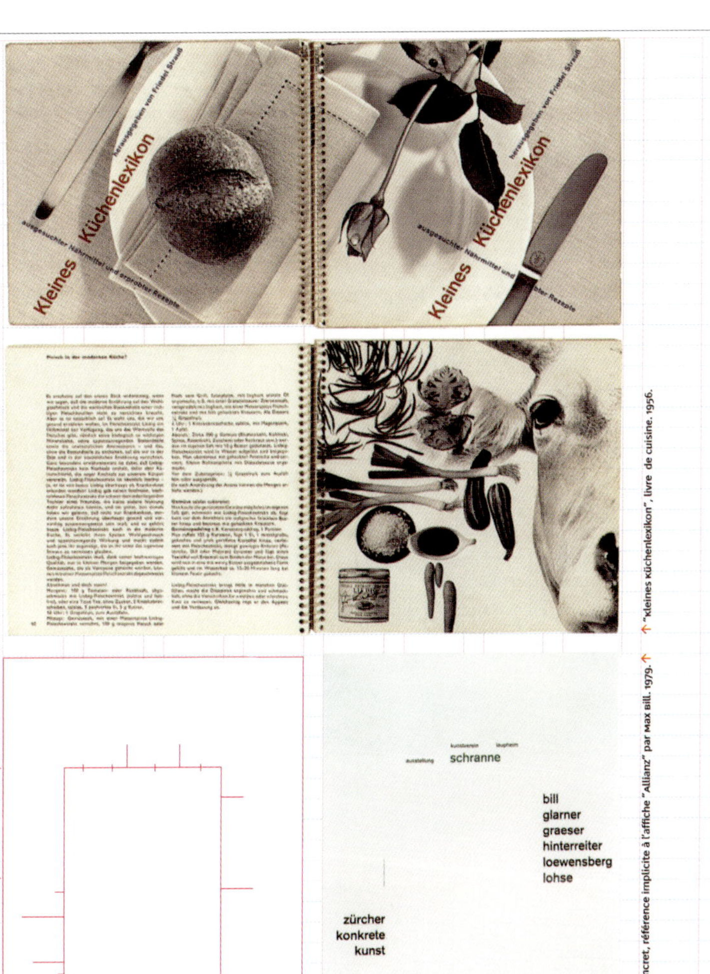

"zürcher konkrete kunst", affiche pour une exposition d'art concret, référence implicite à l'affiche "Allianz" par MAX BILL. 1979. ↑

↑ "kleines küchenlexikon", livre de cuisine. 1956.

comme des recettes appliquées par défaut. Phénomène encore appuyé par la structure des logiciels de PAO, qui recourent au gabarit comme point de départ à l'édition de tout document. L'efficacité radicale de l'abstraction sera quant à elle escamotée au profit d'effets plus spectaculaires et moins préoccupés.

ceci dit, au boulot

Depuis ses débuts de scénographe, Müller-Brockmann a réalisé un grand nombre de travaux, seul ou à la tête de son agence (1965-1984): scénographies d'expositions didactiques ou commerciales, identité, communication et édition (brochures, publicités et stands) d'entreprises pour des fabricants de carton (L + C: lithographie et cartonnage, 1954 et 1955), de machines-outils (Elmag, 1954), de machines à écrire (Addo AG, 1960) pour des fournisseurs de savon (CWS, 1958) de produits alimentaires (Nestlé, de 1956 à 1960) ou pour la chaîne de magasins néerlandais Bijenkorf (1960). En 1962, il décroche d'importants contrats auprès d'entreprises allemandes: Max Weishaupt (systèmes de chauffage) et Rosenthal

:83

WIEDER AUFGENOMMENER RASTER

88. Nähe durch Beschnitt

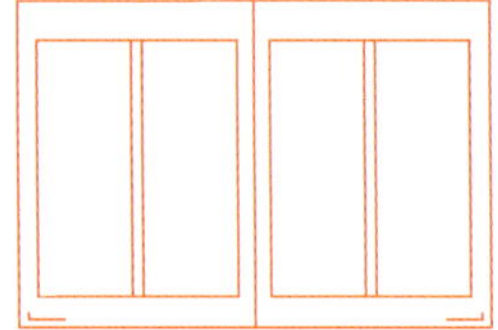

Raster können ein Projekt erdrücken und zu einer zerstörerischen Kraft werden oder sie können einen subtilen Unterbau bilden, der – in den Worten eines Grafikdesigners – „anmutig und logisch, aber nie aufdringlich ist".

Siehe auch Seite
18

PROJEKT
Chuck Close | Work

KUNDE
Prestel Publishing

DESIGN
Mark Melnick

Ein zurückhaltendes grafisches Design präsentiert gekonnt Gemälde großer Persönlichkeiten.

Die Stärke der Titelseite liegt in ihrer Einfachheit. Im Mittelpunkt stehen der Künstler und sein Werk. Das Layout für einen Bucheinband wird später gefaltet und dient dann als Schutzumschlag für das gebundene Buch.

Die Bilder auf den Vorsatzblättern beginnen mit dem Künstler bei der Arbeit und enden mit dem Künstler im Profil.

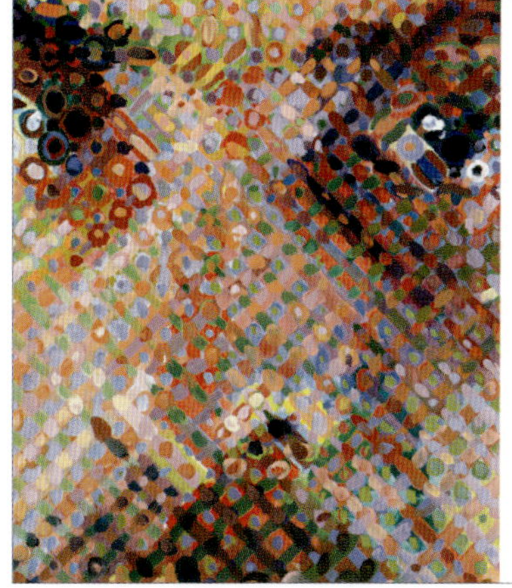

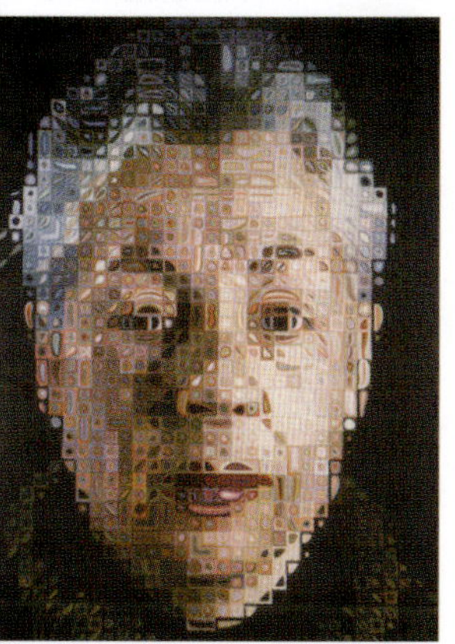

Oben links: Die Titeldoppelseite zeigt eine Vergrößerung des Auges. Der Titel selbst ist wiederum einfach.

Oben rechts: Bild und Titel verweisen deutlich auf den Raster.

Die beiden Bilder in der Mitte: Auch hier ist der Raster vorherrschend.

89. Grenzen verschieben

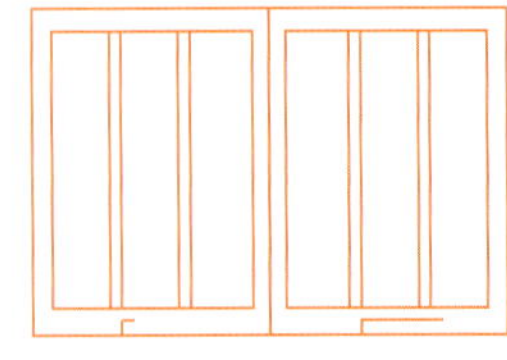

Ergänzende Informationen können so schön sein wie der Haupttext – und die Grenzen zum primären Material verschieben. Ein Nachspann oder Kataloge wie Anhänge, Zeittafeln, Anmerkungen, Bibliografien und Inhaltsverzeichnisse sind oft komplex. Durch die Verteilung solcher Angaben über das ganze Projekt erscheint das Design vollendet. Auch weniger beachtete Seiten werden dadurch interessant.

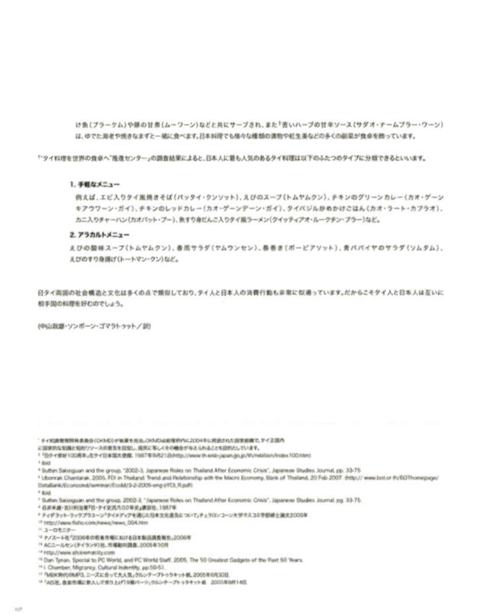

PROJEKT
Ausstellungskatalog *Show Me Thai*

KUNDE
Institut für zeitgenössische Kunst und Kultur, Kultusministerium, Thailand

DESIGN
Practical Studio/Thailand

DESIGN DIRECTOR
Santi Lawrachawee

DESIGNER
Ekaluck Peanpanawate
Montchai Suntives

Dieser Ausstellungskatalog enthält zahlreiche nützliche Raster. Besonders interessant ist die Teilnehmerliste gestaltet.

Appendix C ::

Exhibitions

Linke Seite oben: Einer Seite mit einem einzelnen Foto steht eine Seite mit starker Rasterung gegenüber.

Linke Seite unten: Auf der linken Seite haben die Textspalte und je zwei Bilder dieselbe Breite. Obwohl sich diese große Breite im Allgemeinen nicht empfiehlt, ist das Layout hier gelungen.

Ein dreispaltiger Raster und eine Tabelle erzeugen in kunstvoller Weise ein Gefühl von Ordnung.

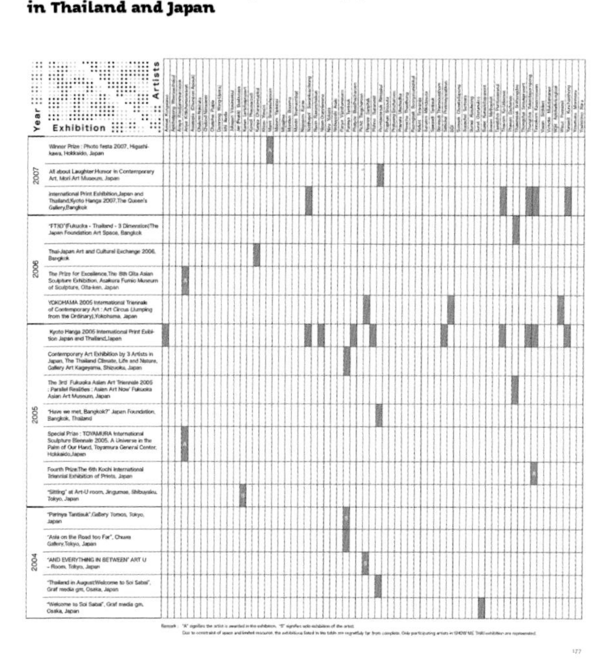
Appendix B ::

Exhibitions participated by Thai and Japanese Artists in Thailand and Japan

Die Tabelle auf dieser Doppelseite ist übersichtlich, ansprechend und interessant gestaltet. Ein Ornament gibt ihr Textur.

90. Vertrauen in das Modul

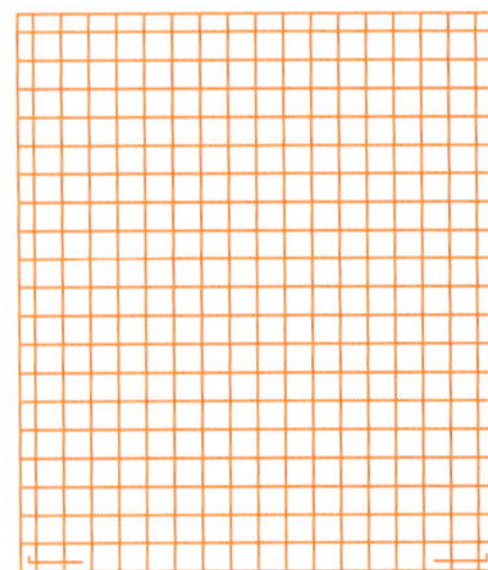

Siehe auch Seite
29

Mit Farbe lässt sich selbst das fast Unmögliche gestalten. Farbe kann Formen und Flächen schaffen. Eine schwache Farbe erzeugt einen Negativraum, eine starke hingegen kann zu einem Teil des Vordergrunds werden. Überschneidungen geben dem Ganzen noch eine andere Dimension. Probieren Sie aus, was sich mit Farbschichten und -formen alles erreichen lässt. Das Ergebnis kann den Goldenen Schnitt verkörpern.

Der Raster in seiner elementaren Form, eines Puzzle, erhält durch Marian Bantjes grafisches Können Tiefe. Die Künstlerin ist gerne geneigt, „die Regeln, die ich kenne, zu durchbrechen und etwas zu versuchen und zu schaffen, das mir Unbehagen bereitet, aber auf angenehme Weise".

PROJEKT
Deckel für das Puzzle *The Guardian's G2*

KUNDE
The Guardian Media Group

DESIGN
Marian Bantjes

Für diesen Deckel der Puzzleversion von *G2* wurden Linien und Quadrate in Schichten angeordnet.

PUZZLE
SPECIAL

91. Mit mehreren Dimensionen arbeiten

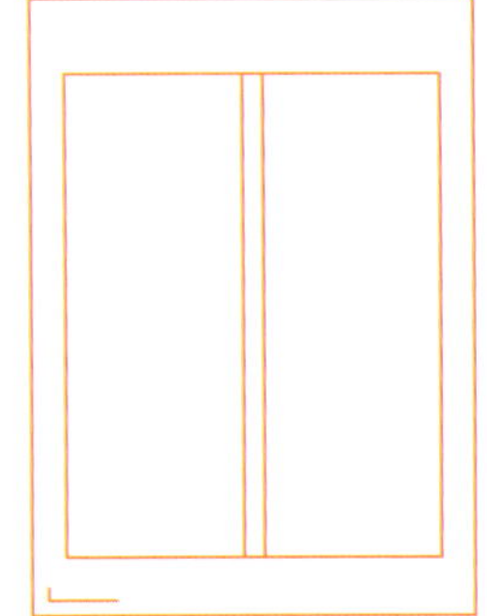

Obwohl die meisten Layouts mit Rastern flach sind, müssen sie die Dimensionen der dargestellten Arbeit einfangen, ganz gleich, ob es sich um eine Druckseite oder einen PC-Bildschirm handelt. Ein Prospekt kann in einem anderen Format produziert werden als ein Buch, eine Broschüre oder eine Einzelseite. Dreidimensional konzipierte, aber flach gestaltete Prospekte erhalten ein zusätzliches Tiefenmoment, wenn sie wie ein Leporello gefalzt werden.

PROJEKT
Ausstellungskatalog für *Stuck*, eine Kunstausstellung über Collagen.

KUNDE
Molloy College

GALLERY DIRECTOR
Dr. Yolande Trincere

CURATOR
Suzanne Dell'Orto

DESIGNER
Suzanne Dell'Orto

Dieser als Ausfalter konzipierte Prospekt einer Collagenausstellung lässt die verspielten Kunstwerke in der Galerie erahnen..

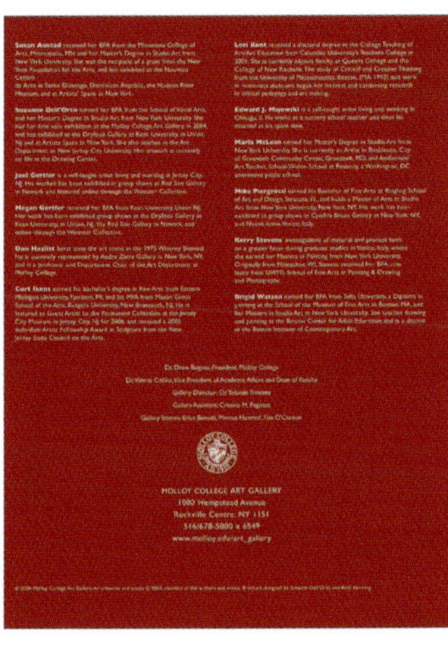

stuck

THE INFLUENCE OF COLLAGE ON 21ST CENTURY ARTISTS

Persistent Provocation: The Enduring Discourse of Collage

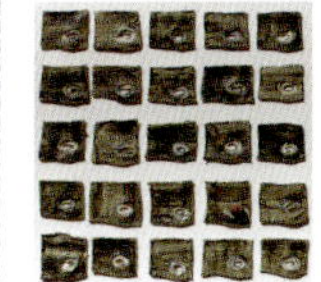

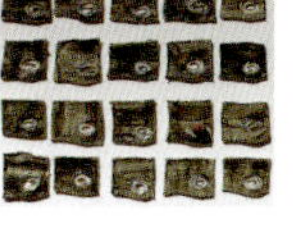

Ein traditioneller Raster bildet die Grundlage für die etwas eigenartigen Collagen einer Ausstellung. Die nüchtere Schrift und die gut geplanten Flächen rahmen die lebendigen Kunstwerke. Der beidseitig bedruckte und wie ein Leporello gefaltete Prospekt wirkt dreidimensional. Die obere Abbildung zeigt die Außenseite, die untere die Innenseite.

Linke Seite: Eine der vier Tafeln auf der Prospektinnenseite zeigt ein zerlegtes Buch über Kunstgeschichte. Das Bild ist ordentlich auf einer Spalte platziert. Die Kombination von stattlicher Gill Sans und verspielter P.T. Bantum erinnert an eine Collage, in der fremde Elemente ebenfalls nebeneinander platziert werden.

92. Global denken

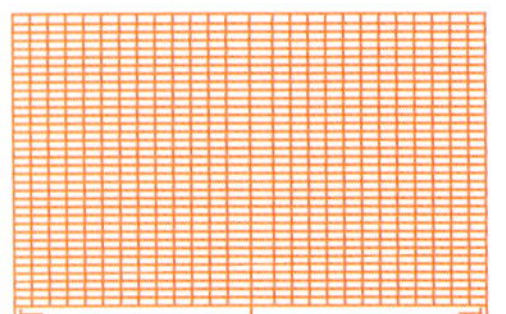

Der Raster als Grundlage kann viele übereinandergelagerte Elemente enthalten. Aber denken Sie daran, dass

- die Typografie für Informationen leserlich sein muss
- freie Flächen für eine gelungene Komposition sehr wichtig sind und
- nicht jedes Pixel und Pica ausgefüllt sein muss.

Auf der äußersten, der buchstäblichen Ebene können Schichten das Interesse des Lesers wecken. Auf einer tieferen Ebene laden sie zum Nachdenken über die Kombination von Elementen ein.

Carbon, Climate and the Race for New Energy Technology

"No other center of learning combines a depth of expertise in fundamental earth science with the ability to apply this knowledge to the solution of the world's wide-ranging social and economic challenges. This is why Columbia's Earth Institute is unique."

— G. Michael Purdy, Director
Lamont-Doherty Earth Observatory

For nearly six decades, Columbia University has played a leading role in understanding the workings of our planet's climate system. Today, Earth Institute scientists stand on the cutting edge of research to understand how human activity may be destabilizing the planet's physical systems and how changes in climate will impact society.

Just as human behavior affects climate, so too does climate influence human activities, including those that are geopolitical and economic in nature. Climate is often the reason why some people are well fed and technologically developed and others not, or why some civilizations flourish and others disappear into history. The more we can understand about the complexities of climate science, the more the world can hope to protect the planet and humankind, particularly the most vulnerable.

Beide Bilder: Auf der Titelseite und einer Innendoppelseite eines Prospekts werden mit Schichten Dimensionen erzeugt, die Mitteilung wird aber gleichzeitig klar gehalten.

PROJEKT
Branding posters

KUNDE
Earth Institute at Columbia University

CREATIVE DIRECTOR
Mark Inglis

DESIGNER
John Stislow

ILLUSTRATOR
Mark Inglis

Fotos in Schichten, Illustrationen und Icons fügen dem Design Tiefe hinzu und implizieren zusätzliche Bedeutungsebenen ebenso wie Interesse am Layout.

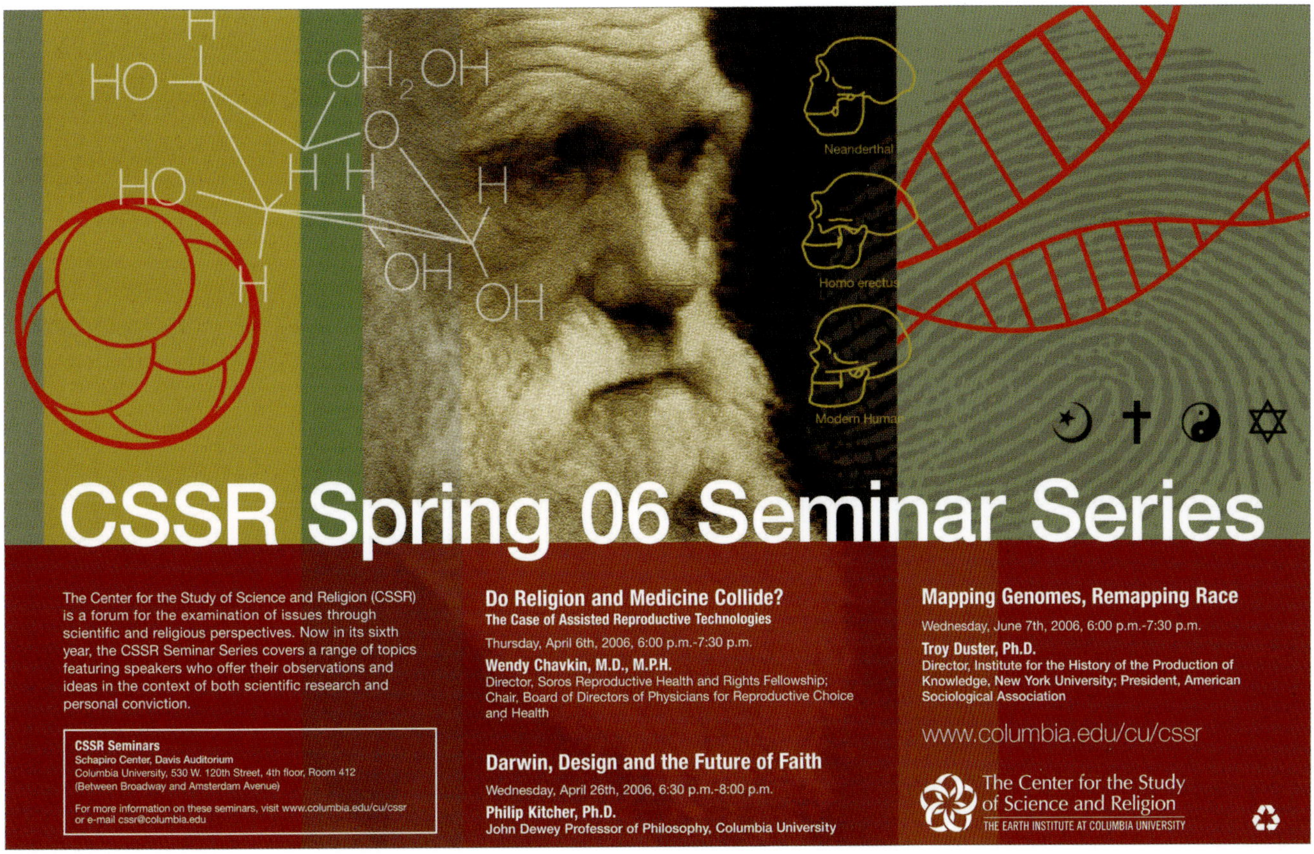
HO
H
CH_2OH
O
HO
H
H
H
H
OH
OH
H
Neanderthal
Homo erectus
Modern Human
CSSR Spring 06 Seminar Series
The Center for the Study of Science and Religion (CSSR) is a forum for the examination of issues through scientific and religious perspectives. Now in its sixth year, the CSSR Seminar Series covers a range of topics featuring speakers who offer their observations and ideas in the context of both scientific research and personal conviction.
CSSR Seminars
Schapiro Center, Davis Auditorium
Columbia University, 530 W. 120th Street, 4th floor, Room 412
(Between Broadway and Amsterdam Avenue)
For more information on these seminars, visit www.columbia.edu/cu/cssr or e-mail cssr@columbia.edu
Do Religion and Medicine Collide?
The Case of Assisted Reproductive Technologies
Thursday, April 6th, 2006, 6:00 p.m.-7:30 p.m.
Wendy Chavkin, M.D., M.P.H.
Director, Soros Reproductive Health and Rights Fellowship; Chair, Board of Directors of Physicians for Reproductive Choice and Health
Darwin, Design and the Future of Faith
Wednesday, April 26th, 2006, 6:30 p.m.-8:00 p.m.
Philip Kitcher, Ph.D.
John Dewey Professor of Philosophy, Columbia University
Mapping Genomes, Remapping Race
Wednesday, June 7th, 2006, 6:00 p.m.-7:30 p.m.
Troy Duster, Ph.D.
Director, Institute for the History of the Production of Knowledge, New York University; President, American Sociological Association
www.columbia.edu/cu/cssr
The Center for the Study of Science and Religion
THE EARTH INSTITUTE AT COLUMBIA UNIVERSITY

Die einem Foto überlagerten Elemente und transparente Farbflächen unterstützen die dreispaltige Typografie.

Accelerating the Fight against AIDS, Tuberculosis, and Malaria:
A European Union Perspective
A Talk by
Poul Nielson
European Union Commissioner for Development and Humanitarian Aid
MONDAY
APRIL
4:30 PM
28
Faculty Room, Low Library, Columbia University
116th Street and Broadway, New York
The Institute for the Study of Europe, SIPA
The Earth Institute
AT COLUMBIA UNIVERSITY

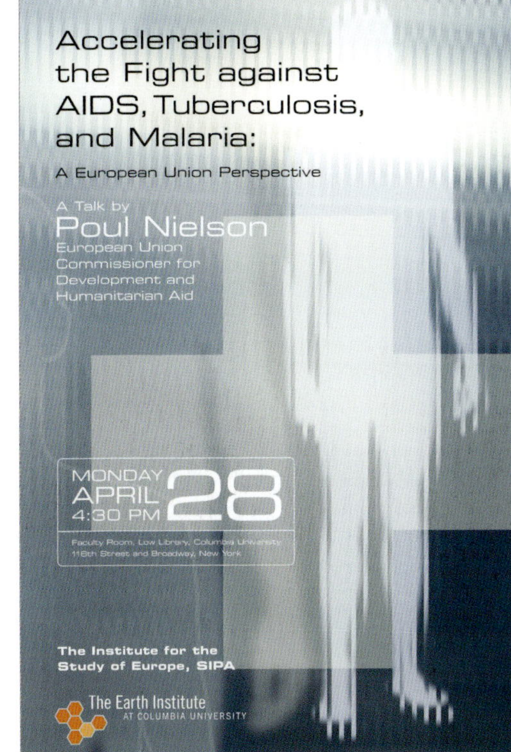

Auf diesem Plakat für einen Vortrag über aktuelle Gesundheitsthemen macht die Typografie nur die alleroberste Schicht aus.

93. Für alle Plattformen

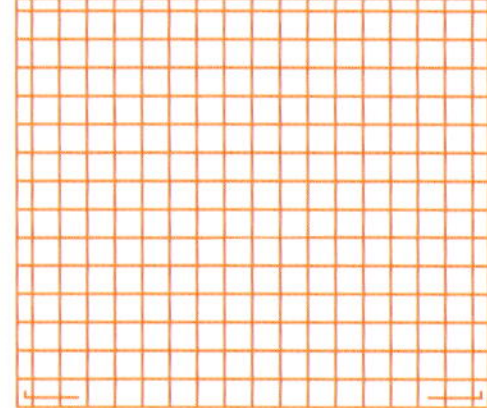

Felder und Farben können in bytegroßen Behältnissen Informationen liefern. Hin und wieder bildet das Logo eines Kunden die Vorlage für Farben und Typografie. Informationskategorien können in Boxen untergebracht und Navigationsleisten um die Webseite herum angeordnet werden. Eine dicht gestaltete Seite erinnert an das Straßennetz einer Metropole: ein klares, aber quirliges Raster, zu dem man in einer virtuellen, schwindelerregenden Fahrt Zugang erlangt.

Schwarze Balken für die Überschriften und taxigelbe Boxen kennzeichnen Design Taxi.

PROJEKT
Webseite

KUNDE
Design Taxi

DESIGN
Design Taxi

DESIGN DIRECTOR
Alex Goh

Die Website von Design Taxi, die aus Singapur grüßen lässt, bringt den Nutzer von einem Raster zum nächsten – in einer hochdichten Digitopolis mit Rahmen, Linien, Boxen, Leitfäden, Farben, Schattierungen, Links und Recherchen, aber keinen Starbucks-Laden.

Die Seite enthält zahlreiche Angebote in gerahmten Feldern und verschiedenen Grautönen. Die virtuelle Fahrt ist zuweilen holprig und das Auffinden der gesuchten HTML-Seite schwierig.

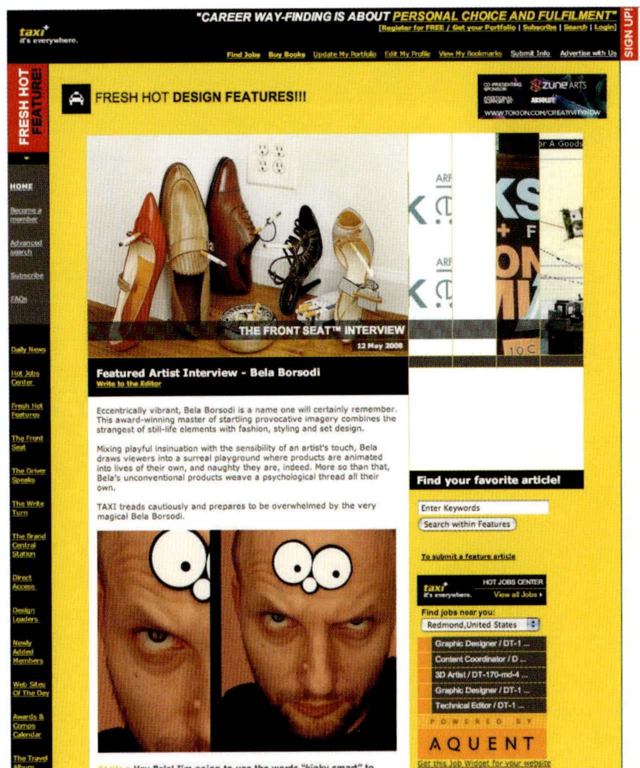

Mit einer einfachen überschaubaren Typografie lassen sich die Informationen ständig und einfach aktualisieren.

94. Gestaffelt

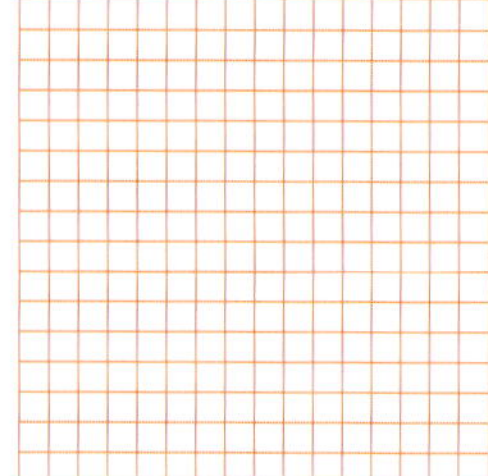

Auf der Grundlage eines Moduls lassen sich Raster staffeln, wiederholen oder auslassen, damit das System der Identität frisch und zukunftsorientiert bleibt.

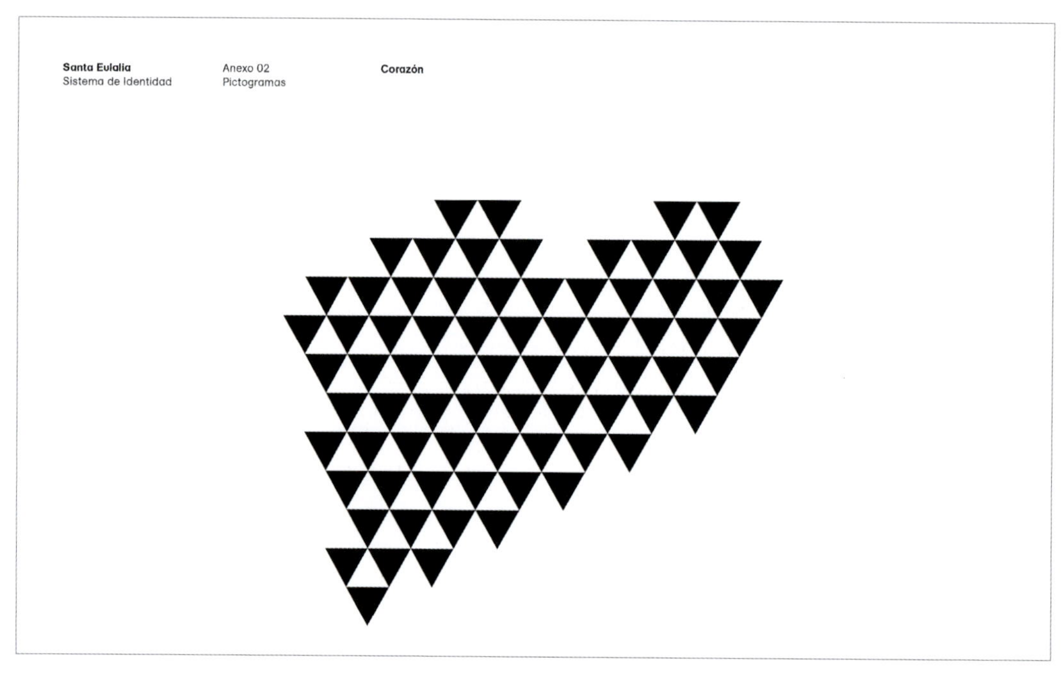

PROJEKT
Santa Eulalia
Identity

KUNDE
Santa Eulalia

DESIGN
Mario Eskenazi Studio

DESIGNER
Mario Eskenazi

Für Santa Eulalia, eine luxuriöse Mehrmarken-Boutique in Barcelona, schuf die Designagentur eine Identität, die ein X suggeriert (in Katalonien steht das X als Symbol für Santa Eulalia, die von den Römern an einem X-förmigen Kreuz gekreuzigt wurde.)

Seit der Entwicklung im Jahr 2006 hat das Studio von Mario Eskenazi auf der Grundlage des ursprünglichen Musters neue Elemente (Zahlen und Piktogramme) hinzugefügt.

Seite Gegenüber, beide Fotos:
Das stilisierte X der Identität von Santa Eulalia findet seinen Kontrast in der schlichten serifenlosen Schrift und dem sauberen Raster für den Markennamen, Daten und das Cover.

Diese Seite, alle Fotos:
Durch die Ausweitung der Grundmuster werden die Raster immer komplexer, erinnern aber immer noch an das X.

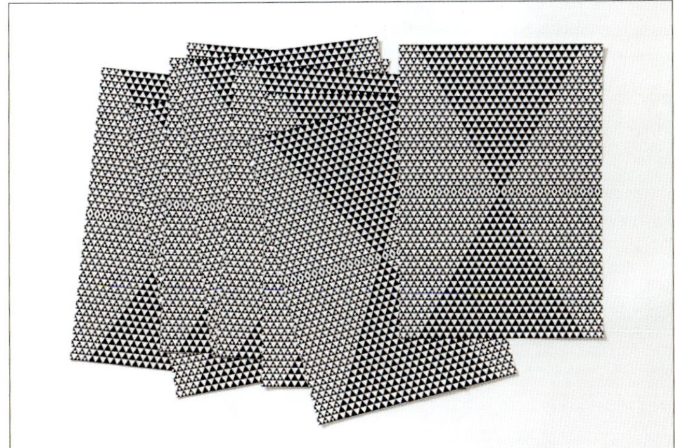

95. Großformat

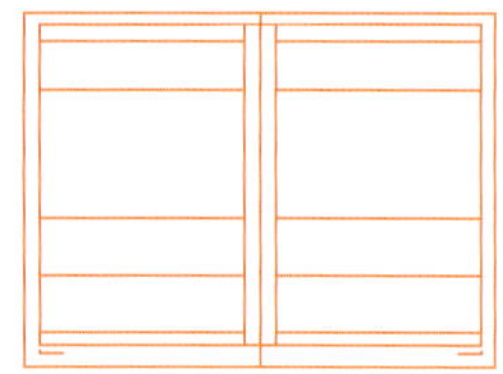

Für großformatige Supergrafiken gelten folgende typografische Regeln:

- Lassen Sie Größen, Stärken und Farben so miteinander agieren, dass dynamische Layouts entstehen.
- Berücksichtigen Sie die Dimension der Buchstaben.
- Achten Sie auf die Dynamik: Verglichen mit der Schrift auf einer Seite, muss Schrift, die sich bewegt, besonders stark spationiert werden, um leserlich zu bleiben.

PROJEKT
Bloomberg Dynamic Digital Displays

KUNDE
Bloomberg LLP

DESIGN
Pentagram, New York

ART DIRECTOR/DESIGNER, ENVIRONMENTAL GRAPHICS
Paula Scher

ART DIRECTOR/DESIGNER, DYNAMIC DISPLAYS
Lisa Strausfeld

DESIGNER
Jiae Kim, Andrew Freeman
Rion Byrd

PROJEKTARCHITEKTEN
STUDIOS Architecture

PROJEKTFOTOGRAFIE
Peter Mauss/Esto

Supergrafiken mit großer, fetter Schrift auf elektronischen Displays mit wechselnden Daten koppeln Informationen mit Marken.

BEIDE SEITEN: Die Supergrafiken kombinieren Inhalt, Statistiken und Stil.

Die dynamischen Zeichen auf den vier horizontalen Paneelen haben unterschiedliche Farben. Größe und Farbe der Buchstaben und Zahlen sind in jeder Grafik anders, wodurch ein bestimmter Blickwinkel erzeugt wird und die Daten hervorgehoben werden.

NIKKEI
VALUE 11276.59 % CHANGE +0.09 LAST UPDATED 4:29

Celsius
HIGH 29° LOW 16° HUMIDITY 81%
Clear

96. Module in Bewegung

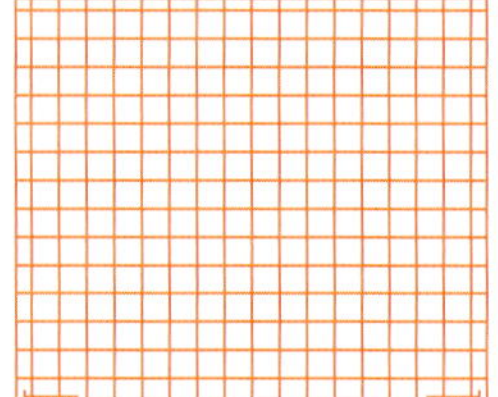

Mit gleichen Modulen lassen sich nicht nur in gedruckten Medien, sondern auch im Internet Inhalte vielseitig aufteilen, u.a. in Bereiche für Videos, die eine Webseite beleben.

FLIESSENDE KONFIGURATIONEN

In der schönen neuen Welt des interaktiven Grafikdesigns stellen veränderliche Raster und Layouts ein Thema dar, das unbedingt Beachtung verdient. Was tut man, wenn die Papiergröße nicht länger wichtig ist? Hält man an willkürlichen Maßen fest und zentriert das Layout auf einem Bildschirm? Oder kreiert man fließende Layouts, die sich automatisch für verschiedene Bildschirmgrößen neu konfigurieren? Webdesigner werden sicherlich Letzterem den Vorzug geben, doch darf man nicht vergessen, dass die technischen Aspekte für den Aufbau solcher Layouts komplexer sind.

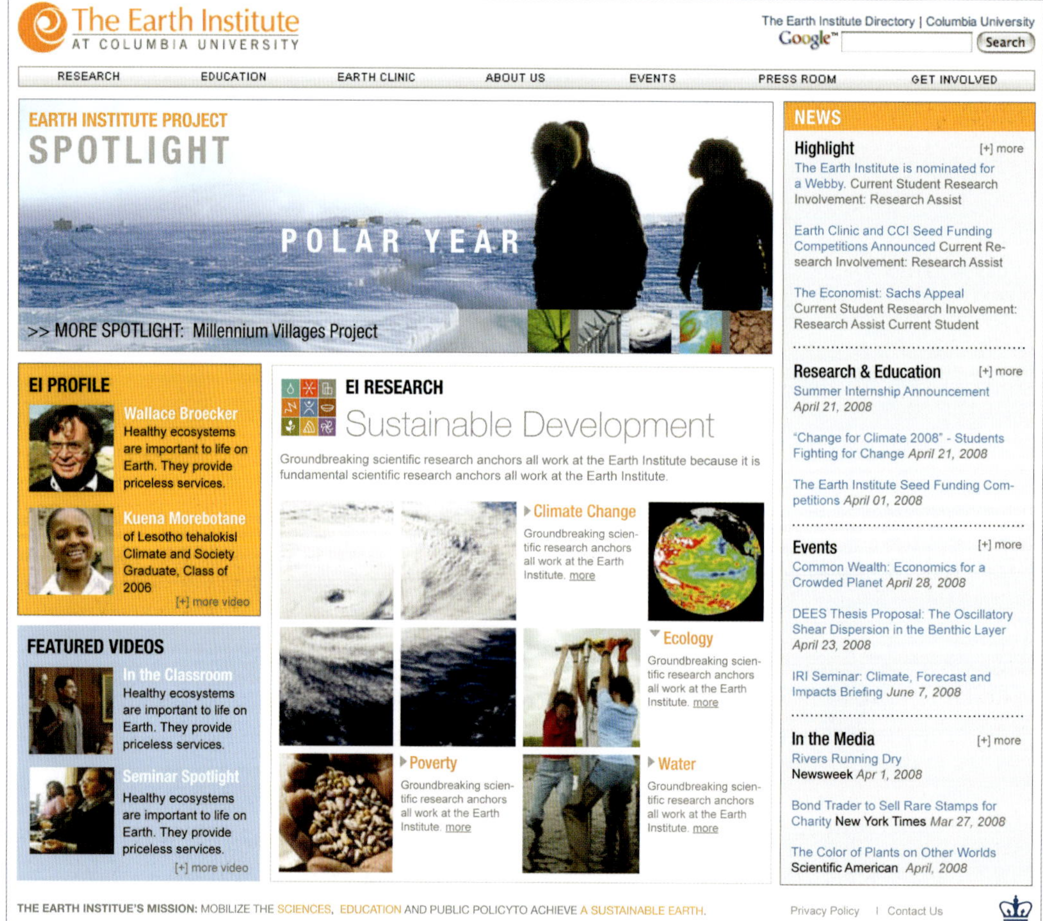

PROJEKT
Webseite

KUNDE
Earth Institute at Columbia University

CREATIVE DIRECTOR
Mark Inglis

DESIGN
Sunghee Kim, John Stislow

Mit Modulen lassen sich umfassende und gänzlich verschiedenartige Informationen ansprechend darstellen.

Beide Seiten: Module einer Homepage, die unterhalb der Hauptnavigationsleiste erscheinen sollen, können vielseitig konfiguriert und kombiniert werden.

- Alle Module, einschließlich der Links, die über die gesamte Breite reichen, dienen als Impressum.
- Ein einzelnes Modul stellt ein einzelnes Thema dar.
- Zwei Module zusammen bilden eine Box.
- Module an der Seite bilden eine lange vertikale Spalte, die als „Schwarzes Brett" für Nachrichten und Ereignisse dient.
- Module können Videos enthalten.

Der Leser kann von der Startseite zu den anderen Seiten gelangen und so sein Leseerlebnis erhöhen.

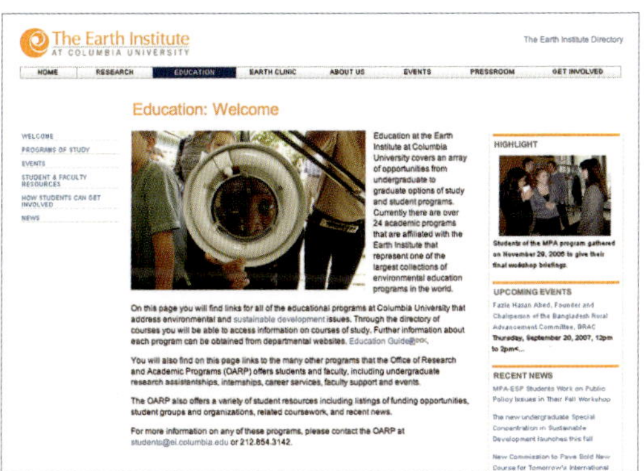

Abhängig von den Anforderungen der Informationen ist der Aufbau der in Module aufgeteilten Unterseiten leicht horizontal.

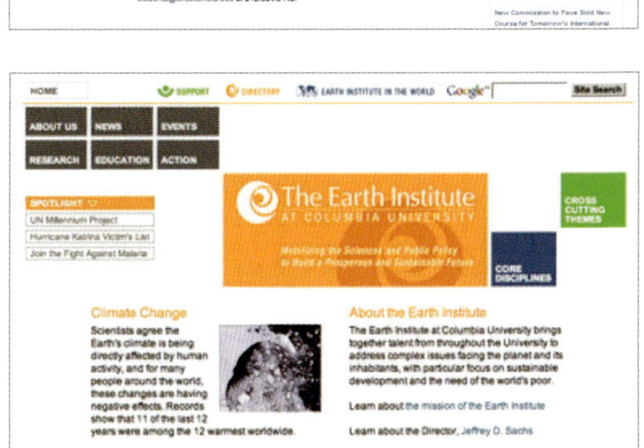

97. Nutzen Sie Ihre Stärken

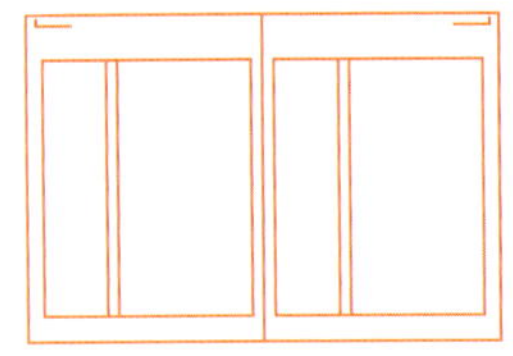

Viele großartige Grafikdesigner behaupten, ohne Raster zu arbeiten. Dennoch zeigen ihre Arbeiten Weite, Textur und Erhabenheit. Ohne sich dessen bewusst zu sein, halten sich die meisten von ihnen an die Prinzipien des guten Designs, um Text und Bilder zur Geltung zu bringen und verständlich zu machen.

Siehe auch Seite
29

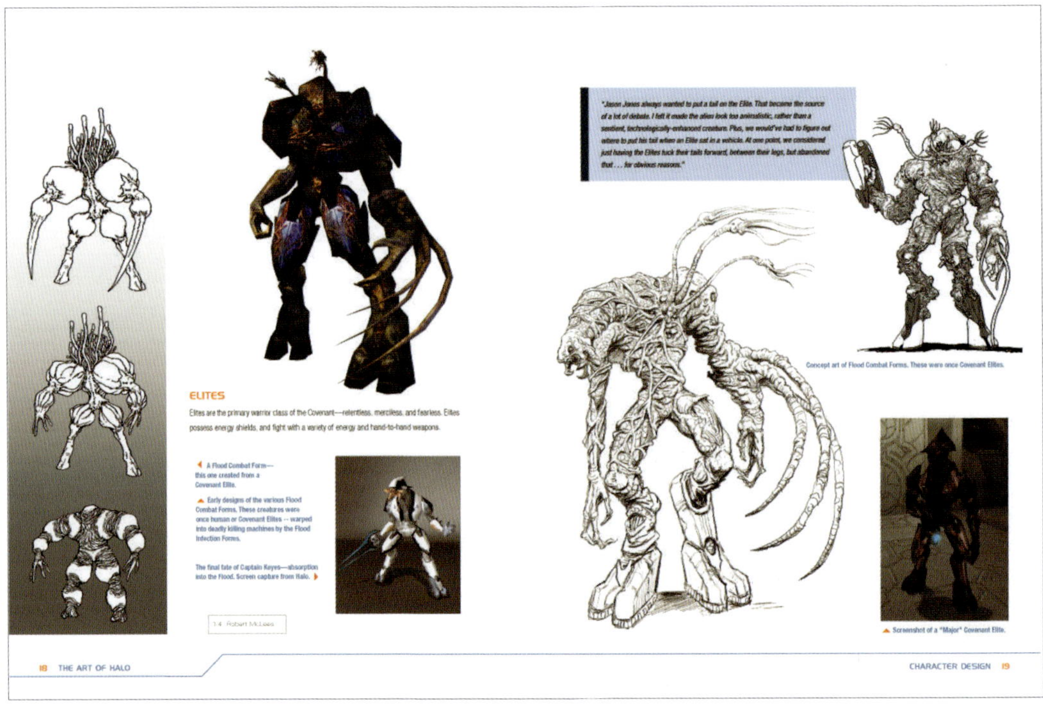

"Jason Jones always wanted to put a tail on the Elite. That became the source of a lot of debate. I felt it made the alien look too animalistic, rather than a sentient, technologically-enhanced creature. Plus, we would've had to figure out where to put his tail when an Elite sat in a vehicle. At one point, we considered just having the Elites tuck their tails forward, between their legs, but abandoned that ... for obvious reasons."

Concept art of Flood Combat Forms. These were once Covenant Elites.

ELITES

Elites are the primary warrior class of the Covenant—relentless, merciless, and fearless. Elites possess energy shields, and fight with a variety of energy and hand-to-hand weapons.

◀ A Flood Combat Form—this one created from a Covenant Elite.

▲ Early designs of the various Flood Combat Forms. These creatures were once human or Covenant Elites -- warped into deadly killing machines by the Flood Infection Forms.

The final fate of Captain Keyes—absorption into the Flood. Screen capture from Halo. ▶

1-4 Robert McLees

▲ Screenshot of a "Major" Covenant Elite.

18 THE ART OF HALO

CHARACTER DESIGN 19

Viele gezeichnete Freisteller zeigen die Entwicklung der Charaktere und sind ein Hinweis auf die Animation des Spieles. Die Figuren basieren auf horizontalen Linien und beleben die Doppelseite mit einer nach unten gerichteten Bewegung.

PROJEKT
The Art of Halo

KUNDE
Random House

DESIGN
Liney Li

Die Hauptfiguren des Spieles The *Art of Halo: Die Erschaffung einer virtuellen Welt* erlangen in diesem Buch doppelte Unsterblichkeit.

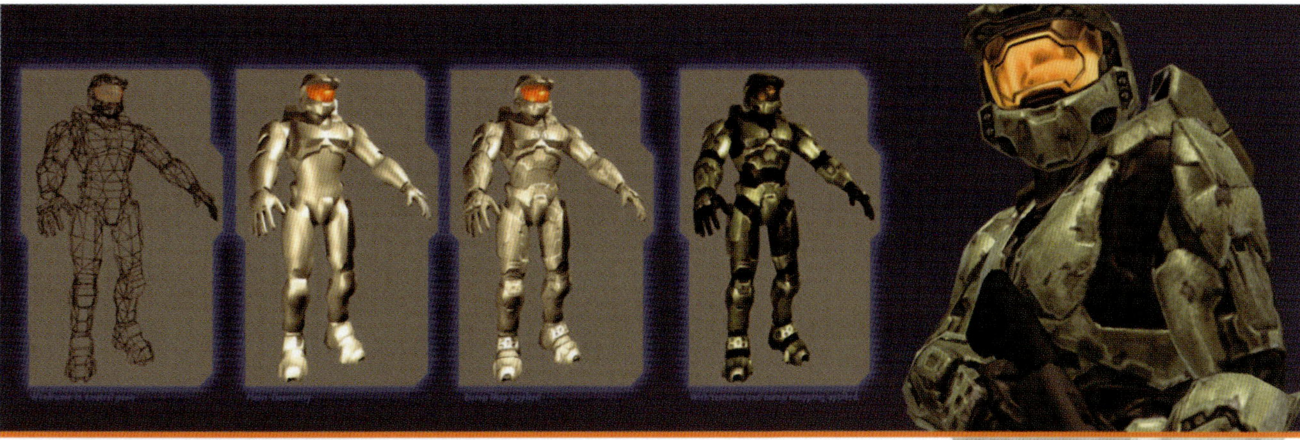

▲ The evolution of Master Chief from wire frame, rendering, and finally clad in his battle armor.

▲ One of the public's first looks at Halo came in the November, 1999 issue of Computer Gaming World.he evolution of Master Chief from wire frame, rendering, and finally clad in his battle armor.

The collaborative process at Bungie wasn't confined to the *Halo* team. There were several Bungie artists and programmers working on other titles during the various stages of *Halo*'s development. "I didn't do a lot on *Halo*—I was assigned to a team working on a different project," said character artist Juan Ramirez. "But most of us would weigh in on what we saw. I like monsters and animals and creatures—plus I'm a *sculptor*, so I did some sculpture designs of the early Elite.

"When I came on, I wasn't really a 'computer guy'—I was more into comics, film, that kind of thing. I try and apply that to my work here—to look at our games as more than just games. Better games equals better entertainment. A lot of that is sold through character design."

4 THE ART OF HALO

THE MASTER CHIEF

Seven feet tall, and clad in fearsome MJOLNIR Mark V battle armor, the warrior known as the Master Chief is a product of the SPARTAN Project. Trained in the art of war since childhood, he may well hold the fate of the human race in his hands.

1-4 Artist Unknown
5 Artist Unknown
6 Artist Unknown
7 Shi Kai Wang

MARCUS LEHTO, ART DIRECTOR: *"At first, Rob [artist Robt. McLees] and I were the only artists working on Halo. After that we hired Sheik [artist ShiKai Wang], who's just great from the conceptual standpoint. I'd do a preliminary version of something, then Sheik would work from that, and really enhance the concept.*

"The Master Chief design sketch that really took hold came after heavy collaboration with ShiKai. One of his sketches—this kind of manga-influenced piece, with ammo bandoliers across his chest, and a big bladed weapon on his back—really caught our imagination.

"Unfortunately, when we got that version into model form, he looked a little too slender, almost effeminate. So, I took the design and tried to make it look more like a modern tank. That's how we got to the Master Chief that appears in the game."

CHARACTER DESIGN 5

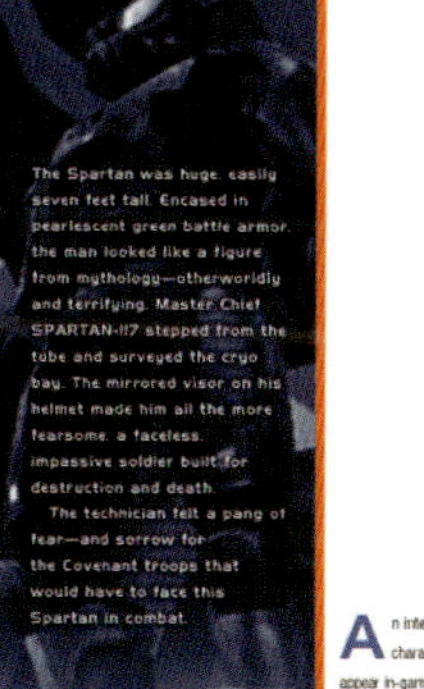

The Spartan was huge, easily seven feet tall. Encased in pearlescent green battle armor, the man looked like a figure from mythology—otherworldly and terrifying. Master Chief SPARTAN-117 stepped from the tube and surveyed the cryo bay. The mirrored visor on his helmet made him all the more fearsome, a faceless, impassive soldier built for destruction and death.

The technician felt a pang of fear—and sorrow for the Covenant troops that would have to face this Spartan in combat.

—Excerpt from Halo: The Flood by William C. Dietz, the novelization of the game.

An integral part of creating a good story is the creation of believable and interesting characters. Bungie's 3-D modelers craft designs of the various characters that appear in-game, which must then be "textured"—telling the game engine how light and shadow react with the model. From there, the models must be rigged so they can be animated. "Overlap is vital, particularly among modelers and animators," says animator William O'Brien. "We depend on each other for the final product to work—and none of us can settle. We always have to up it a notch."

"Our job is to bring the characters to life in the game," said Nathan Walpole, animation lead for *Halo 2*. "It's what we're best at. We don't use motion capture—most of us are traditional 2-D animators, so we prefer to hand-key animation. Motion capture just looks so bad when it's done poorly. We have more control over hand-keyed animation, and can produce results faster than by editing mocap."

Crafting the animations that bring life to the game characters is a painstaking process. "Usually, we start with a thumbnail sketch to build a look or feel," explained Walpole. "Then, you apply it to the 3-D model and work out the timing."

"Sometimes the timing's *so* off, it's hilarious," adds animator Mike Budd. "Everyone comes over and has a good laugh. Working together like we do keeps us fresh. There's such a variety of characters—human and alien. And you work on them in a matter of weeks. You're always working on something new and interesting."

▲ A pair of Grunts prepare to engage the enemt. Screen capture from Halo.

To design the characters' motions, the animators study virtually any source of movement for inspiration—though this can create some challenges for animator William O'Brien: "Just being surrounded by people with good senses of humor makes it easier to do your job. The drawback is, I've always had my own office. To animate a character, I often act out motions and movements; this gives you a sense of what muscle and bone actually do. But now, I have an audience. 'Hey, look at the crazy stuff Bill's doing now!' So now, I tend to do that kind of work on video, in private."

◀ Opposite page: Captions needed for illustrations 1, 2 ,3 , and 4.

1-4 Screenshot from Halo
5 Artist Unknown

2 CHARACTER DESIGN 3

Das Buch kombiniert eine klassische mit einer stilisierten futuristischen Typografie. Bildunterschriften sind durch die Farbe Blau vom Text getrennt. Linien und Richtungsanzeiger (Pfeile und Wörter wie „links" und „rechts") sind leuchtend orange.

Spalten entlang der Außenkante generieren Boxen und grenzen die Charaktere voneinander ab.

98. Seien Sie flexibel

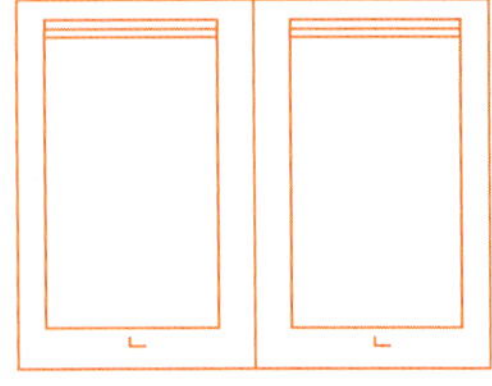

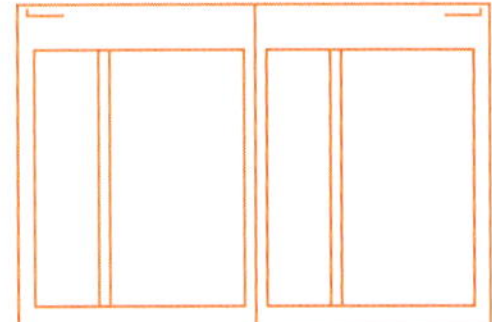

Manchmal müssen formale Gestaltungsaspekte wie breite Ränder, leserliche Schrift und korrekte Kursivschriften beiseitegeschoben werden. Denn in bestimmten Kontexten kann ein „falsches" Design durchaus richtig sein. Ein Durchbrechen der Regeln kann für eine provokative oder visionäre Mitteilung eine perfekte Lösung darstellen.

RECHTS: Elastisch. Überlagert. Faszinierend.

Foreword

With Design and the Elastic Mind, The Museum of Modern Art once again ventures into the field of experimental design, where innovation, functionality, aesthetics, and a deep knowledge of the human condition combine to create outstanding artifacts. MoMA has always been an advocate of design as the foremost example of modern art's ability to permeate everyday life, and several exhibitions in the history of the Museum have attempted to define major shifts in culture and behavior as represented by the objects that facilitate and signify them. Shows like Italy: The New Domestic Landscape (1972), Designs for Independent Living (1988), Mutant Materials in Contemporary Design (1995), and Workspheres (2001), to name just a few, highlighted one of design's most fundamental roles: the translation of scientific and technological revolutions into approachable objects that change people's lives and, as a consequence, the world. Design is a bridge between the abstraction of research and the tangible requirements of real life.

The state of design is strong. In this era of fast-paced innovation, designers are becoming more and more integral to the evolution of society, and design has become a paragon for a constructive and effective synthesis of thought and action. Indeed, in the past few decades, people have coped with dramatic changes in several long-standing relationships—for instance, with time, space, information, and individuality. We must contend with abrupt changes in scale, distance, and pace, and our minds and bodies need to adapt to acquire the elasticity necessary to synthesize such abundance. Designers have contributed thoughtful concepts that can provide guidance and ease as science and technology proceed in their evolution. Design not only greatly benefits business, by adding value to its products, but it also influences policy and research without ever reneging its poietic, nonideological nature—and without renouncing beauty, efficiency, vision, and sensibility, the traits that MoMA curators have privileged in selecting examples for exhibition and for the Museum's collection.

Design and the Elastic Mind celebrates creators from all over the globe—their visions, dreams, and admonitions. It comprises more than two hundred design objects and concepts that marry the most advanced scientific research with the most attentive consideration of human limitations, habits, and aspirations. The objects range from

Schmale Ränder, verschiedene Schriften, verschwindende Seitenzahlen und Kolumnentitel sind Teile eines Planes, das Thema fesselnd, provokativ und reflektierend darzustellen.

PROJEKT
Design and the Elastic Mind

KUNDE
Museum of Modern Art

DESIGN
Irma Boom, Niederlande

UMSCHLAGGESTALTUNG
Daniël Maarleveld

In diesem Katalog für die Ausstellung „Design and the Elastic Mind" meidet der Grafikdesigner herkömmliche formale Gestaltungsaspekte. Das Ergebnis ist so provokativ – und manchmal so irritierend – wie die Ausstellung selbst

sometimes for hours, other times for minutes, using means of communication ranging from the most encrypted and syncopated to the most discursive and old-fashioned, such as talking face-to-face—or better, since even this could happen virtually, let's say nose-to-nose, at least until smells are translated into digital code and transferred to remote stations. We isolate ourselves in the middle of crowds within individual bubbles of technology, or sit alone at our computers to tune into communities of like-minded souls or to access information about esoteric topics.

Over the past twenty-five years, under the influence of such milestones as the introduction of the personal computer, the Internet, and wireless technology, we have experienced dramatic changes in several mainstays of our existence, especially our rapport with time, space, the physical nature of objects, and our own essence as individuals. In order to embrace these new degrees of freedom, whole categories of products and services have been born, from the first clocks with mechanical time-zone crowns to the most recent devices that use the Global Positioning System (GPS) to automatically update the time the moment you enter a new zone. Our options when it comes to the purchase of such products and services have multiplied, often with an emphasis on speed and automation (so much so that good old-fashioned cash and personalized transactions—the option of talking to a real person—now carry the cachet of luxury). Our mobility has increased along with our ability to communicate, and so has our capacity to influence the market with direct feedback, making us all into arbiters and opinion makers. Our idea of privacy and private property has evolved in unexpected ways, opening the door

top: James Powderly, Evan Roth, Theo Watson, and HELL. Graffiti Research Lab. **L.A.S.E.R. Tag.** Prototype. 2007. 60 mW green laser, digital projector, camera, and custom GNU software (L.A.S.E.R. Tag V1.0, using OpenFrameworks)

New forms of communication transcend scale and express a yearning to share opinions and information. This project simulates writing on a building. A camera tracks the beam painter of a laser pointer and software transmits the action to a very powerful projector.

16 Design and the Elastic Mind

17 Paola Antonelli Design and the Elastic Mind

bottom: James Powderly, Evan Roth, Theo Watson, DASK, FOXY LADY, and BENNETT4SENATE. Graffiti Research Lab. **L.A.S.E.R. Tag graffiti projection system.** Prototype. 2007. 60 mW green laser, digital projector, camera, custom GNU software (L.A.S.E.R. Tag V1.0, using OpenFrameworks), and mobile broadcast unit

for debates ranging from the value of copyright to the fear of ubiquitous surveillance.[2] Software glitches aside, we are free to journey through virtual-world platforms on the Internet. In fact, for the youngest users there is almost no difference between the world contained in the computer screen and real life, to the point that some digital metaphors, like video games, can travel backward into the physical world: At least one company, called area/code, stages "video" games on a large scale, in which real people in the roles of, say, Pac Man play out the games on city streets using mobile phones and other devices.

Design and the Elastic Mind considers these changes in behavior and need. It highlights current examples of successful design translations of disruptive scientific and technological innovations, and reflects on how the figure of the designer is changing from form giver to fundamental interpreter of an extraordinarily dynamic reality. Leading up to this volume and exhibition, in the fall of 2006 The Museum of Modern Art and the science publication Seed launched a monthly salon to bring together scientists, designers, and architects to present their work and ideas to each other. Among them were Benjamin Aranda and Chris Lasch, whose presentation immediately following such a giant of the history of science as Benoit Mandelbrot was nothing short of heroic, science photographer Felice Frankel, physicist Keith Schwab, and computational design innovator Ben Fry, to name just a few.[3] Indeed, many of the designers featured in this book are engaged in exchanges with scientists, including Michael Burton and Christopher Woebken, whose work is influenced by nanophysicist Richard A. L. Jones; Elio Caccavale, whose interlocutor is Armand Marie Leroi, a biologist from the Imperial

Bilder, die im Bund verschwinden, sind bei einem normalen, weniger dynamischen Projekt verboten.

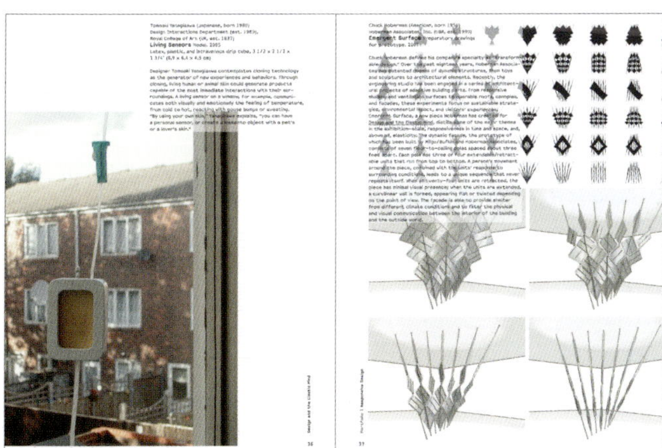

In diesem Buch ist Schrift absichtlich von Abbildungen überlagert.

Textboxen sind über Bilder gedruckt.

99. Die Stimme des Herzens

Den Raster für ein Projekt zu finden ist wie ein Puzzlespiel. Manchmal ist das Konzept selbst ein Rätsel, sowohl thematisch als auch in der Aufgabe, eine Idee zu vermitteln. Eine gute Sache kann starke Arbeiten inspirieren, die eine wichtige Botschaft übermitteln und Herz und Seele erfreuen – das ist wichtig bei Arbeiten, die die Liebe zu einer Gemeinschaft zeigen und diese so beschreiben, dass sie für das Publikum interessant wird.

PROJEKT
Conundrum

DESIGNER
Dayna Iphill,
New York City College
of Technology, 2018

PROFESSOR
Douglas Davis

Conundrum versucht, das Bewusstsein für Autismus zu stärken. Da Autisten auf ein uninformiertes Publikum schnell verwirrend wirken, setzt das Design auf ein typografisches Puzzle. Das Identitätssystem von Conundrum umfasst das Logo, Poster, Broschüren, die Webseite und soziale Medien.

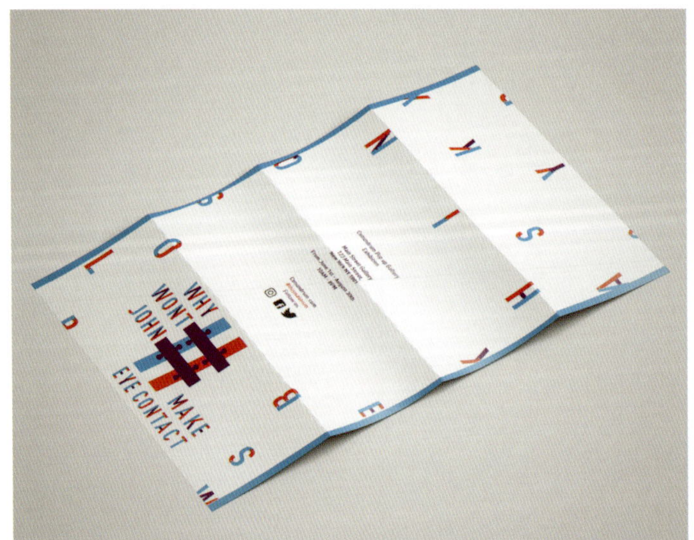

Diese Seite: Als Gegensatz zum Puzzle des Logos von Conundrum setzt die Broschüre auf einen klaren zweispaltigen Raster, der den Leser führt. Durch alle Elemente führen blaue vertikale Linien.

WHAT IS AUTISM ?

Autism is a mental condition that develops from a young age. 1 in 68 children are diagnosed with an autism disorder. Boys are nearly 5 times more likely to have autism than girls. People who have autism exhibit repetitive behaviors such as hand flapping and rocking. They also have difficulty making eye contact as well as communicating and forming relationships. They may have sensory processing difficulties causing them to be sensitive to towuch, smell and noise. Autism comes in different forms, and no two people are the same. Although there have been many theories about the cause of autism such as vaccines or genetic, no exact cause has been confirmed.

GET INVOLVED

WALK FOR A CAUSE
Find a walk near you during Autism awareness month in April. Help raise awareness by walking for autism with your friends, family or neighbors who can also help support the cause.

JOIN AN AUTISM SUPPORT GROUP
Join a community who knows what it is like to live with autism.

ATTEND WORKSHOPS/INFORMATION SESSIONS
Get to learn more about what autism is. Attending workshops/ information sessions will allow you to educate yourself more and understand how to interact with people who have autism.

DONATE
Donate to an autism organization in your area to help support the cause.

ADVOCATE
Become an advocate for persons who are autistic. You can help them by providing support and assistance.

ATTEND AN AUTISM EXHIBITION
Visit the conundrum exhibit to understand what it might feel like to be autistic.

PERSONAL STORIES

Christian Andersen
A 27-year-old from Copenhagen, is a role model for young adults on the spectrum. Andersen has Asperger's Syndrome, and he is currently working at company called "Lundbeck."

Jesse Saperstein
A 32-year-old man with Asperger's Syndrome and an anti-bullying/autism activist. Saperstein used his passion for helping others to become a best-selling author who's written two books.

Seite gegenüber: Wie bei der Broschüre sitzen blaue Linien oben auf der Webseite. Die Bilder sind für die sozialen Medien vereinfacht. Auf den Postern an den Haltestellen wird der Text in das Puzzle des Logos einbezogen.

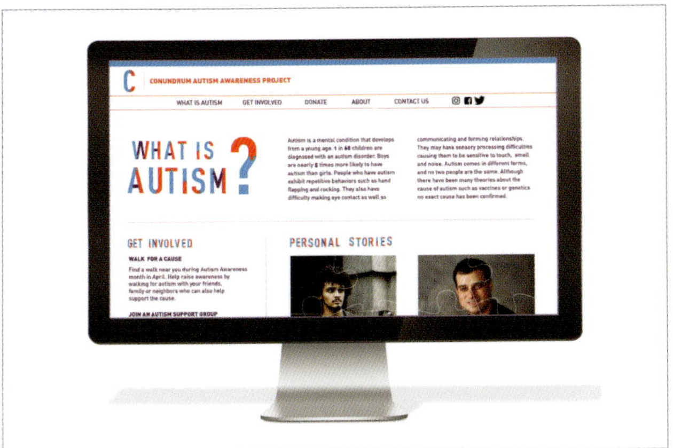
C
CONUNDRUM AUTISM AWARENESS PROJECT
WHAT IS AUTISM
GET INVOLVED
DONATE
ABOUT
CONTACT US
WHAT IS AUTISM ?
GET INVOLVED
PERSONAL STORIES

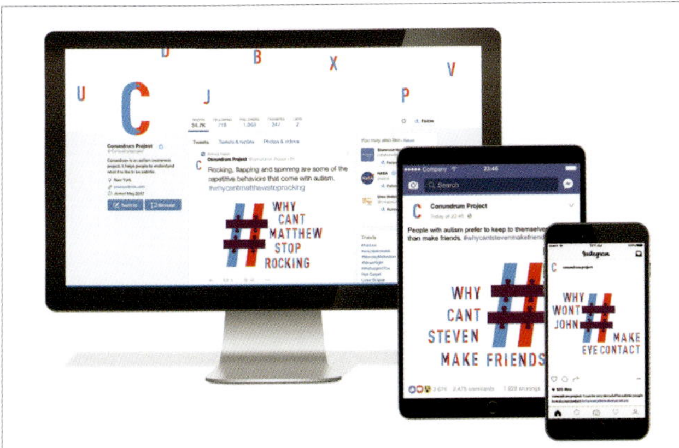
WHY CANT MATTHEW STOP ROCKING
WHY CANT STEVEN MAKE FRIENDS
WHY WONT JOHN MAKE EYE CONTACT

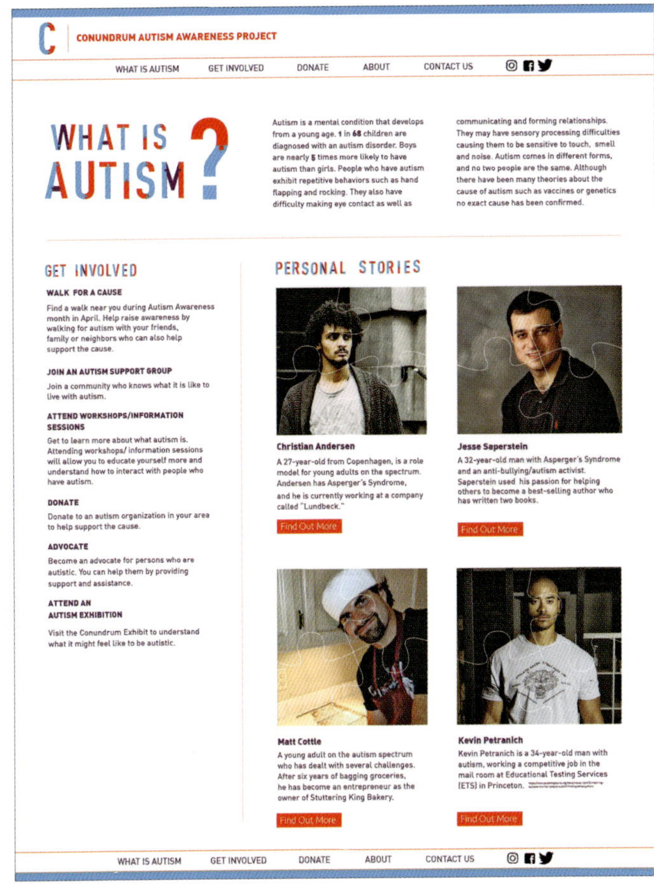
C
CONUNDRUM AUTISM AWARENESS PROJECT
WHAT IS AUTISM
GET INVOLVED
DONATE
ABOUT
CONTACT US
WHAT IS AUTISM ?
Autism is a mental condition that develops from a young age. 1 in 68 children are diagnosed with an autism disorder. Boys are nearly 5 times more likely to have autism than girls. People who have autism exhibit repetitive behaviors such as hand flapping and rocking. They also have difficulty making eye contact as well as
communicating and forming relationships. They may have sensory processing difficulties causing them to be sensitive to touch, smell and noise. Autism comes in different forms, and no two people are the same. Although there have been many theories about the cause of autism such as vaccines or genetics no exact cause has been confirmed.
GET INVOLVED
WALK FOR A CAUSE
Find a walk near you during Autism Awareness month in April. Help raise awareness by walking for autism with your friends, family or neighbors who can also help support the cause.
JOIN AN AUTISM SUPPORT GROUP
Join a community who knows what it is like to live with autism.
ATTEND WORKSHOPS/INFORMATION SESSIONS
Get to learn more about what autism is. Attending workshops/ information sessions will allow you to educate yourself more and understand how to interact with people who have autism.
DONATE
Donate to an autism organization in your area to help support the cause.
ADVOCATE
Become an advocate for persons who are autistic. You can help them by providing support and assistance.
ATTEND AN AUTISM EXHIBITION
Visit the Conundrum Exhibit to understand what it might feel like to be autistic.
PERSONAL STORIES
Christian Andersen
A 27-year-old from Copenhagen, is a role model for young adults on the spectrum. Andersen has Asperger's Syndrome, and he is currently working at a company called "Lundbeck."
Find Out More
Jesse Saperstein
A 32-year-old man with Asperger's Syndrome and an anti-bullying/autism activist. Saperstein used his passion for helping others to become a best-selling author who has written two books.
Find Out More
Matt Cottle
A young adult on the autism spectrum who has dealt with several challenges. After six years of bagging groceries, he has become an entrepreneur as the owner of Stuttering King Bakery.
Find Out More
Kevin Petranich
Kevin Petranich is a 34-year-old man with autism, working a competitive job in the mail room at Educational Testing Services (ETS) in Princeton.
Find Out More
WHAT IS AUTISM
GET INVOLVED
DONATE
ABOUT
CONTACT US

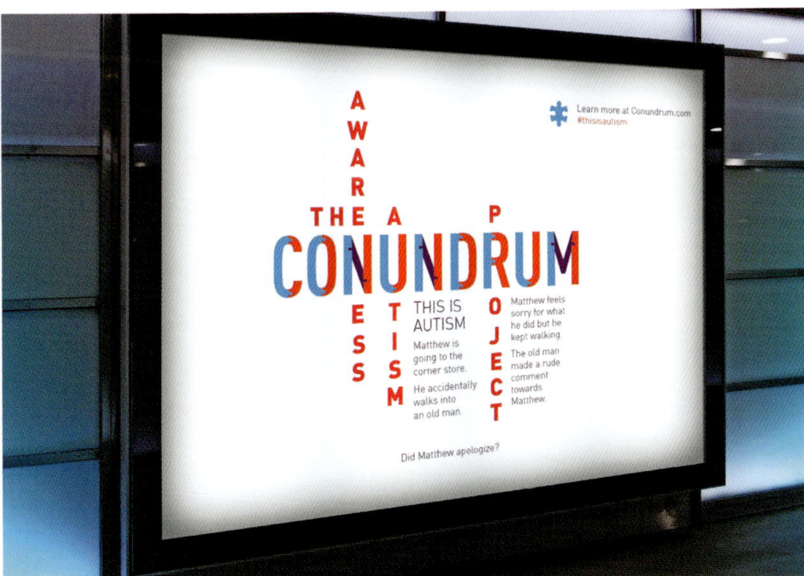
Learn more at Conundrum.com
#thisisautism
THE CONUNDRUM
AWARENESS
AUTISM
PROJECT
THIS IS AUTISM
Matthew is going to the corner store.
He accidentally walks into an old man.
Matthew feels sorry for what he did but he kept walking.
The old man made a rude comment towards Matthew.
Did Matthew apologize?

THE
CONUNDRUM
AWARENESS
AUTISM
PROJECT
THIS IS AUTISM

DEN RASTER AUFBRECHEN

100. Die Regeln ignorieren

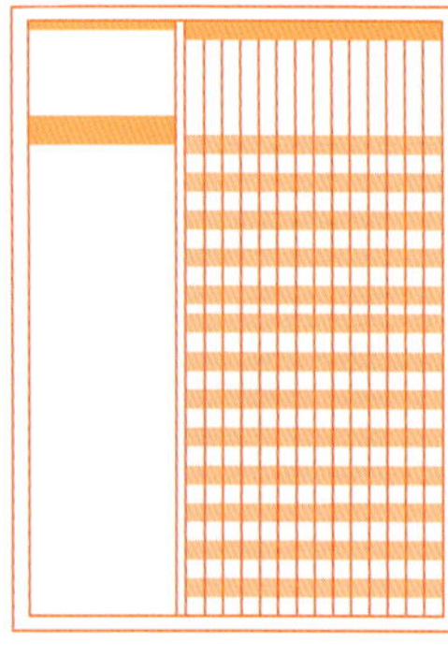

Dieses Buch zeigt verschiedene Kommunikationssysteme, die auf Raster setzen, und beschreibt die Grundkomponenten des Layouts wie Typografie, Raum und Farbe. Wie bereits gesagt, der Raster muss das Design mit dem Thema verbinden. Sie brauchen eine klare Hierarchie der Information und müssen bei der Typografie darauf achten, ob sie klassisch, klar oder ein bunter Mix verschiedener Schriften und Strichstärken ist. Beim Layout zählt das Handwerk. Arbeiten Sie ausgewogen und konsequent.

Lernen Sie aus diesem Buch, aber denken müssen Sie selbst. Sie müssen die formalen Prinzipien beherrschen, aber ab und zu auch die Regeln brechen. Kein Buch und keine Webseite kann alles vermitteln. Beobachten Sie. Fragen Sie. Arbeiten Sie mit anderen zusammen. Lernen Sie von anderen. Bitten Sie notfalls um Hilfe. Bewahren Sie Ihren Humor. Seien Sie flexibel und ausdauernd. Üben Sie. Üben Sie immer weiter. Der Erfolg des Designs liegt in der Überarbeitung und im Spaß an der Arbeit. Ihre Fragen sind stets willkommen.

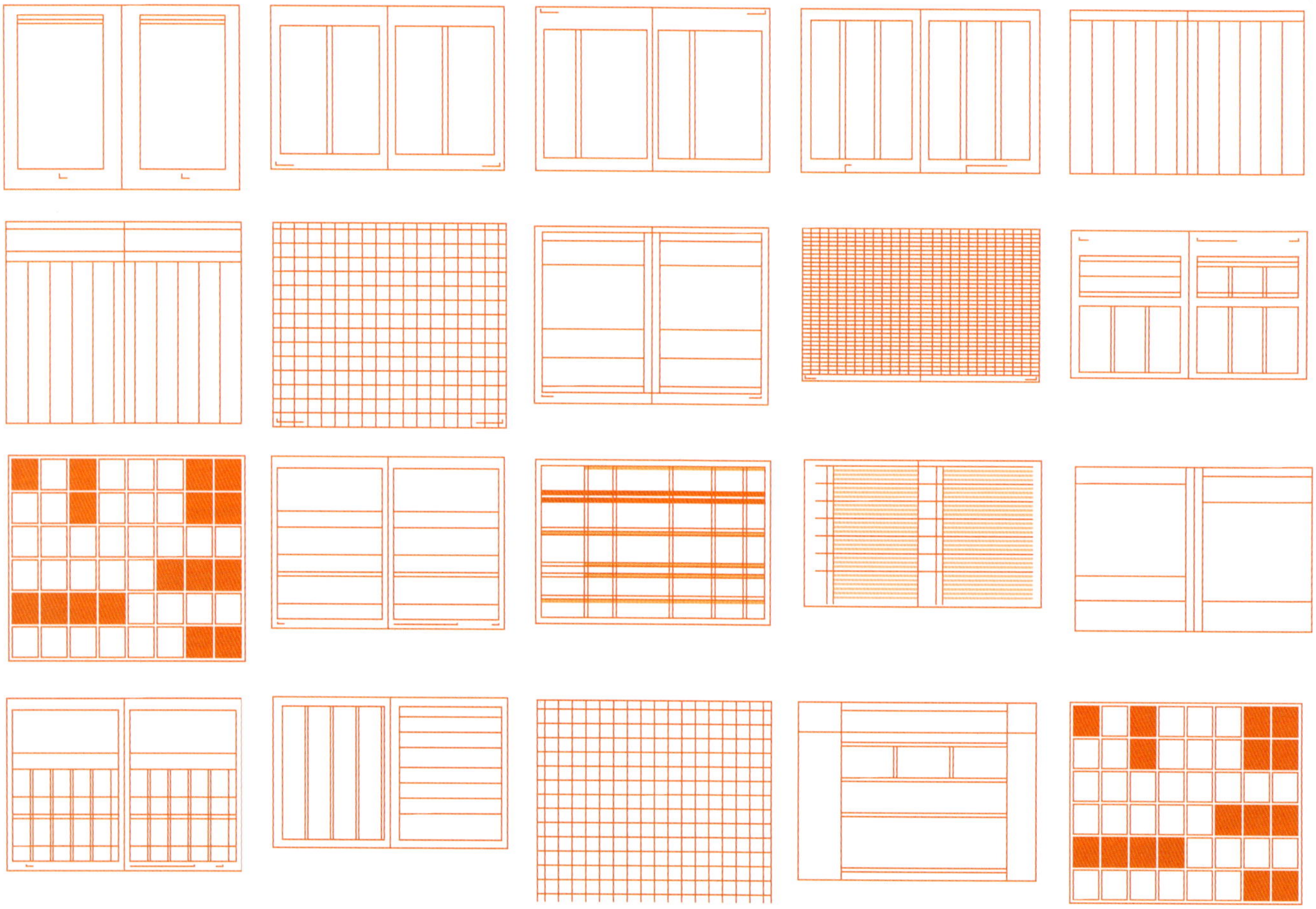

„Lassen Sie sich nicht vom Raster beherrschen, beherrschen Sie ihn. Es ist wie ein Löwenkäfig. Bleibt der Dompteur zu lange drinnen, wird er aufgefressen. Sie müssen wissen, wann Sie den Käfig verlassen müssen und wann Sie den Raster verlassen müssen.“

—MASSIMO VIGNELLI

„Der Raster unterstützt, er garantiert nichts. Er bietet eine Reihe von Möglichkeiten, und jeder Designer kann die für seinen Stil passende Lösung finden. Aber er muss lernen, den Raster anzuwenden. Das ist eine Kunst, sie will geübt sein.“

—JOSEF MÜLLER-BROCKMANN

„Ein Raster ist wie Unterwäsche. Sie tragen sie, aber die anderen sehen sie nicht.”

—MASSIMO VIGNELLI

Glossar

EINE AUSWAHL WICHTIGER FACHBEGRIFFE

Blocksatz – Der Text einer Spalte ist beidseitig bündig ausgerichtet.

CMYK – Abkürzung für *C*yan, *M*agenta, *Y*ellow und Blac*k*. Diese Farben werden beim Vierfarbendruck verwendet.

Fließtext – Nur Text, d. h. in der Regel nicht unterbrochen von Überschriften, Grafiken, Bildern etc.

Font – Digital beinhaltet ein Font Schriftschnitte einer Schrift, wie sie tatsächlich im Layout Verwendung finden. Eine Schriftart ist hingegen eine Sammlung von Zeichen, die dasselbe charakteristische Design aufweisen. Die Begriffe werden häufig verwechselt, weil heutzutage beinahe synonym verwendet. Der Font wird in der Produktion und Praxis am Computer eingesetzt, während man sich unter Schriftart eher das übergeordnete Design vorstellen sollte.

Freisteller – Ein Bild, auf dem ein Objekt oder eine Figur vom Hintergrund freigestellt, ohne diesen dargestellt ist.

Hurenkind – Die letzte Zeile eines Absatzes allein am Anfang einer Spalte oder Seite; Gegenteil: Schusterjunge.

Impressum – Eine Auflistung der Mitwirkenden einer Publikation sowie sonstige Informationen darüber.

JPEG – Steht für *Joint Photographic Experts Group*. Ein Komprimierungsformat, um Bilder im Internet darstellen zu können. JPEG ist nicht unbedingt für den herkömmlichen Druckprozess geeignet.

Klammerbindung – Bindung mit Draht, ähnlich wie Heftklammern.

Klebebindung – Eine klebstoffbasierte Bindetechnik. Die Papierkanten werden auf dem Buchrücken mit Klebstoff versehen und dann mit dem Buchumschlag umgeben. Anschließend wird das Buch samt Umschlag sauber an den anderen drei Seiten beschnitten.

Kopfsteg – Die unbedruckte Fläche zwischen Beschnitt und Satzspiegel am oberen Rand einer Seite.

Layout – Die Anordnung von Text- und Bildelementen auf einer Seite oder dem Bildschirm.

Nachspann – Ergänzende Informationen, die nicht zum eigentlichen Inhalt des Buches gehören, wie etwa Anhänge, Anmerkungen, Bibliografien, Glossare und Inhaltsverzeichnisse; Pendant: Titelei.

Negativraum – Der Raum zwischen Formen und Figuren wird vornehmlich im Bereich von Kunst, Bildhauerei und Musik gebraucht.

Lebender Kolumnentitel – Text am oberen oder unteren Seitenrand, der etwa den Buch- oder Kapiteltitel wiederholt und die Seitenzahl (Pagina) angibt. Diese allein heißt „toter“ Kolumnentitel.

Pica – Angelsächsische Maßeinheit für den Schriftsatz. Eine Pica entspricht dem Schriftgrad von 12 Punkt, bei Postscriptdruckern $^1/_6$ Inch.

Pixel – Ein Pixel ist der kleinste Bildpunkt auf einem Computerbildschirm (steht für *picture elements*).

Punkt – Maßeinheit für Schriftgrößen. Eine Pica entspricht 12 Punkten und 1 Inch etwa 72 Punkten.

Rechtsbündig – Der Text wird (gleichmäßig) am rechten Rand ausgerichtet, wobei am linken Rand die Zeilen mit unterschiedlichen Längen enden, flattern.

RGB (*Rot, Grün und Blau*) – Additive Primärfarben für die Farbdarstellung auf dem Bildschirm. Beim Scannen erhält man in Photoshop Bilder in den additiven Primärfarben. Beim Rollenoffsetdruck müssen Bilder meistens in CMYK-TIFF gedruckt werden.

Rollenoffsetdruck – Bei diesem Druckverfahren kommt das Papier von der Rolle. Es wird indirekt über Gummiband auf das Papier gedruckt.

Sättigung – Eine gesättigte Farbe ist intensiv und enthält einen geringen Grauanteil. Je höher der Sättigungsgrad, desto niedriger der Grauanteil einer Farbe.

Schriftart – Ein Schrifttyp mit charakteristischen Merkmalen. Verschiedene Schriftarten können die gleichen Merkmale aufweisen. Eine Schriftart umfasst ebenso kursive und fett gedruckte Buchstaben, Kapitälchen sowie Duktusvariationen. Die Schriftart ist die künstlerische Ausgestaltung einer Schrift. Siehe auch Font.

Schusterjunge – Erste Zeile eines Absatzes allein am Ende einer Spalte oder Seite. Gegenteil: Hurenkind.

Spalte – Ein vertikales Feld, das Text und/oder Bilder beinhaltet. Die Spaltentexte werden horizontal gemessen.

Spec – Ursprünglich *specification*. Anleitungen für den Schriftsatz. Mittlerweile die Bezeichnung für die Seitengestaltungsfunktion von Layoutprogrammen.

Tagline – Ein Slogan oder einige Zeilen, die aus dem Haupttext herausgenommen wurden.

TIFF (*Tagged Image Filed Format*) – Ein Format für die elektronische Speicherung und Übermittlung von Bitmaps sowie von Schwarz-Weiß- und Farbbildern. TIFF ist das traditionelle Druckformat.

Titelei – Steht in einem Buch oder einer Publikation vor dem eigentlichen Text wie etwa Schmutz- und Innentitel, Impressum oder Inhaltsverzeichnis; Pendant: Nachspann.

Typografie – Stil, Anordnung und Form der Schriften; auch die Kunst, Schriften auszuwählen und zu gestalten.

Übereinanderdruck – Die Farben/Farbschichten werden nacheinander aufgebracht.

Weißfläche – Freie Flächen auf der Seite oder dem Bildschirm ohne Text oder Bild.

Weiterführende Literatur und Medien

BÜCHER

Antonelli, Paola. *Design and the Elastic Mind.* Museum of Modern Art, 2008.

Bierut, Michael. *Wie man als Grafikdesigner Produkte erfolgreicher verkauft, Dinge besser erklärt, Sachen schöner macht, Leute zum Lachen bringt (oder zum Weinen) - und manchmal sogar die Welt verbessert.* niggli Verlag, 2015

Birdsall, Derek. *Notes on Book Design.* Yale University Press, 2004.

Bringhurst, Robert. *The Elements of Typographic Style.* Hartley & Marks Publishers, 1992, 1996, 2002.

Heller, Steven, und Fili, Louise. *Stylepedia. A Guide to Graphic Design Mannerisms, Quirks, and Conceits.* Chronicle Books, 2007.

Heller, Steven. Jedes seiner 150 oder mehr Bücher. Kidd, Chip, *Work: 1986–2006; Book One.* Rizzoli International Publications, Inc., 2005.

Lawson, Alexander. *Anatomy of a Typeface.* David R. Godine Publisher, Inc., 1990.

Leborg, Christian. *Visual Grammar.* Princeton Architectural Press, 2004.

Lee, Marshall. *Bookmaking: Editing, Design, Production.* Third Edition. W. W. Norton & Co., 2004.

Lidwell, William; Holden, Kristina; Butler, Jill. *Universal Principles of Design.* Rockport Publishers, 2003.

Lupton, Ellen. *Thinking with Type.* Princeton Architectural Press, 2004.

Müller-BROCKMANN Rastersysteme für die visuelle Gestaltung - Grid systems in Graphic Design. Niggli. Zweisprachige Ausgabe, 1996.

Rand, Paul. *Design Form and Chaos.* Yale University Press, 1993.

Samara, Timothy. *Making and Breaking the Grid.* Zweite Ausgabe. Rockport Publishers, 2017.

Spiekermann, Erik, and E. M. Ginger. *Stop Stealing Sheep & Find Out How Type Works.* Peachpit Press, 2003.

Stevenson, George A., Revised by William A. Pakan. *Graphic Arts Encyclopedia.* Design Press, 1992.

Updike, Daniel Berkeley. *Printing Types; Their History Forms, and Use.* Volumes I and II. Harvard University Press, 1966.

WEBSEITEN ODER ARTIKEL IM WEB

Haley, Allan. „They're not fonts!" http://www.aiga.org/content.cfm/theyre-not-fonts

Vinh, Khoi. „Grids are Good (Right)?" Blog Entry on subtraction.com

PODCASTS

The Observatory. Design Observer

Design Matters with Debbie Millman

Beiträge

FETTE SEITENANGABEN BEZIEHEN SICH AUF PRINZIPIEN

Prinzip **8**, 17
Sean Adams, Burning Settlers Cabin

Prinzipien **7**, 16; **20**, 40/41
AdamsMorioka, Inc.
Sean Adams, Chris Taillon, Noreen Morioka, Monica Shlaug

Prinzipien **32**, 64/65; **54**, 108/109; **73**, 146/147; **90**, 180/181
Marian Bantjes
Marian Bantjes, Ross Mills, Richard Turley

Prinzipien **4**, 13; **5**, 14; **11**, 22/23; **12**, 24; **16**, 32/33; **94**, 188/189
BTDnyc

Prinzipien **28**, 56/57; **40**, 80/81; **64**, 128/129
Carapellucci Design

Prinzipien **17**, 34/35; **65**, 130/131
The Cathedral Church of Saint John the Divine

Prinzip **60**, 120/121
Collins
Brian Collins, John Moon, Michael Pangilnan

Prinzipien **18**, 36/37; **46**, 92/93; **76**, 152/153
Croissant
Seiko Baba

Prinzip **34**, 68/69
Design Institute, University of Minnesota
Janet Abrams, Sylvia Harris

Prinzip **84**, 168/169
Design within Reach/Morla Design, Inc.
Jennifer Morla, Michael Sainato, Tina Yuan, Gwendolyn Horton

Prinzip **93**, 186/187
Design Taxi

Prinzipien **28**, 56/57; **71**, 142/143; **91**, 182/183
Suzanne Dell'Orto

Prinzip **70**, 140/141
Andrea Dezsö

Prinzipien **22**, 44/45; **25**, 50/51; **35**, 70/71; **43**, 86/87
Barbara deWilde

Prinzip **72**, 144/145
Simon & Schuster, Inc.
Jason Heuer

Prinzipien **37**, 74/75; **38**, 76/77; **56**, 112/113; **58**, 116/117; **94**, 188/189
Mario Eskenazi Studio
Mario Eskenazi, Gemma Villegas, Marc Ferrer Vives, Dani Rubio

Prinzipien **42**, 84/85; **48**, 96/97; **62**, 124/125; **92**, 184/185; **96**, 192/193
The Earth Institute of Columbia University
Mark Inglis, Sunghee Kim

Prinzip **19**, 38/39
Heavy Meta
Barbara Glauber, Hilary Greenbaum

Prinzip **85**, 170/171
Cindy Heller

Prinzipien **49**, 98/99, **75**, 150/151,
Drew Hodges
Naomi Mizusaki

Prinzip **21**, 42/43
Katie Homans

Prinzipien **98**, 198/199
Dayna Iphill

Prinzip **82**, 164/165
INDUSTRIES stationery
Drew Souza

Prinzip **29**, 58/59; **31**, 62/63; **39**, 78/79; **53**, 106/107
Kurashi no techno/Everyday Notebook
Shuzo Hayashi, Masaaki Kuroyanagi

Prinzip **97**, 194/195
Liney Li

Prinzipien **26**, 52/53; **59**, 118/119; **79**, 158/159
Bobby C. Martin Jr.

Prinzipien **9**, 18; **71**, 142/143; **88**, 176/177
Mark Melnick Graphic Design

Prinzipien **5**, 14; **25**, 50/51
Martha Stewart Omnimedia

Prinzip **60**, 120/121
The Martin Agency
Mike Hughes, Sean Riley, Raymond McKinney, Ty Harper

Prinzipien **44**, 88/89; **58**, 116/117
Memo Productions
Douglas Riccardi

Prinzipien **27**, 54/55
Metroplis magazine
Criswell Lappin

Prinzipien **13**, 26/27
Fritz Metsch Design

Prinzip **98**, 196/197
The Museum of Modern Art
Irma Boom

Prinzip **47**, 94/95
New York City Center
Andrew Jerabek, David Saks

Prinzip **30**, 60/61
The New York Times
Design Director: Tk

Prinzip **28**, 56/57
New York University School of Medicine

Prinzip **86**, 172/173
Nikkei Business Publications, Inc.
Prinzip **28**, 56/57
New York University School of Medicine

Prinzip **55**, 110/111
Noom Studio
Punyapol "Noom" Kittayarak

Prinzipien **10**, 19; **12**, 25;
42, 84/85, **50**, 84/85
OCD, Original Champions of Design
Bobby C. Martin Jr. and Jennifer Kinon

Prinzipien **41**, 82/83;
83, 166/167
Pentagram, Emily Oberman, Partner
Creative Direction, Emily Oberman
Christina Hogan, Elizabeth Goodspeed, Joey Petrillo, Anna Meixler

Prinzipien **3**, 12; **23**, 46/47;
37, 74/75; **38**, 76/77
Open, Scott Stowell

Prinzip **65**, 130/131; **95**, 190/191
Pentagram Design
Paula Scher, Lisa Strausferd, Jiae Kim, Andrew Freeman, Rion Byrd, Peter Mauss/Esto

Prinzip **24**, 48/49
The Pew Charitable Trusts
IridiumGroup

Prinzip **72**, 144/145
Picador
Henry Sene Yee, Adam Auerbach, Julyanne Young

Prinzip **89**, 178/179
Practical Studio/Thailand
Santi Lawrachawee, Ekaluck Peanpanawate, Montchai Suntives

Prinzip **67**, 134/135
Rebecca Rose

Prinzip **57**, 114/115
SpotCo
Gail Anderson, Frank Gargialo, Edel Rodriguez

Prinzipien **14**, 28/29;
51, 102/103
Studio RADIA

Prinzip **68**, 136/137
Jacqueline Thaw Design
Jacqueline Thaw

Prinzip **50**, 101
ThinkFilm

Prinzipien **52**, 104/105, **77**, 154/155,
Threaded
Kyra Clarke, Fiona Grieve, Reghan Anderson
Phil Kelly, Desna Whaanga/ Schollum, Karyn Gibbons, Te Raa Nehua

Prinzipien **15**, 30/31; **17**, 34/35;
45, 90/91
Tsang Seymour Design
Patrick Seymour, Laura Howell, Susan Brzozowski

Prinzipien **66**, 132/133
Anna Tunick
The New York Times and Scholastic
Creative Direction: Judith Christ/Lafund
Photo Illustration: Leslie Jean/ Bart

Prinzipien **6**, 15; **61**, 122/123;
74, 148/149; **78**, 156/157;
87, 174/175
Anna Tunick
Pyramyd/*étapes* magazine

Prinzipien **63**, 126/127, **65**, 130/131
Richard Turley

Prinzip **36**, 72/73
Two Twelve Associates
New Jersey Transit Timetables. Principals: David Gibson, Ann Harakawa; Project Manager: Brian Sisco; Designers: Laura Varacchi, Julie Park; Illustrator: Chris Griggs; Copywriter: Lyle Rexer

Prinzip **33**, 66/67
The Valentine Group
Robert Valentine

Prinzipien **80**, 160/161
Veenman Drukkers, Kuntsvlaai/ Katya van Stipout, Photo Beth Tondreau

Prinzip **81**, 162/163
Vignelli Associates
Massimo Vignelli, Dani Piderman

Prinzipien **69**, 138/139
Saima Zaidi

Bilder auf Seite 13:
Oben von *Astronomy 365*, erschienen bei Harry N. Abrams, Inc. © 2006 Jerry T. Bonnel und Robert J. Nemiroff. Mit frdl. Gen.
Unten von *Symbols of Power*, erschienen bei Harry N. Abrams, Inc. © 2007 American Federation of Arts and Les Arts Décoratifs. Mit frdl. Gen.

Bilder auf Seite 14:
Links von *Symbols of Power*, erschienen bei Harry N. Abrams, Inc. © 2007 American Federation of Arts and Les Arts Décoratifs. Mit frdl. Gen.

Bilder auf Seite 24:
Von *Sauces*, erschienen bei John Wiley & Sons, © 2008 by James Peterson. Nachdruck mit frdl. Gen. John Wiley & Sons, Inc.

Schnellstart

1

DAS MATERIAL BEURTEILEN

- ❑ Worum geht es?
- ❑ Gibt es viel Fließtext?
- ❑ Gibt es viele Elemente? Kapitelüberschriften? Zwischentitel? Kolumnentitel? Tabellen? Diagramme? Bilder?
- ❑ Wurde die Informationshierarchie bereits von der Redaktion festgelegt oder gehört das zu Ihren Aufgaben?
- ❑ Muss noch Design gestaltet oder fotografiert werden?
- ❑ Wird das Projekt traditionell gedruckt oder online verbreitet?

2

VORABPLANUNG UND TECHNISCHE SPEZIFIKATION

- ❑ In welchem Verfahren wird gedruckt?
- ❑ Ein-, Zwei- oder Vierfarbendruck?
  - ● ● ● Beim herkömmlichen Druck müssen Sie im TIFF-Format mit 300 dpi als Reproduktionsgröße arbeiten bzw. rechnen.
  - ● ● ● 72 dpi als JPEG ist nicht für den Druck, sondern nur für das Internet geeignet.
- ❑ Sind es viele Elemente? Kapitelüberschriften? Zwischentitel? Kolumnentitel? Tabellen? Diagramme? Bilder?
- ❑ Wird das Projekt traditionell gedruckt oder online verbreitet?
- ❑ Wie groß ist der Beschnittbereich?
- ❑ Ist bei dem Projekt eine Seitenzahl vorgegeben oder gibt es Spielraum?
- ❑ Sind von Ihren Kunden oder der Druckerei Mindestmaße bei den Stegen vorgegeben?

3

FORMAT, STEGE UND FONTS FESTLEGEN

- ❑ Wählen Sie das beste Format für die zu gestaltenden Seiten/ Bildschirmansichten.
  - ● ● ● Bei wissenschaftlichem Material oder auf großformatigen Seiten sind Zwei- oder Mehrspalter sinnvoll.
- ❑ Bestimmen Sie die Maße der Stege. Für Anfänger erweist sich das oft als der schwierigste Teil. Lassen Sie sich Zeit und probieren Sie aus. Bedenken Sie, dass Fläche, auch wenn viel Material auf der Seite untergebracht werden muss, Design bedeutet.
- ❑ Legen Sie anhand des Themas, das Sie im 1. Schritt beurteilt haben, die Fonts fest. Verlangt das Material lediglich einen Font mit verschiedenen Stärken oder verschiedene?
  - ● ● ● Die meisten Computer verfügen über diverse Fonts. Machen Sie sich mit Schriften und Schriftfamilien vertraut. Wagen Sie Schlichtheit, es müssen nicht die verrückten Fonts sein.
- ❑ Setzen Sie sich mit Schriftgrad und Zeilenabstand auseinander. Visualisieren Sie Ihr Projekt, machen Sie einen Entwurf und ziehen Sie Text in das Dokument, um zu sehen, wie es passt.

4
TYPOGRAFISCHE REGELN UND SCHRIFTSATZ

- ❑ Beim Satz wird nach einem Punkt nur ein Leerzeichen gesetzt.
- ● ● ● Das Arbeiten mit Layoutprogrammen unterscheidet sich von Textverarbeitungsprogrammen: Sie erstellen hier richtigen Schriftsatz. Doppelte Leerzeichen zur Simulation des Schriftbildes einer Schreibmaschine sind Vergangenheit.
- ❑ Verwenden Sie innerhalb eines Absatzes Soft Return nur, wenn Sie einen Zeilenumbruch wegen zu vieler Bindestriche oder unansehnlicher Absätze machen.
- ❑ Benutzen Sie sich öffnende und schließende Anführungszeichen und keine 'senkrechten Anführungszeichen' (die geraden Zeichen für Inches und Feet).
- ❑ Nutzen Sie die Rechtschreibprüfung.
- ❑ Prüfen Sie, ob kursive und fette Buchstaben tatsächlich in dem Font gesetzt sind, mit dem Sie arbeiten. Lassen Sie sich keinesfalls dazu verleiten, den Text manuell kursiv oder fett zu setzen.
- ❑ Achten Sie auf unschöne Zeilenumbrüche wie getrennte Namen, mehr als zwei Bindestriche in einer Zeile oder einen Gedankenstrich nach einem Bindestrich am Ende einer Zeile.
- ❑ **Striche unterscheiden:**
  - **Bindestrich** Verbindet Wörter und Redewendungen, trennt Wörter am Zeilenende. (Bindestrichtaste)
  - **Gedankenstrich** Für grammatikalische Pausen und Denkpausen des Lesers. So breit wie der Buchstabe ‚n' im jeweiligen Font. (Alt + Bindestrich)
  - **Geviertstrich** Bei Zeit- oder Nummernangaben. Die Hälfte des Geviertstrichs entspricht der Breite des Buchstabens ‚m' im jeweiligen Font. (Shift + Alt + Bindestrich)

„SENKRECHTE ANFÜHRUNGSZEICHEN" VERMEIDEN

"senkrechte Anführungszeichen"
„öffnende und schließende Anführungszeichen"

"senkrechte Anführungszeichen"
„öffnende und schließende Anführungszeichen"

SONDERZEICHEN UND AKZENTE (APPLE-MACINTOSH)

SONDERZEICHEN

–	Alt + Bindestrich	Gedankenstrich
—	Alt + Shift + Bindestrich	Geviertstrich
…	Alt + .	Auslassungspunkte (werden nicht getrennt)
■	n *(in der ZapfDingbats)*	gefülltes Quadrat
□	n *(in der ZapfDingbats, outlined)*	leeres Quadrat
°	Shift + ^	Gradzeichen
™	Shift + Alt + d	
©	Alt + g	
®	Alt + r	
¢	Alt + $	

AKZENTE (EINFACHE ÜBER AKZENTE-TASTE)

˜	Shift + Alt + 8	Tilde
ˆ	Shift + Alt + k	Accent grave

5
DIE REGELN EINES GUTEN SEITENAUFBAUS

SEITENAUFBAU

- ❑ Vermeiden Sie Hurenkinder und Schusterjungen (siehe auch Glossar).
- ❑ Vorhergehende Seiten berücksichtigen, aber nicht kopieren.
- ❑ Geben Sie mit der Datei auch Fonts und Bilder (sowohl in QuarkXPress wie auch in InDesign) mit, wenn Sie das Projekt zum Drucken geben.

Danksagung

Dieses Buch war ein Abenteuer und eine Erfahrung für mich. Ich danke Steven Heller. Ich danke auch Judith Cressy für ihre Geduld und David Martinell für sein diplomatisches Geschick. .

Die Designer, deren Projekte in diesem Buch gezeigt werden, haben sich Zeit genommen, das Material zusammenzustellen, meine Fragen zu beantworten und ihre Arbeiten großzügig zur Verfügung gestellt. Bei ihnen allen bedanke ich mich herzlich, auch dafür, dass ich so von ihnen lernen und ihrem Talent profitieren durfte.

Vielen Dank auch an Donna David für die Gelegenheit zu unterrichten und für einige in diesem Buch verwendete Fachbegriffe.

Mit diesem Buch habe ich festgestellt, dass Grafikdesign Teamarbeit ist. Janice Carapellucci hat mich, wie schon bei der ersten Ausgabe, unterstützt. George Garrastegui, Jr., er überarbeitete die erste Ausgabe, machte kluge Bemerkungen und gab mir Tipps, wie diese Ausgabe für Studenten und erfahrene Grafiker gleichermaßen nützlicher werden kann.

Ein großes Dankeschön auch an Punyapol „Noom" Kittayarak und besonders Patricia Chang, der nichts entgeht. Sie ist begabt, zuverlässig, verantwortungsvoll und die perfekte Mitarbeiterin.

Mein liebster Mitarbeiter Pat O'Neill war wieder unverwechselbar freundlich und witzig. So soll auch das Buch sein.